Contemporary Physics

Based on Lectures
at the University of Maryland
College Park

Editor-in-Chief
Edward F. Redish
(University of Maryland, College Park)

Editorial Board
(At the University of Maryland, College Park)

Alex J. Dragt
Richard A. Ferrell
Oscar W. Greenberg
Richard E. Prange
Philip G. Roos
Guarang Yodh

Contemporary Physics

R.N. Mohapatra: **Unification and Supersymmetry: The Frontiers of Quark-Lepton Physics**

R.E. Prange and S.M. Girvin (eds.): **The Quantum Hall Effect**

Rabindra N. Mohapatra

Unification and Supersymmetry

The Frontiers of Quark–Lepton Physics

With 49 Illustrations

Springer-Verlag
New York Berlin Heidelberg
London Paris Tokyo

Rabindra N. Mohapatra
Department of Physics and Astronomy
University of Maryland
College Park, MD 20742
U.S.A.

Editor in Chief

Edward F. Redish
Department of Physics and Astronomy
University of Maryland
College Park, MD 20742
U.S.A.

Library of Congress Cataloging in Publication Data
Mohapatra, R. N. (Rabindra Nath)
Unification and supersymmetry.
(Contemporary physics)
Includes bibliographies and index.
1. Grand unified theories (Nuclear physics)
2. Supersymmetry. 3. Particles (Nuclear physics)
I. Title. II. Series: Contemporary physics
(New York, N.Y.)
QC794.6.G7M64 1986 530.1 86-1814

Typeset by Asco Trade Typesetting Ltd., Hong Kong.
Printed and bound by R. R. Donnelley & Sons, Harrisonburg, Virigina.
Printed in the United States of America.

9 8 7 6 5 4 3 2 1

ISBN 0-387-96285-9 Springer-Verlag New York Berlin Heidelberg
ISBN 3-540-96285-9 Springer-Verlag Berlin Heidelberg New York

To
Manju,
Pramit and Sanjit

Preface

The theoretical understanding of elementary particle interactions has undergone a revolutionary change during the past one and a half decades. The spontaneously broken gauge theories, which in the 1970s emerged as a prime candidate for the description of electro-weak (as well as strong) interactions, have been confirmed by the discovery of neutral weak currents as well as the W- and Z-bosons. We now have a field theory of electro-weak interactions at energy scales below 100 GeV—the Glashow–Weinberg–Salam theory. It is a renormalizable theory which enables us to do calculations without encountering unnecessary divergences. The burning question now is: What lies ahead at the next level of unification? As we head into the era of supercolliders and ultrahigh energy machines to answer this question, many appealing possibilities exist: left–right symmetry, technicolor, compositeness, grand unification, supersymmetry, supergravity, Kaluza–Klein models, and most recently superstrings that even unify gravity along with other interactions. Experiments will decide if any one or any combination of these is to be relevant in the description of physics at the higher energies. As an outcome of our confidence in the possible scenerios for elementary particle physics, we have seen our understanding of the early universe improve significantly. Such questions as the origin of matter, the creation of galaxies, and the puzzle of the cosmic horizon are beginning to receive plausible answers in terms of new ideas in particle physics. Although a final solution is far from being at hand, reasonable theoretical frameworks for carrying out intelligent discussions have been constructed. Even such difficult questions as the "birth" and "death" of the universe have been discussed.

This book, based on advanced graduate courses offered by me at CCNY (City College of New York) and in Maryland, attempts to capture these exciting developments in a coherent chapter-by-chapter account, in the hope that the frontiers of our understanding (or lack of it) in this exciting field of science can be clearly defined for students as well as beginning researchers. The emphasis has been on physical, rather than technical and calculational, aspects although some necessary techniques have been included at various points. Extensive references are provided to original works, which the readers are urged to consult in order to become more proficient in the techniques. The prerequisites for this book are an advanced course in quantum field theory (such as a knowledge of Feynman diagrams, the renormalization program, Callan–Symanzik equations, etc.), group theory (Lie groups at the level of the books by Gilmore or Georgi), basic particle theory such as the quark model, weak interaction and general symmetry principles (at the level of the books by Marshak, Riazuddin and Ryan or by Commins), and familiarity with spontaneously broken gauge theories (at the level of the books by J. C. Taylor or Chris Quigg).

The book is divided into two parts: the first eight chapters deal with the introduction to gauge theories, and their applications to standard $SU(2)_L \times U(1) \times SU(3)_C$ models and possible extensions involving left–right symmetry, technicolor, composite models, quarks and leptons, strong and weak CP-violation and grand unification. In the last eight chapters, we discuss global and local supersymmetry, its application to particle interactions, and the possibilities beyond $N = 1$ supergravity. Interesting recent developments in the area of superstrings are only touched on but, unfortunately, could not be discussed as extensively as they ought to be.

I would like to acknowledge many graduate students at CCNY and in Maryland over the past seven years, as well as colleagues at both places who attended the lectures, and helped to sharpen the focus of presentation by their comments. I am grateful to many of my colleagues and collaborators for generously sharing with me their insight into physics. I wish to thank E. F. Redish for suggesting that I compile the lectures into book form, and C. S. Liu and the members of the editorial board of this lecture series for their support at many crucial stages in the production of this book. Finally, I wish to thank Mrs. Rachel Olexa for her careful and prompt typing of the manuscript, and J. Carr for reading several chapters and suggesting language improvements.

I wish to acknowledge the support of the U.S. National Science Foundation during the time this book was written.

College Park Rabindra N. Mohapatra
March, 1986

Contents

"…. Where tireless striving stretches its arms towards perfection:

Where the clear stream of reason has not lost its way into the dreary desert sand of dead habit, …. "

GITANJALI, RABINDRA NATH TAGORE

CHAPTER 1

Important Basic Concepts in Particle Physics

§1.1. Introduction

Forces observed in nature can be classified into four categories according to their observed strength at low energies: strong, electromagnetic, weak, and gravitational. Their strengths, when characterized by dimensionless parameters, are roughly of the following orders of magnitude:

		Orders of magnitude
Strong	$g_{NN\pi}^2/4\pi \simeq 15$	10^1
Electromagnetic	$e^2/4\pi \simeq 1/137 \cdot 035982$	10^{-2}
Weak	$G_F \simeq 1.16632 \times 10^{-5}$ GeV^{-2}	10^{-5}
Gravitational	$G_N \simeq 6 \cdot 70784 \times 10^{-39}$ GeV^{-2}	10^{-40}

Furthermore, the range of electromagnetic and gravitational interactions is (to a very good approximation) infinite, whereas the weak and strong interactions are known to have very short range. Since, in these lectures, we will study the nature of these forces [1] as they act between known "elementary" (only at our present level of understanding) particles, we will start our discussion by giving a broad classification of the known "elementary" particles (Fig. 1.1) according to first, their spin statistics, and then, whether they participate in strong interactions or not. Leptons and nonhadronic bosons are supposed to participate only in weak and electromagnetic interactions, whereas baryons and mesons have strong and weak (as well as electromagnetic) interactions. Another important point of difference between the

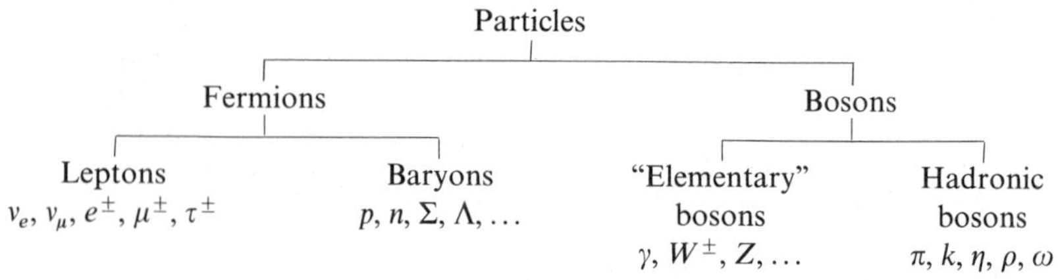

Figure 1.1

nonhadronic (leptons, γ, $w^\pm$, $z \ldots$) and hadronic (baryons and mesons) particles is that all observed nonhadronic particles exhibit pointlike structure to a good approximation (to the extent that experimentalists can tell), whereas the hadronic particles exhibit finite-size effects (or form factors). It is, therefore, convenient to assume that the former are more elementary than the latter, and this has led to the postulate [2] that there exists a more elementary layer of matter (called quarks), of which baryons and mesons are made. Thus, according to present thinkings, there are really two sets of basic building blocks of matter: the basic fermions (leptons and quarks, on which the forces like weak, electromagnetic, strong, etc., act) and a basic set of "elementary" bosons such as photon, W, Z, color gluons (V) which are the carriers of the above forces. According to these ideas, the quarks participate in strong, as well as weak, interactions and it is the binding force of strong interactions mediated by a color octet of gluons that generates the baryons and mesons.

There are, of course, certain other differences between strong interactions and other forms of interactions, the most important being the existence of manifest approximate global symmetries associated with strong interactions like, isospin, SU(3), etc., ..., which are shared by neither the leptons nor the nonhadronic bosons. Lack of availability of any reliable way to solve the bound-state problems in relativistic physics has given significance to the above symmetries in studying the particle spectra and interactions among the baryons and mesons. The basic strong interactions of the quarks must, of course, respect these symmetries. As far as the nonstrong interactions are concerned, although there does not exist any manifest global symmetry, strong hints of underlying hidden symmetries were detected as early as the late 1950s and early 1960s. Gauge theories [3], in a sense, have opened the door to this secret world of nonmanifest symmetries which have been instrumental in building mathematically satisfactory theories for weak and electromagnetic interactions and strong interactions.

Understanding the symmetries of the quark–lepton world is crucial to our understanding the nature of the various kinds of forces. To understand symmetries, we need to identify the basic degrees of freedom on which the symmetries operate. It is currently believed that the quarks carry two independent degrees of freedom (known as flavor and color), the flavor being a manifest degree of freedom which is responsible for the variety and richness of spectrum in the baryon–meson world; whereas color [4] is a hidden "coordinate"

Table 1.1

	Color			Lepton
F	u_1	u_2	u_3	ν_e
L	d_1	d_2	d_3	e^-
A	c_1	c_2	c_3	ν_μ
V	s_1	s_2	s_3	μ^-
O	t_1	t_2	t_3	ν_τ
R	b_1	b_2	b_3	τ^-

which is responsible for the binding of quarks and antiquarks in appropriate combination to baryons and mesons, as well as for the way in which the baryons and mesons interact. Leptons, on the other hand, carry only a flavor degree of freedom. It is, however, a remarkable fact (perhaps indicative of a deeper structure) that the flavor degrees of freedom of quarks and leptons are identical—a fact which is known as quark–lepton symmetry [5]. The existence of a new kind of quark, the charm quark, and a new set of hardrons called charmhadrons was first inferred on the basis of this symmetry. In Table 1.1, we display these quark–lepton degrees of freedom (known to date).

The current experience is that the electro-weak interactions operate along the flavor degree of freedom, whereas the strong interactions operate along the color degree of freedom. A more ambitious program is, of course, to introduce new kinds of interactions operating between quarks and leptons [6], thus providing a framework for the unified treatment of quarks and leptons. No evidence for such new interactions has yet been uncovered, but when that happens we would have taken a significant new stride in our understanding of the subatomic world.

To study the structure and forces of the quark–lepton world, we will start by giving a mathematical formulation of symmetries and their implications in the sections which follow. We will work with the assumption that all physical systems are described by a Lagrangian, which is a local function of a set of local fields $\phi_i(x)$ and their first derivative $\partial_\mu \phi_i$, respecting invariance under proper Lorentz transformations. The requirement of renormalizability further restricts the Lagrangian to contain terms of total mass dimension 4 or less. (To count dimensions, we note that bosonic fields have dimension 1 ($d = 1$); fermionic fields, $d = 3/2$; momentum, $d = 1$; and angular momentum, $d = 0$; etc.)

§1.2. Symmetries and Currents

Associated with any given symmetry is a continuous or discrete group of transformations whose generators commute the Hamiltonian. Examples of symmetries abound in classical and quantum mechanics, e.g., momentum and angular momentum, parity, symmetry, or antisymmetry of the wave function for many body systems, etc. In the following paragraphs, we will discuss the implications of continuous symmetries within the framework of Lagrangian field theories. To describe the behavior of a dynamical systems, we will start by writing down the action S as a functional of the fields ϕ_i, $\partial_\mu \phi_i$

$$S = \int d^4x \, \mathscr{L}(\phi_i, \partial_\mu \phi_i) \tag{1.2.1}$$

($\mathscr{L}$ being the Lagrange density). Under any symmetry transformation

$$\phi \to \phi + \delta\phi, \qquad \partial_\mu \phi \to \partial_\mu \phi + \delta(\partial_\mu \phi)$$

and

$$\mathscr{L} \to \mathscr{L} + \delta\mathscr{L}. \tag{1.2.2}$$

First, we will show that associated with any symmetry transformation there is a current, called Noether current, J_μ, which is conserved if the symmetry is exact, i.e.,

$$\delta\mathscr{L} = 0. \tag{1.2.3}$$

It is sufficient to have $\delta\mathscr{L} = \partial_\mu 0_\mu$ (i.e., four-divergence of a four-vector) so that the action is invariant under symmetry, i.e., $\delta S = 0$. This is actually the case for supersymmetry transformations (see Chapter 9). To obtain J_μ, we look at the variation of $\mathscr{L}$ under the transformations given in eq. (1.2.2):

$$\delta\mathscr{L} = \sum_i \left[\frac{\delta}{\delta\phi_i} \delta\phi_i + \frac{\delta}{\delta(\partial_\mu \phi_i)} \delta(\partial_\mu \phi_i) \right]. \tag{1.2.4}$$

Physical fields are those that satisfy the Euler–Lagrange equations resulting from extremization of the action

$$\frac{\delta\mathscr{L}}{\delta\phi_i} = \partial_\mu \frac{\delta\mathscr{L}}{\delta(\partial_\mu \phi_i)}. \tag{1.2.5}$$

Substituting this in eq. (1.2.4), we get (assuming $\delta(\partial_\mu \phi_i) = \partial_\mu(\delta\phi_i)$)

$$\delta\mathscr{L} = \sum_i \partial_\mu \left[\frac{\delta}{\delta(\partial_\mu \phi_i)} \delta\phi_i \right] \tag{1.2.6}$$

$$\equiv \sum_i \partial_\mu(\varepsilon_a J_\mu^a), \tag{1.2.7}$$

where ε_a is the parameter characterizing the transformation. J_μ^a is the Noether current we were looking for and it satisfies the equation

$$\varepsilon_a \partial_\mu J_\mu^a = \delta\mathscr{L}. \tag{1.2.8}$$

Thus, if $\delta\mathscr{L} = 0$, since ε is arbitrary, $\partial_\mu J_\mu^a = 0$. It is obvious that the corresponding charge

$$Q^a = -i \int d^3x \, J_0(x) \tag{1.2.9}$$

satisfies the equation

$$\frac{dQ}{dt} = 0 \quad \text{if } \delta\mathscr{L} = 0$$

leading to conservation laws for the corresponding charge.

Let us give some examples of symmetries present in the world of elementary particles. The simplest one is the conservation of electric charge Q, which is a symmetry associated with arbitrary phase transformation of complex field. To give an example, consider the Dirac field ψ transforming as $\psi \to e^{iq_i\theta}\psi$ that keeps the following Lagrangian invariant:

$$\mathscr{L} = -[\bar{\psi}\gamma_\mu\partial^\mu\psi + m\bar{\psi}\psi]. \tag{1.2.10}$$

The corresponding conserved current is $J_\mu = i\bar{\psi}\gamma_\mu\psi$. The associated charge operator which generates these symmetries is

$$Q = -i\int d^3x\, J_4(x) = \int d^3x\, \psi^\dagger\psi. \tag{1.2.11}$$

Using equal-time commution relations between ψ and $\psi^\dagger$, we can verify that $e^{iQ\theta}$ indeed generates the phase transformations on the Dirac field ψ.

Isospin

Charge independence of the nuclear forces (pp, np, nn) led Heisenberg to introduce the concept of isospin, according to which proton and neutron are two states (isospin up and down) of the same object, the nucleon. An immediate implication of this hypothesis is that, if we define an internal space, where p and n form the coordinates, any rotation in that space (which, of course, mixes proton and neutron) will leave the nuclear forces invariant. In the language of the Lagrangian, the Lagrangian describing a system of neutrons and protons, interacting only via nuclear forces, will remain invariant under any rotation in that space. So there must be Noether currents, and the charges and conservation laws associated with them. To discuss this question we write a Lagrangian

$$\mathscr{L} = \mathscr{L}_0(\psi, \partial_\mu\psi),$$

where

$$\psi = \begin{pmatrix} P \\ n \end{pmatrix}. \tag{1.2.12}$$

A general rotation in this space is given by

$$\psi \to e^{i\boldsymbol{\varepsilon}\cdot\boldsymbol{\tau}}\psi$$

or

$$\delta\psi = \frac{i}{2}\boldsymbol{\varepsilon}\cdot\boldsymbol{\tau}\psi + O(\varepsilon^2)$$

for an infinitesimal parameter ε_a, which is a constant independent of space and time. From (1.2.7) we can easily read off the corresponding conserved Noether currents (of which there are now three, corresponding to the three Pauli matrices, τ^i)

$$J^i_\mu = \frac{i}{2}\bar{\psi}\gamma_\mu \tau^i \psi. \tag{1.2.13}$$

Again, using equal-time commutation relations between the fields ψ, we can verify that the corresponding charges Q_i satisfy the algebra

$$[Q_i, Q_j] = i\varepsilon_{ijk} Q_K, \tag{1.2.14}$$

which is the same as the familiar angular momentum algebra of SU(2). The Casimir operator of SU(2), which will be used to designate irreducible representations of SU(2) algebra, has eigenvalues $I(I + 1)$ where $t = 0, 1/2, 1, 3/2, \ldots$, etc.

Isospin is actually an approximate symmetry rather than an exact one, since proton and neutron differ in their electric charge, and electromagnetic interactions do not respect isospin invariance. To see this, let us write the electromagnetic interaction of the proton–neutron system in isospin notation

$$\mathscr{L}_{\mathrm{em}}(p, n, A_\mu) = ie\bar{p}\gamma_\mu p A_\mu = \frac{ie}{2}\bar{\psi}\gamma_\mu(1 + \tau_3)\psi. \tag{1.2.15}$$

Note the appearance of the factor τ_3, which makes $\mathscr{L}_{\mathrm{em}}$ noninvariant under isospin. A more realistic Lagrangian for the proton–neutron system includes both $\mathscr{L}_0$ and $\mathscr{L}_{\mathrm{em}}$.

It is worth pointing out that, even though the isospin symmetry is not respected by $\mathscr{L} \equiv \mathscr{L}_0 + \mathscr{L}_{\mathrm{em}}$, the charges Q_i (which are now time-dependent), satisfy the SU(2) algebra

$$[Q_i(t), Q_j(t)] = i\varepsilon_{ijk} Q_K(t). \tag{1.2.16}$$

It was pointed out by Gell-Mann [7] that important physical information can be gained by assuming the algebra of time-dependent charges (the so-called current algebra) even if the symmetry is broken. This idea extended to the chiral symmetry group SU(2) × SU(2) and was exploited by Adler and Weisberger [8] to give a satisfactory explanation of the axial vector coupling constant renormalization (g_A/g_V) in β-decay.

Equation (1.2.15) contains the germs of another important concept in particle physics, i.e., a formula relating electric charge Q to isospin

$$Q = I_3 + \frac{B}{2}, \tag{1.2.17}$$

where B is the baryon number and I_3 is the third component of isospin. The discovery of V-particles by Rochester and Butler (and its subsequent analysis in terms of a new quantum number, the strangeness (S)) added a new dimension to the discussion of internal symmetries such as electric charge and isospin.

The formula for electric charge now becomes

$$Q = I_3 + \frac{B + S}{2}, \tag{1.2.18}$$

this is the Gell-Mann–Nishijima formula. Gell-Mann and Neeman combined isospin and strangeness (or hypercharge $Y \equiv B + S$) into the larger symmetry SU(3), which now describes both nonstrange and strange baryons and mesons. In the language of quarks SU(3) is the transformation that mixes u, d, and s quarks. With the quark spectrum given in Table 1.1, the new approximate global flavor symmetry is SU(6) and this symmetry is not only broken by electromagnetic interactions but also by different masses of the quarks.

In subsequent sections, we will note that broken symmetries are not the "monopoly" of the hadronic world alone; they exist in the leptonic world too. It is in the way they are hidden that makes them less obviously discernible. The symmetries of the hadronic world can be detected by looking at the level spectrum of particles and approximate mass degeneracies, but such studies do not reveal any symmetry for leptons since $\nu_e, e^-, \mu^-, \ldots$, etc. have so vastly different masses in their scale. Their symmetries, on the other hand, are manifest in their interactions and are consequences of local symmetries introduced in the next section.

§1.3. Local Symmetries and Yang–Mills Fields

In Section 1.2 we discussed global symmetries, where symmetry transformation on the fields is identical at all space–time points. We now introduce the concept of local symmetries, where the symmetry transformations can be arbitrarily chosen at different space–time points. Aesthetically, local symmetries appear much more plausible than global symmetries, where symmetry transformation must be implemented in an identical manner at all points no matter how far their separation is or even whether they are causally connected. The local symmetries are implemented by making the group parameters dependent on space–time. Let G be an arbitrary symmetry group whose elements are expressed in terms of its generators θ_a as follows:

$$G = e^{i\varepsilon^a \theta^a}, \tag{1.3.1}$$

where θ^a satisfy the Lie algebra of the group, i.e.,

$$[\theta^a, \theta^b] = if^{abc}\theta^c, \tag{1.3.2}$$

where f^{abc} are the structure constants. If $\varepsilon_a(x)$ is a function of space–time x, then we call G a local symmetry. We see below that demanding invariance of a Lagrangian under the local symmetry $G(x)$ requires the introduction a new set of spin 1 fields, A_μ^a as dynamical variables, and furthermore, its coupling to other fields in the Lagrangian is uniquely fixed by the requirement of gauge invariance. The fields A_μ^a will be called gauge fields. Thus, local symmetries dictate dynamics and therefore provide a more powerful theoretical tool for studies of particle interactions. Moreover, in contrast with global symmetries, the current in the case of local symmetries participates in the interactions and is therefore a physical quantity which, in principle, can be measured (e.g., the electric charge). To state it more pedagogically, a particle carrying a charge of a local symmetry has a field surrounding it (e.g., the electric field), whereas no such thing happens for particles carrying global symmetry charge.

To construct the Lagrangian invariant under an arbitrary local symmetry transformation [9, 10], $G(x)$, we consider a set of spin 1/2 matter fields $\psi(x)$ transforming as an irreducible representation of $G(x)$ as follows:

$$\psi(x) \to S(x)\psi(x). \tag{1.3.3}$$

Under this transformation, the kinetic energy term in the Lagrangian changes as follows:

$$\bar{\psi}\gamma_\mu \partial_\mu \psi \to \bar{\psi}\gamma_\mu \partial_\mu \psi + \bar{\psi}\gamma_\mu S^{-1}\partial_\mu S\psi. \tag{1.3.4}$$

To construct an invariant Lagrangian, we must therefore introduce a spin 1 field $B_\mu(x)$ represented as a matrix in the space of the column vector $\psi(x)$ and transforming under $G(x)$ as follows:

$$B_\mu(x) \to SB_\mu S^{-1} + S\partial_\mu S^{-1}. \tag{1.3.5}$$

If we now write a modified kinetic energy term in the Lagrangian as

$$\mathscr{L} = \bar{\psi}\gamma_\mu(\partial_\mu + B_\mu)\psi, \tag{1.3.6}$$

it is invariant under the transformations of the group $G(x)$.

To make B_μ into a dynamical field that can propagate, we have to write a G-invariant kinetic energy term. To do this we note that the function

$$F_{\mu\nu} = \partial_\mu B_\nu - \partial_\nu B_\mu + [B_\mu, B_\nu] \tag{1.3.7}$$

transforms covariantly under $G(x)$, i.e., $F_{\mu\nu} \xrightarrow{G} SF_{\mu\nu}S^{-1}$. Then the quantity $T_r F_{\mu\nu}^2$ is G-invariant. We can therefore write full gauge invariant Lagrangians involving gauge and matter fields as

$$\mathscr{L} = +\frac{1}{4g^2}\mathrm{Tr}\, F_{\mu\nu}F_{\mu\nu} - \bar{\psi}\gamma_\mu(\partial_\mu + B_\mu)\psi. \tag{1.3.8}$$

To make connection with the gauge field A_μ^a, we have to project out the independent components from the gauge field matrix B_μ. To count the number of independent A_μ^a note that there is one gauge field for each group parameter; therefore, the number of gauge fields is the same as the number of group parameters. We can therefore write the G_μ matrix as follows:

$$B_\mu = -ig\theta^a A_\mu^a. \tag{1.3.9}$$

Using eqns. (1.3.2) and (1.3.9), eqn. (1.3.8) can be rewritten as:

$$\mathscr{L} = -\tfrac{1}{4} f^a_{\mu\nu} f^a_{\mu\nu} - \bar{\psi}\gamma_\mu(\partial_\mu - ig\theta^a A_\mu)\psi, \tag{1.3.10}$$

where

$$f^a_{\mu\nu} = \partial_\mu A^a_\nu - \partial_\nu A^a_\mu + g f^{abc} A^b_\mu A^c_\nu. \tag{1.3.11}$$

Here we have assumed θ^a to be normalized generators satisfying the condition $\mathrm{Tr}\, \theta^a \theta^b = \delta^{ab}$.

A feature of Yang–Mills theories of great physical significance can be gleaned immediately from eqn. (1.3.10), i.e., there is only one coupling constant describing interactions of all matter fields with the gauge field A^a_μ as well as the self-interactions of gauge fields. Thus, Yang–Mills theories are potentially strong candidates for describing interactions such as weak interactions, which, as we will see, have the feature that their interactions respect universality of coupling. There is, however, a potential hurdle in the path of such applications since gauge invariance demands that the guage fields A^a_μ be massless.

While the Lagrangian obtained in the above manner is, of course, correct, we present an alternative derivation for this which emphasizes the role of the Noether current as a dynamical current participating in the gauge interactions. To see this, we write the gauge transformations for the ψ and A^a_μ for infinitesimal values of the group parameters $\varepsilon_a(x)$

$$\psi \to \psi + i\boldsymbol{\theta}\boldsymbol{\varepsilon}(x)\psi,$$

$$\mathbf{A}_\mu \to \mathbf{A}_\mu + \frac{1}{g}\partial_\mu \boldsymbol{\varepsilon} + \mathbf{A}_\mu \times \boldsymbol{\varepsilon} + O(\boldsymbol{\varepsilon}^2). \tag{1.3.12}$$

We have denoted the adjoint representation of group G by a vector symbol (boldface) for simplicity, and $\mathbf{A} \times \boldsymbol{\varepsilon}$ stands for $f^{abc} A^b_\mu \varepsilon^c$. Let us consider an arbitrary Lagrangian $\mathscr{L}(\psi, \partial_\mu\psi, \mathbf{A}_\mu, \partial_\mu \mathbf{A}_\nu)$ and try to determine its form by requiring invariance under eqn. (1.3.12). The variation of the Lagrangian under eqn. (1.3.12) is (using eqn. (1.3.12))

$$\delta\mathscr{L} = \frac{\delta\mathscr{L}}{\delta\psi} i\boldsymbol{\theta}\boldsymbol{\varepsilon}\psi + \frac{\delta\mathscr{L}}{\delta(\partial_\mu\psi)} i\boldsymbol{\theta}\partial_\mu(\boldsymbol{\varepsilon} \times \psi) + \frac{\delta\mathscr{L}}{\delta\mathbf{A}_\mu}\left(\frac{1}{g}\partial_\mu\boldsymbol{\varepsilon} + \mathbf{A}_\mu \times \boldsymbol{\varepsilon}\right)$$
$$+ \frac{\delta\mathscr{L}}{\delta(\partial_\mu \mathbf{A}_\mu)}\left(\frac{1}{g}\partial_\mu\partial_\nu\boldsymbol{\varepsilon} + \partial_\nu \mathbf{A}_\mu \times \boldsymbol{\varepsilon} + \mathbf{A}_\mu \times \partial_\nu\boldsymbol{\varepsilon}\right). \tag{1.3.13}$$

To obtain the Noether current generating this symmetry transformation, we can choose a particular form for the gauge parameter functions $\varepsilon(x)$, i.e., $\varepsilon(x) = \text{const.}$ and get

$$-\mathbf{J}_\mu = i\frac{\delta\mathscr{L}}{\delta(\partial_\mu\psi)}\boldsymbol{\theta}\psi + \frac{\delta\mathscr{L}}{\delta(\partial_\mu \mathbf{A}_\nu)} \times \mathbf{A}_\nu, \tag{1.3.14}$$

$\mathbf{J}_\mu$ is conserved, i.e., $\partial_\mu \mathbf{J}_\mu = 0$. Using this, we can rewrite eqn. (1.3.13) as

$$\delta\mathscr{L} = -\partial_\mu\boldsymbol{\varepsilon}\mathbf{J}_\mu + \frac{1}{g}\frac{\delta\mathscr{L}}{\delta\mathbf{A}_\mu}\partial_\mu\boldsymbol{\varepsilon} + \frac{1}{g}\frac{\delta\mathscr{L}}{\delta(\partial_\nu \mathbf{A}_\mu)}\partial_\mu\partial_\nu\boldsymbol{\varepsilon}. \tag{1.3.15}$$

For $\delta\mathscr{L} = 0$, we must have

$$\mathbf{J}_\mu = \frac{1}{g}\frac{\delta\mathscr{L}}{\delta\mathbf{A}_\mu} \tag{1.3.16}$$

and

$$\frac{\delta\mathscr{L}}{\delta(\partial_\nu\mathbf{A}_\mu)} = -\frac{\delta\mathscr{L}}{\delta(\partial_\mu\mathbf{A}_\nu)}. \tag{1.3.17}$$

We conclude from eqn. (1.3.16) that the Lagrangian must be a function of the symmetry current in the case of local symmetries. Equation (1.3.17) dictates an antisymmetric form for the kinetic energy term in agreement with our previous observations.

To obtain the exact form for $\mathbf{f}_{\mu\nu}$ derived earlier, we can write eqn. (1.3.17) to imply the following general form

$$\frac{\delta\mathscr{L}}{\delta(\partial_\mu\mathbf{A}_\nu)} = \partial_\mu\mathbf{A}_\nu - \partial_\nu\mathbf{A}_\mu + \mathbf{g}_{\mu\nu}(\mathbf{A}_\alpha). \tag{1.3.18}$$

Using eqns. (1.3.16) and (1.3.14), we get

$$\frac{1}{g}\frac{\delta\mathscr{L}}{\delta\mathbf{A}_\mu} = -(\partial_\mu\mathbf{A}_\nu - \partial_\nu\mathbf{A}_\mu + \mathbf{g}_{\mu\nu}) \times \mathbf{A}_\nu. \tag{1.3.19}$$

We can solve for $\mathbf{g}_{\mu\nu}$ by iteration. Let us assume $\mathbf{g}_{\mu\nu} = 0$; then eqn. (1.3.19) implies, along with (1.3.18), that

$$\mathscr{L} = -\tfrac{1}{2}g(\partial_\mu\mathbf{A}_\nu - \partial_\nu\mathbf{A}_\mu)\mathbf{A}_\nu \times \mathbf{A}_\mu - \tfrac{1}{4}(\partial_\mu\mathbf{A}_\nu - \partial_\nu\mathbf{A}_\mu)^2. \tag{1.3.20}$$

Equations (1.3.18) and (1.3.20) then determine the following unique form for $\mathscr{L}$ to be

$$\mathscr{L} = -\tfrac{1}{4}(\partial_\mu\mathbf{A}_\nu - \partial_\nu\mathbf{A}_\mu + g\mathbf{A}_\mu \times \mathbf{A}_\nu)^2. \tag{1.3.21}$$

Thus, for local symmetries, the Noether theorem is even powerful enough to fix the dynamics.

§1.4. Quantum Chromodynamic Theory of Strong Interactions: An Application of Yang–Mills Theories

In the previous section, we saw how powerful the implications of exact local symmetries are. We would therefore like to search for areas of physics where it may be applicable. It is well known that electromagnetic interactions are invariant under a local U(1) symmetry associated with phase transformation of complex fields. The associated massless spin 1 boson is the photon. The only other force in nature which is known to be long range is gravitation, but to consistently describe the properties of gravity we need a massless spin 2 boson, the graviton, which, as we will see in Chapter 14, can be thought of as the gauge field associated with local translation symmetry. So if we want to apply the beautiful idea of non-abelian local symmetries to nature, either: (i) we must find a way to give mass to the gauge bosons without spoiling the local symmetry of the Lagrangian; or (ii) we must explain the absence of long-range forces arising from the exchange of massless gauge particles. The first method will be chosen in applying gauge theories to the study of electroweak interactions, whereas the second alternative will be chosen in applications to strong interactions, and this is the focus of this section.

Let us briefly discuss the features of strong interactions that suggest its description in terms of a non-abelian gauge theory. The first important step in the understanding of nuclear forces was the success of SU(3) symmetry for hadrons, which led Gell-Mann and Zweig to introduce a quark picture of hadrons. According to this picture, the quarks are assumed to have spin 1/2 and baryon number 1/3 and transform as SU(3) triplets denoted by (u, d, s); (u, d) carry isospin and s carries strangeness. This picture provides a clear understanding of the spectroscopy of low-lying mesons and baryons in terms simple S- and P-wave bound states of quarks (u, d, s) or quark and antiquarks. The baryon spectroscopy is understood in terms of three-quark bound states whereas meson spectroscopy arises from nonrelativistic quark–antiquark bound states. Accepting quarks as the constituents of hadrons, we have to search for a field theory which provides the binding force between the quarks.

In trying to understand the Fermi statistics for baryons (such as Ω^-), it became clear that if they are S-wave bound states then the space part of their wave function is totally symmetric; since a particle such as Ω^- consists of three strange quarks, and has spin 3/2, the spin part of its wave function is symmetric. If there was no other degree of freedom, this would be in disagreement with the required Fermi statistics. A simple way to resolve this problem is to introduce [11] a threefold degree of freedom for quarks, called color (quarks being color triplets) and assume that all known baryons are singlet under this new SU(3). Since an $SU(3)_C$-singlet constructed out of three triplets is antisymmetric in the interchange of indices (quarks), the total baryon wave function is antisymmetric in the interchange of any two constituents as required by Fermi statistics.

It is now tempting to introduce strong forces by making $SU(3)_C$ into a local symmetry. In fact, if this is done, we can show that exchange of the associated gauge bosons provides a force for which the $SU(3)_C$ color singlet is the lowest-lying state; and triplet, sextet, and octet states all have higher mass. By choosing this mass gap large, we can understand why excited states corresponding to the color degree of freedom have not been found.

While this argument in favor of an $SU(3)_C$ gauge theory of strong interaction was attractive, it was not conclusive. The most convincing argument in favor of $SU(3)_C$ gauge theory came from the experimental studies of deep inelastic neutrino and electron scattering off nucleons. These experiments involved the scattering of very high energy (E) electrons and neutrinos with the exchange of very high momentum transfers (i.e., q^2 large). It was found that the structure functions, which are analogues of form factors for large q^2 and E, instead of falling with q^2, became scale invariant functions depending only on the ratio ($q^2/2mE$). This was known as the phenomenon of scaling [12]. Two different theoretical approaches were developed to understand this problem. The first was an intuitive picture called the parton model suggested by Feynman [13] and developed by Bjorken and Paschos [14], where it was assumed that, at very high energies, the nucleon can be thought of as consisting of free pointlike constituents. The experimental results also showed that these pointlike constituents were spin 1/2 objects, like quarks, and the scaling function was simply the momentum distribution function for the partons inside the nucleon. These partons could be identified with quarks, thus providing a unified description of the nucleon as consisting of quarks at low, as well as high, energies. The main distinction between these two energy regimes uncovered by deep inelastic scattering experiments is that at low energies the forces between the quarks are strong, whereas at high energies the forces vanish letting the quarks float freely inside the nucleons. If g described the strong interaction coupling constant for the quarks

$$g(Q^2) \xrightarrow[Q^2 \to \infty]{} 0,$$

where

$$g(Q^2) \xrightarrow[Q^2 \to 0]{} \text{large}. \tag{1.4.1}$$

The same picture emerged in trying to understand the scaling phenomena from a field-theoretic point of view. It was noted by Wilson and others [15] that the scaling region, i.e., $Q^2 \to \infty$, $E \to \infty$ with $x = Q^2/2mE =$ finite, corresponded in coordinate space to the behavior of current operator products on the light cone, and revealed that only if current products on the light cone have free field singularities [15], would we get the observed scaling phenomena. Again, this implied that the strong interaction coupling among quarks must vanish for large Q^2 values.

The big challenge to theoretical physics of the early 1970s was to find a field theory which had this property of $g(Q^2) \to 0$ for large Q^2 or "asymptotic freedom." The discovery of the Callan–Symanzik equation [16] in 1970–1971 had given a way to probe the behavior of coupling constants at different momentum transfers. It is given by the following equation

$$\frac{dg}{dt} = \beta(g), \tag{1.4.2}$$

where $t = \ln Q^2$ and $\beta(g)$ is a function of the coupling constants which can be obtained by studying the coupling constant renormalization in a field theory. Asymptotic freedom would result if $\beta(g) \xrightarrow[g \to 0]{} 0$ with $d\beta/dg < 0$. It was discovered by Gross, Wilczek, and Politzer [17] that only non-abelian gauge theories have this property if the number of fermion flavors are less than 16. More quantitatively, they calculated the function $\beta(g)$ for an SU(N) non-abelian gauge theory with N_f fermion species each transforming as N-dimensional representation under SU(N) and found

$$\beta(g) = -\frac{1}{16\pi^2}\left[\tfrac{11}{3}N - \tfrac{2}{3}N_f\right]. \tag{1.4.3}$$

Thus, in the $SU(3)_C$ theory of quarks, the coupling constant has the desired property of asymptotic freedom and no other theory (such as scalar, pseudoscalar interactions of quarks, etc.) has this attractive property. Thus, unbroken $SU(3)_C$ gauge theory became the accepted theory of strong interactions and came to be known as quantum chromodynamics (QCD). The associated $SU(3)_C$ gauge bosons are called gluons. In this picture, all observed nuclear forces are residual effects of gluon exchanges between the quarks which are bound inside the nucleon.

An immediate problem was to understand why long-range forces resulting from single gluon exchange have not been seen. Similarly, since gluons are massless they should be copiously produced in all strong interaction processes. Furthermore, since at high energies quarks become free, they should have been seen since their masses are expected to be less than the nucleon mass. To resolve this conundrum a new principle, known as "confinement," was introduced [18]. According to this principle, all unbroken non-abelian gauge theories with the property of asymptotic freedom should confine the gauge nonsinglet states, and only observable sectors of the Hilbert space correspond to states which are singlets under this local symmetry. This would then explain why quarks and gluons, and indeed any other nonsinglet bound hadron states, have not been seen. In the last 10 years, many plausible arguments and techniques have been developed to prove confinement and it has, now-a-days, become an accepted principle.

We close this subsection by writing down the QCD Lagrangian for different quark flavors:

$$\mathscr{L}_{\text{QCD}} = \sum_{a=flavors} -\bar{q}_a\gamma_\mu(\partial_\mu - ig_2\boldsymbol{\lambda}_c\mathbf{V}_\mu)q_a - \tfrac{1}{4}f^i_{\mu\nu}f^i_{\mu\nu}, \tag{1.4.4}$$

where $q_a = (q_a^1, q_a^2, q_a^3)$, for three colors.

§1.5. Hidden Symmetries of Weak Interactions

In the preceding section we presented arguments for a non-abelian gauge theory of strong interactions. In this section we would like to look for clues that will enable us to apply the ideas of gauge theories in the domain of weak interactions. Weak interaction processes known to date are of three kinds: (a) those that involve only leptons such as ν_e, e^-, ν_μ, μ^- (to be called leptonic); (b) those that involve only hadrons such as Σ, Λ, p, π^-, K (to be called hadronic), etc.; and (c) those that involve both hadrons and leptons (to be called semileptonic). (For a detailed study of weak interactions prior to the advent of gauge theories, we refer to several excellent texts [1].) Some typical weak processes are (the lifetime is given in parenthesis)

$$\begin{aligned}
\mu^- &\to e^-\bar{\nu}_e\nu_\mu && (2.19 \times 10^{-6}\ \text{s}),\\
n &\to pe^-\bar{\nu}_e && (898\ \text{s}),\\
K^+ &\to \pi^0e^+\nu_e && (1.2371 \times 10^{-8}\ \text{s}),\\
\Lambda &\to p\pi^- && (\sim 3.7 \times 10^{-10}\ \text{s}),\\
\Sigma &\to p\pi^- && (\sim 1.6 \times 10^{-10}\ \text{s}),\\
K^0 &\to 2\pi && (0.89 \times 10^{-10}\ \text{s}).
\end{aligned}$$

The first important point to note is that even though the various decay half-lives vary by orders of magnitude of from 10^3 s to 10^{-10} s), if we try to understand them in terms of a Hamiltonian involving four fermion operators, i.e.,

$$H_{\text{wk}} = \frac{G_F}{\sqrt{2}}\bar{\psi}_1 O\psi_2\bar{\psi}_3 O\psi_4, \tag{1.5.1}$$

we find that for β-decay as well as μ-decay the associated couplings G_β and G_μ are almost the same ($\sim 10^{-5}/m_p^2$), and for the decays involving strange particles the coupling strength is $G_\mu \sin\theta_C$ where θ_C is known as the Cabibbo angle whose value is determined to be [19]

$$\sin\theta_C = 0.231 \pm 0.003. \tag{1.5.2}$$

So even though the particles taking part in weak interactions have widely varying masses (and would thus not indicate any obvious symmetry) their interaction strength is universal. Since gauge theories lead to universal coupling strengths, as we saw in Section 1.3, this may be the first reason to suspect the relevance of gauge symmetries to weak interactions.

To proceed further, we need to know the form of O in eqn. (1.5.1). When Fermi originally wrote down the interaction in eqn. (1.5.1), he chose the most general form for O (i.e., $1, \gamma_5, \gamma_\mu, \gamma_\mu\gamma_5, \sigma_{\mu\nu}$) consistent with Lorentz invariance and parity symmetry. In 1956, in order to understand the $\tau - \theta$ puzzle, Lee and Yang proposed that weak interactions may not conserve parity, an idea that was experimentally confirmed by Wu, Ambler, Hayward, and Hobson in the decay of polarized Co^{60}. Then in 1957, Marshak, Sudarshan, Feynman, Gell-Mann, and Sakurai proposed the $V - A$ theory of weak interaction that uniquely gave the form for the operator O to be

$$O = \gamma_\mu(1 + \gamma_5). \tag{1.5.3}$$

The fact that the operator involved in (1.5.1) is a vector-axial vector type implies that currents are responsible for weak interactions. This is another indication that gauge theories may be relevant to weak interactions since, as we argued in Section 1.3, gauge symmetries imply that the interaction Lagrangian *must* involve the Noether current that generates the symmetry.

Having said this much, we can now look for the kind of symmetries by simply looking at the current involved in weak processes, and its equal-time commutation relations to get the Lie algebra of the weak symmetries.

To make this discussion more transparent we will depart from the historical order of events and proceed to express the weak Hamiltonian and the weak currents in terms of the basic constituents of matter, the quarks (rather than $p, n, \Lambda, \Sigma \ldots$) and leptons. The weak Hamiltonian $H_{\mathrm{wk}}^{\mathrm{cc}}$ (cc stands for charged current) can be written in terms of weak charged currents as

$$H_{\mathrm{wk}}^{\mathrm{cc}} = \frac{G_F}{\sqrt{2}} J_\mu J_\mu^+, \tag{1.5.4}$$

where

$$J_\mu = \bar{u}\gamma_\mu(1 + \gamma_5)(\cos\theta_C d + \sin\theta_C s) + \bar{\nu}_e\gamma_\mu(1 + \gamma_5)e^- + \bar{\nu}_\mu\gamma_\mu(1 + \gamma_5)\mu^-. \tag{1.5.5}$$

We will soon generalize this to include all generations. To uncover the hidden symmetry of weak interactions, we go to the limit of $\theta_C = 0$ and ignore the muon and electron pieces and look at the $\bar{u}\gamma_\mu(1 + \gamma_5)d$ piece alone. The charge corresponding to this current is

$$Q_L^- = \int d^3x\, u^+(1 + \gamma_5)d, \tag{1.5.6}$$

and using equal-time canonical commutation relations for quark fields we get

$$[Q_L^-, Q_L^+] = 2Q_L^3, \tag{1.5.7}$$

where

$$Q_{3L} = \int d^3x\, \{u^+(1 + \gamma_5)u - d^+(1 - \gamma_5)d\}. \tag{1.5.8}$$

Thus, the weak charges have the potential to generate a weak symmetry; however, if this symmetry is to be local, then the Q_{3L} piece should also be experimentally measured with the same strength as charged currents, i.e., the general form of the full weak Hamiltonian should be

$$H_{\mathrm{wk}} = 4\frac{G_F}{\sqrt{2}}[J_{\mu L}^+ J_{\mu L}^- + J_{\mu L}^3 J_{\mu L}^3], \tag{1.5.9}$$

where

$$\begin{aligned} J_{\mu L}^3 = \tfrac{1}{4}[&\bar{u}\gamma_\mu(1 + \gamma_5)u - \bar{d}\gamma_\mu(1 + \gamma_5)d + \bar{\nu}_e\gamma_\mu(1 + \gamma_5)\nu_e - \bar{e}\gamma_\mu(1 + \gamma_5)e \\ &+ \bar{\nu}\gamma_\mu(1 + \gamma_5)\nu_\mu - \bar{\mu}\gamma_\mu(1 + \gamma_5)\mu + \cdots]. \end{aligned} \tag{1.5.10}$$

Again, ignoring the chronological order of events, the processes arising from the neutral current piece of the Hamiltonian are those where charge does not change at each current vertex, i.e.,

$$\nu + p \to \nu + p,$$
$$\nu_\mu + e \to \nu_\mu + e,$$
$$\nu_e + e \to \nu_e + e. \tag{1.5.11}$$

These processes have been observed [20] and they confirm eqn. (1.5.9) as the correct structure for the weak Hamiltonian up to a piece, which we denote by H'_{wk}:

$$H'_{\mathrm{wk}} = -\frac{4G_F}{\sqrt{2}} k\bar{\nu}\gamma_\mu(1 + \gamma_5)\nu J_\mu^{\mathrm{em}}. \tag{1.5.12}$$

If we temporarily ignore eqn. (1.5.12) we find that the weak interaction Hamiltonian has all the right properties to incorporate an underlying local weak $SU(2)_L$ symmetry. Under this weak $SU(2)_L$ symmetry, the left-handed fermions must transform as doublets as follows:

$$\begin{pmatrix} \nu_{e_L} \\ e_L^- \end{pmatrix} \begin{pmatrix} u_L \\ d_L \cos\theta_C + s_L \sin\theta_C \end{pmatrix}$$
$$\begin{pmatrix} \nu_{\mu_L} \\ \mu_L^- \end{pmatrix}. \tag{1.5.13}$$

If we write down eqn. (1.5.9) using the doublets given in eqn. (1.5.13) the charged current piece of the weak Hamiltonian is correctly reproduced. But as far as the neutral current piece goes, however, it gives rise to interactions that change strangeness, i.e.,

$$H_{\Delta S=1}^{\Delta Q=0} = \frac{G_F}{\sqrt{2}} \cos\theta_C \sin\theta_C \bar{d}\gamma_\alpha(1 + \gamma_5)s\bar{\mu}\gamma_\alpha(1 + \gamma_5)\mu + \cdots. \tag{1.5.14}$$

This will lead to the process $K_L^0 \to \mu^+\mu^-$ with a lifetime comparable to $K^+ \to \mu^+\nu$; but, experimentally, it is known that

$$\frac{\Gamma(K_L^0 \to \mu^+\mu^-)}{\Gamma(K^+ \to \mathrm{all})} \simeq (9.1 \pm 1.9) \times 10^{-9}. \tag{1.5.15}$$

This implies that the strangeness changing neutral current piece must be absent from the weak Hamiltonian. It was suggested by Glashow, Illiopoulos, and Maiani [5] that a fourth quark (called charm quark, c) with properties similar to the up-quark should be introduced and that it must belong to a new weak doublet

$$\begin{pmatrix} c_L \\ -d_L \sin\theta_C + s_L \cos\theta_C \end{pmatrix}.$$

Then the $\Delta S = 1$, $\Delta Q = 0$ piece of the Hamiltonian disappears and the $K_L \to \mu^+\mu^-$ puzzle is solved. In fact, in 1974, the charm quark was discovered [21] and it is now well established that the weak interactions of the charm quark are given by the doublet structure given above.

With the discovery of two new leptons (ν_τ, τ^-) [22] and a new quark b [23], the complete weak $SU(2)_L$ doublet structure, including the effects of all three generations, can be written as follows:

$$\begin{pmatrix} u_L \\ d'_L \end{pmatrix}\begin{pmatrix} c_L \\ s'_L \end{pmatrix}\begin{pmatrix} t_L \\ b'_L \end{pmatrix},$$
$$\begin{pmatrix} \nu_{eL} \\ e_L^- \end{pmatrix}\begin{pmatrix} \nu_{\mu L} \\ u_L^- \end{pmatrix}\begin{pmatrix} \nu_{\tau L} \\ \tau_L^- \end{pmatrix}, \tag{1.5.16}$$

where

$$\begin{pmatrix} d'_L \\ s'_L \\ b'_L \end{pmatrix} = U_{\mathrm{KM}}(\theta_1, \theta_2, \theta_3, \delta)\begin{pmatrix} d_L \\ s_L \\ b_L \end{pmatrix}. \tag{1.5.17}$$

U_{KM} is a unitary matrix, known as the Kobayashi–Maskawa [24] matrix, and is parametrized by three real angles and a phase.

$$U_{\mathrm{KM}} = \begin{pmatrix} c_1 & -s_1 c_3 & -s_1 s_3 \\ s_1 c_2 & c_1 c_2 c_2 - s_2 s_3 e^{i\delta} & c_1 c_2 s_3 + s_2 c_3 e^{i\delta} \\ s_1 s_2 & c_2 s_2 c_3 + c_2 s_3 e^{i\delta} & c_1 s_2 s_3 - c_2 c_3 e^{i\delta} \end{pmatrix}. \tag{1.5.18}$$

As of this writing, the t-quark remains to be discovered. Its discovery will complete the third generation of quarks and leptons.

Before closing this section we note that, while having convincingly argued for describing weak interactions by a local $SU(2)_L$ symmetry theory, we realize that these interactions are short range, and therefore, we must give mass to the gauge bosons without destroying the gauge invariance of the Lagrangian. Note that we cannot use confinement arguments, as in the case of QCD, since ν, e^- are observed particles. This will be dealt with in the next chapter where we seek different realizations of the symmetries present in a Lagrangian.

References

[1] For a discussion of weak interactions, see
R. E. Marshak, Riazuddin, and C. Ryan, *Theory of Weak Interactions in Particle Physics*, Wiley, New York, 1969;
E. D. Commins, *Weak Interactions*, McGraw-Hill, New York, 1973;
J. J. Sakurai, *Currents and Mesons*, University of Chicago Press, Chicago, 1969.

[2] M. Gell-Mann, *Phys. Lett.* **8**, 214 (1964);
G. Zweig, CERN preprint 8182/Th. 401 (1964), reprinted in *Developments in Quark Theory of Hadrons*, Vol. 1, 1964–1978 (edited by S. P. Rosen and D. Lichtenberg), Hadronic Press, MA., 1980.

[3] Some general references on recent developments in gauge theories are
J. C. Taylor, *Gauge Theories of Weak Interactions*, Cambridge University Press, Cambridge, 1976;
C. Quigg, *Gauge Theories of the Strong, Weak, and Electromagnetic Interactions*, Benjamin-Cummings, New York, 1983;
R. N. Mohapatra and C. Lai, *Gauge Theories of Fundamental Interactions*, World Scientific, Singapore, 1981;
A. Zee, *Unity of Forces in Nature*, World Scientific, Singapore, 1983;
M. A. Beg and A. Sirlin, *Phys. Rep.* **88**, 1 (1982); and *Ann. Rev. Nucl. Sci.* **24**, 379 (1974);
E. S. Abers and B. W. Lee, *Phys. Rep.* **9C**, 1 (1973);
T. P. Cheng and L. F. Li, *Gauge Theory of Elementary Particle Physics*, Oxford University Press, New York, 1984;
P. Langacker, *Phys. Rep.* **72C**, 185 (1981);
G. G. Ross, *Grandunified Theories*, Benjamin-Cummings, 1985.
L. B. Okun, *Leptons and Quarks*, North-Holland, Amsterdam, 1981.
R. N. Mohapatra, *Fortsch. Phys.* **31**, 185 (1983).
A. Masiero, D. Nanopoulos, C. Kounas and K. Olive, *Grandunification and Cosmology*, World Scientific, Singapore, 1985.

[4] O. W. Greenberg, *Phys. Rev. Lett.* **13**, 598 (1964);
M. Y. Han and Y. Nambu, *Phys. Rev.* **139**, B1006 (1965);
H. Fritzsch, M. Gell-Mann, and H. Leutweyler, *Phys. Lett.* **478**, 365 (1973).

[5] A. Gamba, R. E. Marshak, and S. Okubo, *Proc. Nat. Acad. Sci.* (*USA*) **45**, 881 (1959);
J. D. Bjorken and S. L. Glashow, *Phys. Lett.* **11**, 255 (1965);
S. L. Glashow, J. Illiopoulos, and L. Maiani, *Phys. Rev.* **D2**, 1285 (1970).

[6] J. C. Pati and A. Salam, *Phys. Rev.* **D10**, 275 (1974);
H. Georgi and S. L. Glashow, *Phys. Rev. Lett.* **32**, 438 (1974).

[7] M. Gell-Mann, *Physics* **1**, 63 (1964).

[8] S. Ll Adler, *Phys. Rev. Lett.* **14**, 1051 (1965);
W. I. Weisberger, *Phys. Rev. Lett.* **14**, 1047 (1965).

[9] C. N. Yang and R. L. Mills, *Phys. Rev.* **96**, 191 (1954).

[10] R. Shaw, Problem of Particle Types and Other Contributions to the Theory of Elementary Particles Ph.D. Thesis, Cambridge University, 1955.

[11] O. W. Greenberg, *Phys. Rev. Lett.* **13**, 598 (1964);
M. Y. Han and Y. Nambu, *Phys. Revs.* **139**, B1006 (1965).
For a review and references on the subject, see
O. W. Greenberg and C. A. Nelson, *Phys. Rep.* **32**, 69 (1977);
W. Marciano and H. Pagels, *Phys. Rep.* **36C**, 137 (1978).

[12] J. D. Bjorken, *Phys. Rev.* **179**, 1547 (1969).

[13] R. Feynman, *Photon–Hadron Interactions*, Benjamin, Reading, MA., 1972.

[14] J. D. Bjorken and E. A. Paschos, *Phys. Rev.* **185**, 1975 (1969).

[15] K. Wilson, *Phys. Rev.* **179**, 1499 (1969);
R. Brandt and G. Preparata, Nucl. Phys. **27B**, 541 (1971);
H. Fritzsch and M. Gell-Mann, *Proceedings of the Coral Gables Conference*, 1971;
Y. Frishman, *Phys. Rev. Lett.* **25**, 966 (1970).

[16] C. G. Callan, *Phys. Rev.* **D2**, 1541 (1970);
K. Symanzik, *Comm. Math. Phys.* **18**, 227 (1970).

[17] D. Gross and F. Wilczek, *Phys. Rev. Lett.* **30**, 1343 (1973);
H. D. Politzer, *Phys. Rev. Lett.* **30**, 1346 (1973).

[18] K. Wilson, *Phys. Rev.* **D3**, 1818 (1971).

[19] Most recent determination of $\sin\theta_C$ is by
M. Bourquin *et al.*, *Z. Phys.* **C** (to appear).

For a review, see
L. L. Chau, *Phys. Rep.* **95C**, 1 (1983).

[20] F. Hasert *et al.*, *Phys. Lett.* **46B**, 121 (1973);
B. Aubert *et al.*, *Phys. Rev. Lett.* **32**, 1457 (1974);
A. Benvenuti *et al.*, *Phys. Rev. Lett.* **32**, 800 (1974);
B. Barish *et al.*, *Phys. Rev. Lett.* **32**, 1387 (1974).

[21] J. J. Aubert *et al.*, *Phys. Rev. Lett.* **33**, 1404 (1974);
J.-E. Augustin *et al.*, *Phys. Rev. Lett.* **33**, 1406 (1974);
G. Goldhaber *et al.*, *Phys. Rev. Lett.* **37**, 255 (1976);
I. Peruzzi *et al.*, *Phys. Rev. Lett.* **37**, 569 (1976);
E. G. Cazzoli *et al.*, *Phys. Rev. Lett.* **34**, 1125 (1975).

[22] For a recent survey, see
M. Perl, *Physics in Collison*, Vol. 1 (edited by W. P. Trower and G. Bellini), Plenum, New York, 1982.

[23] S. W. Herb *et al.*, *Phys. Rev. Lett.* **39**, 252 (1977);
W. R. Innes *et al.*, *Phys. Rev. Lett.* **39**, 1240 (1977);
C. W. Darden *et al.*, *Phys. Lett.* **76B**, 246 (1978);
Ch. Berger *et al.*, *Phys. Lett.* **76B**, 243 (1978).

[24] M. Kobayashi and T. Maskawa, *Prog. Theor. Phys.* **49**, 652 (1973).

CHAPTER 2

Spontaneous Symmetry Breaking, Nambu–Goldstone Bosons, and the Higgs Mechanism

§2.1. Symmetries and Their Realizations

A Lagrangian for a physical system may be invariant under a given set of symmetry [1] transformations; but how the symmetry is realized in nature depends on the properties of the ground state. In field theories the ground state is the vacuum state. We will, therefore, have to know how the vacuum state responds to symmetry transformations.

Let U(ε) denote the unitary representations of the symmetry group in the Fock space of a given field theory. Under the symmetry transformations by parameters ε, a field ϕ transforms as follows:

$$\phi_a \to \phi'_a = \mathrm{U}(\varepsilon)\phi_a \mathrm{U}^{-1}(\varepsilon) = C_{ab}\phi_b \tag{2.1.1}$$

and

$$\mathscr{L}(\phi', \partial_\mu \phi') = \mathscr{L}(\phi, \partial_\mu \phi). \tag{2.1.2}$$

The vacuum state $|0\rangle$ may or may not be invariant under U(ε). Let us consider both these cases.

Case (i): $\mathrm{U}(\varepsilon)|0\rangle = |0\rangle$. This is known as the Wigner–Weyl mode of symmetry realization. We can show that, in this case, there exists mass degeneracy between particles in a supermultiplet as well as relations between coupling constants and scattering amplitudes. We will illustrate the technique of obtaining mass degeneracy.

The mass of a particle is defined as the position of the pole in the variable p^2 in the propagation function $\Delta_{aa}(p)$ for the corresponding field ϕ_a defined as follows:

$$\Delta_{aa}(p) = i \int e^{ip\cdot x}\, d^4x\, \langle 0|\, T(\phi_a(x)\phi_a(0))\,|0\rangle. \tag{2.1.3}$$

We can represent the function near its pole as

$$\Delta_{aa}(p) = \frac{Z_a}{p^2 + m_a^2} + \text{terms regular at } p^2 = -m_a^2. \tag{2.1.4}$$

Similarly, for the field ϕ_b we can write

$$\Delta_{bb}(p) = \frac{Z_b}{p^2 + m_b^2} + \text{terms regular at } p^2 = -m_b^2, \tag{2.1.5}$$

where $Z_{a,b}$ denote the wave function renormalization. If ϕ_a and ϕ_b belong to the same supermultiplet they can be transformed into one another by a symmetry transformation $U_{ab}(\varepsilon)$, i.e.,

$$U_{ab}(\varepsilon)\phi_a U_{ab}^{-1}(\varepsilon) = \phi_b. \tag{2.1.6}$$

Since $U(\varepsilon)|0\rangle = |0\rangle$, this implies $\Delta_{aa}(p) = \Delta_{bb}(p)$, i.e., the two functions are identical. This in turn implies that $m_a = m_b$ and $Z_a = Z_b$.

Similar techniques can be used to relate coupling constants and scattering amplitudes which are derived by looking at the poles of three- and four-point functions of the fields ϕ_a.

Case (ii): $U(\varepsilon)|0\rangle \neq |0\rangle$. This is known as the Nambu–Goldstone realization of the symmetry. We will show that, in this case, the spectrum of particles in the theory must contain a zero-mass particle, known as the Nambu–Goldstone boson.

To prove this we choose an infinitesimal symmetry transformation $U(\varepsilon)$ which can be expanded as follows:

$$U(\varepsilon) = 1 + i\varepsilon_j Q_j, \tag{2.1.7}$$

Q_j are the generators of symmetry. Noninvariance of vacuum then implies

$$Q_j|0\rangle \neq 0 \qquad \text{(for at least one } j\text{)}. \tag{2.1.8}$$

Noether's theorem tells us that

$$\frac{dQ_j}{dt} = 0. \tag{2.1.9}$$

Equation (2.1.8) implies that there exists at least one state $|m\rangle$ in the Hilbert space for which

$$\langle m|Q_j|0\rangle \neq 0. \tag{2.1.10}$$

Equation (2.1.9) leads to

$$\langle m|\frac{dQ_j}{dt}|0\rangle = 0. \tag{2.1.11}$$

Using translation invariance we can rewrite eqn. (2.1.11) as

$$E_m\delta^3(\mathbf{p}_m)\langle m|Q_j|0\rangle = 0. \tag{2.1.12}$$

Equation (2.1.12) implies that

$$\lim_{\mathbf{p}_m \to 0} E_m = 0. \tag{2.1.13}$$

This means that there is a massless, spin 0 particle in the theory. Let us now recapitulate the three conditions required to prove this fundamental theorem of particle physics:

(a) Conserved current $\partial_\mu J_\mu = 0$, corresponding to an exact symmetry of the Lagrangian.

(b) $Q_j|0\rangle \neq 0$. This condition can also be written as

$$\langle 0|[Q, \phi]|0\rangle \neq 0, \tag{2.1.14}$$

where ϕ is a local bosonic operator in the theory or as $\langle\phi\rangle \neq 0$.

(c) Lorentz invariance or locality. For instance, without the locality property, we cannot obtain eqn. (2.1.12) from eqn. (2.1.11).

We will not go into any more detailed discussion of the Goldstone theorem, except to note that the number of Goldstone bosons is precisely the same as the number of charges that do not annihilate the vacuum (or the number of broken symmetry generators).

Example. We can give classical, quantum mechanical, as well as field-theoretical examples of systems for which the Lagrangian respects a symmetry whereas the ground state does not. A classical example is a bead sliding on a frictionless vertical circular wire rotating about its vertical diameter. Above a certain angular velocity the equilibrum (or ground state) corresponds to the bead in a position with azimuthal angle $\theta \neq 0$. This breaks the discrete symmetry corresponding to $\theta \to -\theta$. We leave it as an exercise to the reader to work out the Hamiltonian for this system and prove our assertion.

A well-known quantum mechanical example is the configuration of the ground state of the ozone (O^3) molecule. Instead of the three oxygen atoms occupying the vertices of an equilateral triangle, they occupy those of an isosceles triangle.

Let us now discuss a field-theoretical example, where a continuous U(1) symmetry is broken. Consider a scalar field ϕ with a nonzero U(1) charge

$$\phi \xrightarrow{\mathrm{U}(1)} e^{i\theta}\phi. \tag{2.1.15}$$

The Lagrangian

$$\mathscr{L} = -(\partial_\mu\phi)(\partial_\mu\phi^*) + \mu^2\phi^*\phi - \lambda(\phi^*\phi)^2. \tag{2.1.16}$$

is invariant under the above U(1) transformations. Let us choose $\mu^2 > 0$; $\lambda > 0$ as required for the Hamiltonian to have a lower bound. The minimum of the Hamiltonian corresponds to

$$\frac{\partial V}{\partial \phi} = 0, \tag{2.1.17}$$

which implies

$$\langle\phi\rangle^2 = \frac{\mu^2}{2\lambda} \neq 0. \tag{2.1.18}$$

This implies that the vacuum state is not invariant under the U(1) symmetry. Therefore, there must be a zero mass particle in the theory. If we write

$$\phi = \sqrt{\frac{\mu^2}{2\lambda}} + R + iG, \tag{2.1.19}$$

we can easily check that

$$\begin{aligned} M_G &= 0, \\ M_R &= \sqrt{2}\mu = 2\sqrt{\lambda}\langle\phi\rangle. \end{aligned} \tag{2.1.20}$$

Therefore, $G \equiv \operatorname{Im} \phi$ is the Nambu–Goldstone boson corresponding to spontaneous breaking of the U(1) symmetry.

§2.2. Nambu–Goldstone Bosons for an Arbitrary Non-abelian Group

Let G be an n-parameter group with generators θ_A, $A = 1, \ldots, n$. Let ϕ be an irreducible representation of G. Let $V(\phi)$ be the potential for a physical system which is invariant under the group G. To obtain the ground state, we have to look for minima of the potential

$$\left.\frac{\partial V}{\partial \phi_i}\right|_{\phi_i=\lambda_i} = 0 \tag{2.2.1}$$

and

$$\left.\frac{\partial^2 V}{\partial \phi_i\, \partial \phi_j}\right|_{\phi=\lambda} \equiv M_{ij}^2, \tag{2.2.2}$$

such that M^2 has positive or zero eigenvalues. The minima generally consist of a manifold of points in the ϕ-space obtained by application of the symmetry transformation G, i.e., $\lambda' = G\lambda$. Invariance of V under the group transformations imply

$$\frac{\partial V}{\partial \phi_i}(\theta_A)_{ij}\phi_j = 0. \tag{2.2.3}$$

Equation (2.2.3) is valid for all values of the field ϕ. Differentiating this equation with respect to k and setting $\phi = \lambda$, and using eqns. (2.2.1) and (2.2.2) we obtain

$$M_{ki}^2(\theta_A)_{ij}\lambda_j = 0. \tag{2.2.4}$$

This is the key equation in the study of properties of Nambu–Goldstone bosons. We see that, for those generators for which $(\theta_A)_{ij}\lambda_j \neq 0$, the M^2 matrix has a zero eigenvalue. Thus, the Goldstone boson eigenstates are given by $(\theta_A)_{ij}\lambda_j$. Those generators which satisfy $(\theta_A)_{ij}\lambda_j = 0$ represent the unbroken part of the symmetry group G and generate a symmetry group H which leaves both the vacuum and the Lagrangian invariant. The broken generators belong to the coset space G/H. (Coset space consists of those elements c of group G which satisfy $g = h_i c_j$ for $g \subset G$, $h_i \subset H$. What we have given is the right coset space; we could also define a left coset space in an obvious manner.)

We give a simple O(3) example that illustrates the connection between the zero-mass particles and the broken generators. The generators of the O(3) group can be written as

$$(\theta_A)_{ij} = i\varepsilon_{Aij}. \tag{2.2.5}$$

Consider a triplet representation of O(3), i.e.,

$$\boldsymbol{\phi} = \begin{pmatrix} \phi_1 \\ \phi_2 \\ \phi_3 \end{pmatrix}. \tag{2.2.6}$$

Choosing the potential $V(\varepsilon)$ as follows:

$$V(\boldsymbol{\phi}) = -\mu^2\boldsymbol{\phi}^2 + \lambda(\boldsymbol{\phi}^2)^2, \tag{2.2.7}$$

we see that for $\mu^2 > 0$ the vacuum state corresponds to $\langle\phi\rangle \neq 0$. In general, we could choose

$$\langle\phi\rangle = \begin{pmatrix} v_1 \\ v_2 \\ v_3 \end{pmatrix}. \tag{2.2.8}$$

But as we remarked before, all $\langle\phi\rangle$'s obtained from (2.2.8) by applications of O(3) transformations correspond to minima of the potential and are physically equivalent. So we can choose

$$\langle\phi\rangle = \begin{pmatrix} 0 \\ 0 \\ v \end{pmatrix}. \tag{2.2.9}$$

Using eqn. (2.2.5) we can now conclude that

$$\theta_1\langle\phi\rangle \neq 0 \quad \text{and} \quad \theta_2\langle\phi\rangle \neq 0,$$

but

$$\theta_3\langle\phi\rangle = 0. \tag{2.2.10}$$

Thus, there are two broken generators which reduce the O(3) group to O(2). Working out the mass matrix it is easy to see that $M_{\phi_1} = M_{\phi_2} = 0$; thus, ϕ_2-

and ϕ_1-fields are the Nambu–Goldstone boson eigenstates corresponding to the broken generators θ_1 and θ_2.

§2.3. Some Properties of Nambu–Goldstone Bosons

It was generally believed for a long time that there may be no real Nambu–Goldstone bosons in nature for the reason that their existence would give rise to long-range forces in classical physics, and would lead to new effects in scattering and decay processes. However, starting in 1980 when it was pointed out [2] that any possible nonrelativistic classical long-range forces arising from the existence of massless Goldstone particles are spin dependent and would therefore be extremely difficult to observe [3], the trend of thinking, on the physical reality of Goldstone bosons, has changed and serious studies have been conducted to detect long-range forces in nature [4]. In this section we present some general properties of the coupling of Nambu–Goldstone bosons to matter.

We would like to discuss two specific properties: Decoupling of Goldstone bosons from low-energy particles, and spin dependence of long-range forces. We will study the situation when the scale of symmetry breaking is much larger than the highest scales of low-energy physics ($\sim m_W$), i.e., $\Lambda \gg m_W$. (It is also possible [5] to have theories where the symmetry breaking scale is much lower than m_W, without conflicting with observations.)

To study the physics of decoupling phenomena we consider a simple model with a U(1)′ symmetry under which bosons (ϕ) and fermions (ψ_i) transform nontrivially. The Noether current corresponding to this symmetry is given by

$$J_\mu = i(\phi^* \partial_\mu \phi - \phi \partial_\mu \phi^*) + \bar{\psi}_i \gamma_\mu O_{ij} \psi_j. \tag{2.3.1}$$

We assume that the symmetry is spontaneously broken, i.e., $\langle \phi \rangle = \Lambda \neq 0$. We can then write

$$J_\mu = +\Lambda \partial_\mu G + \bar{\psi}_i \gamma_\mu O_{ij} \psi_j + \cdots, \tag{2.3.2}$$

where $G = -\sqrt{2}\ \mathrm{Im}\ \phi$ is the Nambu–Goldstone boson. Using the fact that $\partial_\mu J_\mu = 0$ and taking the fermion matrix element of (2.3.1), we find that the coupling of fermions to G, i.e., $f_{\psi\psi G}$ is given by

$$f_{\psi\psi G} \simeq \frac{q_\mu}{\Lambda} \bar{u}_i \gamma_\mu O_{ij} u_j. \tag{2.3.3}$$

Since $\bar{u}_i \gamma_\mu \mathrm{O}_{ij} u_j q_\mu$ is a low-energy parameter and is at most of order m_W, we find that $f_{\psi\psi G} \approx m_W/\Lambda \ll 1$. This is the phenomenon of decoupling, i.e., as $\Lambda \to \infty, f_{\psi\psi G} \to 0$.

As a specific aspect of the decoupling property we may ask the following question: The potential $V(\phi)$ involves all scalar fields and since the Goldstone boson G is made out of scalar fields, is it not obvious that G will not appear in $V(\phi)$ with an enhanced coupling to other physical scalar bosons of the theory? Below we will show that in the simple case of U(1) symmetries, $V(\phi)$ is indeed independent of G [6].

First, we consider the simple case discussed in eqn. (2.1.16). Let us parametrize the complex scalar field ϕ as follows:

$$\phi = \frac{1}{\sqrt{2}}(\Lambda + \rho)\, e^{iG/\Lambda}, \tag{2.3.4}$$

where $\Lambda = \sqrt{\mu^2/2\lambda}$. We then see that the potential V is independent of G, proving our assertion.

Let us now study a more complicated situation which involves a U(1) symmetric theory with several scalar fields ϕ_i with U(1) charges Q_i. The Lagrangian in this case can be written as

$$\mathscr{L} = \sum_i - (\partial_\mu \phi_i^*)(\partial_\mu \phi_i) - V(\phi_i), \tag{2.3.5}$$

$V(\phi_i)$ consists of products of monomials of fields ϕ_i. Let us consider a typical term

$$V_x = \prod_i \phi_i^{X_i}, \tag{2.3.6}$$

where X_i are numbers satisfying the relation following from group invariance

$$\sum_i Q_i X_i = 0. \tag{2.3.7}$$

Let the symmetry be broken by various fields ϕ_i acquiring vacuum expectation values (v.e.v.) Λ_i. If we parametrize $\phi_i = (1/\sqrt{2})(\Lambda_i + \rho_i)\exp(i\xi i/\Lambda_i)$, then eqn. (2.2.4) implies that the Nambu–Goldstone boson G is given by

$$G = \sum_i \Lambda_i \xi_i Q_i \Big/ \left(\sum_i \Lambda_i^2\right)^{1/2}. \tag{2.3.8}$$

Substituting ϕ_i in eqn. (2.3.6) we find that

$$V_x = \prod_i \frac{(\rho_i + \Lambda_i)}{\sqrt{2}} \exp\left(i \sum_j X_j \xi_j / \Lambda_j\right). \tag{2.3.9}$$

Using (2.3.7) it is easy to see that the combinations of ξ_i present in V_x are orthogonal to G. This holds for each term in the potential. Thus, V is independent of G.

Exercise

Prove that V is independent of G for a simple non-abelian group such as SU(2).

Let us now discuss the spin dependence of long-range forces resulting from the exchange of real Goldstone bosons. To study this let us look at eqns.

(2.3.2) and (2.3.3). We can see using the Dirac equation that

$$f_{\psi_i\psi_jG} = \frac{1}{\Lambda}\bar{\psi}_i(M_iO_{ij} - \varepsilon O_{ij}M_j)\psi_j, \tag{2.3.10}$$

where $\varepsilon = +1$ if γ_μ commutes with O (i.e., it has no γ_5) and -1 if it anti-commutes with O (i.e., it has γ_5 in it). We see that, if $\varepsilon = +1$, only the off-diagonal part of O_{ij} survives, i.e., the Nambu–Goldstone boson G couples two different fermions. If $\varepsilon = -1$ we can have diagonal couplings of G which could lead to long-range forces between matter. The off-diagonality, or the γ_5 nature of the Nambu–Goldenstone boson coupling to fermions, could be inferred from other considerations, too; for instance, note that Nambu–Goldstone bosons satisfy the property that $G \to G + C$ (where C is a constant) keeps the Lagrangian invariant. Therefore, the only way they can appear is through $\bar{\psi}_i\gamma_5\psi_jG$ $(i \neq j)$ coupling since, on translation $G \to G + C$, each of these interactions contribute a four-divergence to the Lagrangian and therefore vanish.

As we noted the diagonal couplings arise when $\varepsilon = -1$ corresponding to γ_5 couplings and it is well known from standard textbooks that the nonrelativistic limit of γ_5 gives spin-dependent forces as follows:

$$V_G(r) = \frac{f_{\psi\psi G}^2}{m^2r^3}(\boldsymbol{\sigma}_1 \cdot \boldsymbol{\sigma}_2 - 3(\boldsymbol{\sigma}_1 \cdot \hat{r})(\boldsymbol{\sigma}_2 \cdot \hat{r})). \tag{2.3.11}$$

The importance of the spin dependence is that for big "chunks" of matter the net spin is infinitesimal compared to the number of nucleons ($\sim$ Avagadro number) leading to tiny unobservable corrections to gravitational effects. The best bound on $f_{\psi\psi G}$ comes from atomic physics where shifts of energy levels can occur due to the presence of these long-range forces with large strength. From an experiment of Ramsey and Code [7] we deduce a bound on $f_{\psi\psi G} \lesssim 10^{-3}$.

§2.4. Phenomenology of Massless and Near-Massless Spin 0 Bosons

Since 1980 several interesting theoretical models have been proposed that contain massless [2, 4, 8] or near-massless [9] particles. One of the ways they could be detected is through the long-range forces generated by them. If they are Nambu–Goldstone bosons then their low-energy effect would be to induce spin-dependent forces. On the other hand, combining with CP-violation, the γ_5-couplings could, in principle, change to pure scalar interactions which would then lead to spin-independent, long-range forces. It may therefore be useful to summarize the present experimental limits on various long-range forces [3]:

Table 2.1. Limits on the strengths of new long-range forces from Eotvos, Cavendish, and atomic experiments are summarized in Ref. [3].

N	Limit on λ^{SI}	Limit on Λ^{SD}	Limit on λ^{T}
1	10^{-45} (Ref. [10])		10^{-16} (Ref. [7])
	10^{-47} (Ref. [11])		
2	10^{-23} (Ref. [10])	10^{-8} (Ref. [12])	10^{-11} (Ref. [7])
	10^{-20} (Ref. [11])		
3	10^{-2} (Ref. [10])	10^{6} (Ref. [12])	10^{-6} (Ref. [7])
	10^{7} (Ref. [11])		
	10^{-12} (Ref. [12])		
4			10^{-1} (Ref. [7])
5			10^{4} (Ref. [7])

Spin-Independent Forces

In this case we parametrize the potentials as follows:

$$V^{\mathrm{SI}} = \frac{\lambda_N^{\mathrm{SI}}}{r}\left(\frac{r_0}{r}\right)^{N-1}, \qquad r_0 = 200\ \mathrm{MeV}^{-1}. \tag{2.4.1}$$

Spin-Dependent Forces

$$V^{\mathrm{SD}} = \frac{\lambda_N^{\mathrm{SD}}}{r}\left(\frac{r_0}{r}\right)^{N-1} \mathbf{S}_1 \cdot \mathbf{S}_2. \tag{2.4.2}$$

Tensor Forces

$$V^{\mathrm{T}} = \frac{\lambda_N^{\mathrm{T}}}{r}\left(\frac{r_0}{r}\right)^{N-1} (\mathbf{S}_1 \cdot \mathbf{S}_2 - 3\mathbf{S}_1 \cdot \hat{r}\mathbf{S}_2 \cdot \hat{r}). \tag{2.4.3}$$

Finally, we wish to point out that the existence of Nambu–Goldstone bosons or ultralight bosons can effect astrophysical considerations by contributing new mechanisms for energy loss from stars. Consider, for instance, a typical Nambu–Goldstone boson with diagonal γ_5-coupling. Then the Feynman diagram in Fig. 2.1 provides a new mechanism for energy loss. It has

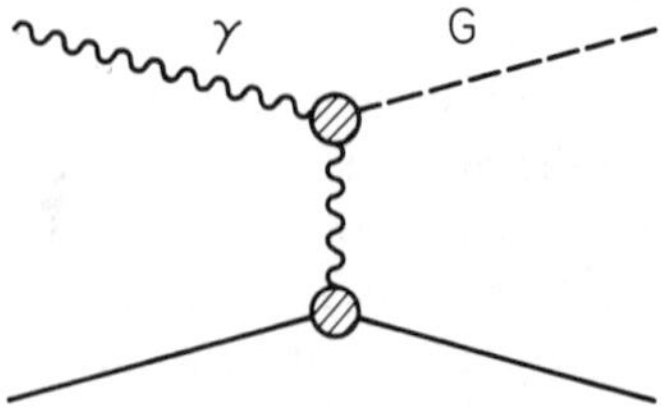

Figure 2.1

been argued [13] that the effective $\psi\psi G$ coupling must be limited by a cosmological known rate of energy loss from red giant stars [14] to be $f_{\psi\psi G} \leq 10^{-12}$ where $\psi = u, d, e$.

§2.5. The Higgs–Kibble Mechanism in Gauge Theories

In this section, we wish to study field theories where local symmetries are spontaneously broken by vacuum. We will see that, in these cases, the zero-mass particles do not appear in the physical spectrum of states, but rather provide longitudinal modes to the gauge fields which then become massive. This is, therefore, an extremely welcome situation where two kinds of massless particles, undesirable for physical purposes, combine to give a massive vector–meson state which may mediate short-range forces such as the weak forces. This is known as the Brout–Englert–Higgs–Guralnik–Hagen–Kibble mechanism and it forms the basis for the construction of unified gauge theories.

Before proceeding to study the physics of this mechanism, let us illustrate it with the help of a U(1) model. We consider a Lagrangian for a complex scalar field ϕ with nonzero U(1) charge invariant under local U(1) transformations

$$\phi \to e^{ie\theta(x)}\phi. \tag{2.5.1}$$

As we discussed in Chapter 1, we need to introduce a real spin 1 field A_μ which transforms under the gauge transformations as follows:

$$A_\mu \to A_\mu + \partial_\mu \theta. \tag{2.5.2}$$

The invariance Lagrangian is then given by

$$\mathscr{L} = -\tfrac{1}{4}F_{\mu\nu}F_{\mu\nu} - (D_\mu\phi)^*(D_\mu\phi) + \mu^2\phi^*\phi - \lambda(\phi^*\phi)^2, \tag{2.5.3}$$

where

$$\begin{aligned} F_{\mu\nu} &= \partial_\mu A_\nu - \partial_\nu A_\mu, \\ D_\mu\phi &= (\partial_\mu - ieA_\mu)\phi. \end{aligned} \tag{2.5.4}$$

If we look at the potential in eqn. (2.5.3) we see that the ground state breaks the U(1) symmetry, i.e.,

$$\langle\phi\rangle = \frac{v}{\sqrt{2}} = \sqrt{\frac{\mu^2}{2\lambda}} \neq 0. \tag{2.5.5}$$

If we parametrize

$$\phi = \frac{1}{\sqrt{2}}(v + \rho)\, e^{iG/V}, \tag{2.5.6}$$

then,

$$D_\mu \phi = \frac{1}{\sqrt{2}} e^{iG/v}(\partial_\mu \rho - ie(\rho + v)B_\mu), \tag{2.5.7}$$

where

$$B_\mu = A_\mu - \frac{1}{ev}\partial_\mu G. \tag{2.5.8}$$

We can rewrite $\mathscr{L}$ as

$$\mathscr{L} = -\tfrac{1}{4}B_{\mu\nu}B_{\mu\nu} - \tfrac{1}{2}(\partial_\mu \rho)^2 - \tfrac{1}{2}e^2(\rho^2 + 2\rho v)B_\mu^2 - \tfrac{1}{2}M_B^2 B_\mu^2 - V(\rho), \tag{2.5.9}$$

where $B_{\mu\nu} = \partial_\mu B_\nu - \partial_\nu B_\mu$.

It is clear, from eqn. (2.5.9), that the massless fields A_μ and G have disappeared and a new massive vector field B_μ has appeared in their place. This phenomenon occurs in more general cases with non-abelian symmetries.

At this stage we digress to understand the physics behind this mechanism. We saw in Section 2.1 that there are three conditions needed to prove the Goldstone theorem, i.e., current conservation, vacuum noninvariance, and Lorentz invariance. Once we couple the gauge fields, to study the theory, we must fix gauge and, of course, there are various ways to fix gauge: e.g., radiation gauge, i.e., $\nabla \cdot \mathbf{A} = 0$ or Lorentz gauge $\partial_\mu A_\mu = 0$, etc. If we work in radiation gauge then gauge fixing destroys explicit Lorentz invariance. Therefore, strictly, the third condition needed to prove the Goldstone theorem does not exist and thus no Nambu–Goldstone bosons appear. On the other hand, if we work in Lorentz gauge, all three conditions for the Goldstone theorem are satisfied. Therefore, there must be a massless particle in the theory. However, in this case, the Hilbert space contains more states than are physically acceptable, for instance, states with zero or negative norm. So, the physical subspace of the Hilbert space has to be projected out. It can be shown that the zero mass particle, as well as the massless gauge fields, remain in the unphysical sector of the Hilbert space; only their linear combination (which gives B_μ) has excitations in the physical subspace.

We illustrate this with the following very simple model involving a real scalar field ϕ and gauge field A_μ:

$$\mathscr{L} = -\tfrac{1}{4}F_{\mu\nu} - \tfrac{1}{2}(\partial_\mu \phi - gA_\mu)^2. \tag{2.5.10}$$

This Lagrangian is invariant under the following local symmetry:

$$\begin{aligned} \phi &\to \phi + g\lambda(x), \\ A_\mu &\to A_\mu + \partial_\mu \lambda(x). \end{aligned} \tag{2.5.11}$$

The current that generates this symmetry is

$$J_\mu = \partial_\mu \phi - gA_\mu, \tag{2.5.12}$$

which is conserved as a result of the field equations. Since the canonical momentum $\pi_\phi = iJ_4$, equal-time canonical commutation relations imply that

$$[Q, \phi] = 1, \tag{2.5.13}$$

i.e., the $\langle 0|[Q, \phi]|0\rangle = 1$ and the symmetry is spontaneously broken. The Higgs phenomena at the Lagrangian level is obtained by introducing the new vector–meson field, $B_\mu = (A_\mu - (1/g)\partial_\mu\phi)$, which now has mass g.

To illustrate the disappearance of the Goldstone boson in Lorentz gauge, we add Lagrangian multiplier terms to $\mathscr{L}$, i.e.,

$$\mathscr{L}' = -A_\mu\partial_\mu B - \tfrac{1}{2}B^2. \tag{2.5.14}$$

The field equations are

$$A_\mu\text{-variation:}\quad \partial_\mu F_{\mu\nu} = \partial_\nu B - g(\partial_\nu\phi - gA_\nu), \tag{2.5.15}$$

$$\phi\text{-variation:}\quad \partial_\mu(\partial_\mu\phi - gA_\mu) = 0, \tag{2.5.16}$$

$$B\text{-variation:}\quad \partial_\mu A_\mu = B. \tag{2.5.17}$$

From these equations we obtain

$$\Box B = 0 \tag{2.5.18}$$

and

$$\Box\phi = gB. \tag{2.5.19}$$

Equation (2.5.18) says that B is a free field and therefore we can make plane wave decomposition and the Gupta–Bleuler gauge condition can be imposed as

$$B^{(+)}|\psi_{\text{phys}}\rangle = 0. \tag{2.5.20}$$

Thus, eqn. (2.5.20) defines the physical subspace of the Hilbert space. An operator that does not commute with B will lead to excitations in the unphysical part of the Hilbert space. All we have to show now is that A_μ and the Goldstone boson field ϕ do not commute with B but that $B_\mu = A_\mu - (1/g)\partial_\mu\phi$ does.

To do this, let us write down the canonical commutation relations for various fields

$$\pi_{A_i} = F_{0i}, \qquad \pi_{A_0} = B, \qquad \pi_\phi = J_0. \tag{2.5.21}$$

From the commutation (equal-time) relations

$$[A_0(x), B(y)]_{x_0=y_0} = i\delta^3(x - y) \tag{2.5.22}$$

and

$$[\pi_{A_i}(x), B(y)]_{x_0=y_0} = 0, \tag{2.5.23}$$

it follows that

$$[\partial_t A_i(x), B(y)]_{x_0=y_0} = i\partial_i\delta^3(x - y). \tag{2.5.24}$$

This then implies that at arbitrary times

$$[B(x), A_\mu(y)] = -i\partial_\mu D(x-y). \tag{2.5.25}$$

Using eqn. (2.5.17) it follows that

$$[B(x), B(y)] = 0. \tag{2.5.26}$$

This condition is important if the physical subspace has to be "stable."

Furthermore,

$$[\pi_\phi(x), \phi(y)]_{x_0=y_0} = i\delta^3(x-y). \tag{2.5.27}$$

This implies that (using (2.5.15))

$$\frac{1}{g}[\partial_0 B(x), \phi(y)]_{x_0=y_0} = -i\delta^3(x-y). \tag{2.5.28}$$

For eqn. (2.5.19) to be consistent with eqn. (2.5.23) we require

$$[B(x), \phi] = -igD(x-y). \tag{2.5.29}$$

We therefore see that neither A_μ nor ϕ are physical operators but that $B_\mu \equiv A_\mu - (1/g)\partial_\mu\phi$ commutes with B and is, therefore, in the physical subspace of states. This proves our assertion.

§2.6. Group Theory of the Higgs Phenomenon

In this section we briefly discuss the Higgs mechanism for the case of general non-abelian local symmetries. Consider the local symmetry group to be G with N generators θ_A. We can choose a basis so that θ_A are imaginary and antisymmetric, i.e.,

$$\theta_A = -\theta_A^{\mathrm{T}} = -\theta_A^* \tag{2.6.1}$$

and satisfy the following commutation relations:

$$[\theta_A, \theta_B] = if_{ABC}\theta_C. \tag{2.6.2}$$

Let V_μ^A be the gauge fields associated with this group and let ϕ be a scalar multiplet used to implement spontaneous symmetry breaking. In the basis we use, ϕ is real. The Lagrangian invariant under gauge transformations can be written as (see Chapter 1)

$$\mathscr{L} = -\tfrac{1}{4}f_{\mu\nu,A}f_{\mu\nu,A} - \tfrac{1}{2}D_\mu\phi^{\mathrm{T}}D_\mu\phi - V(\phi^{\mathrm{T}}\phi), \tag{2.6.3}$$

where

$$D_\mu\phi = (\partial_\mu - i\theta_A V_{\mu,A})\phi.$$

By appropriate choice of $V(\phi)$ we can have a minimum of V which breaks the symmetry G

$$\langle\phi\rangle \neq 0. \tag{2.6.4}$$

Writing

$$\phi = \langle\phi\rangle + \phi' \tag{2.6.5}$$

and substituting it in eqn. (2.6.3) we see that the gauge fields acquire a mass matrix

$$(M^2)_{AB} = +(\langle\phi\rangle^{\mathrm{T}}, \theta_A\theta_B\langle\phi\rangle). \tag{2.6.6}$$

Note that for those generators that annihilate $\langle\phi\rangle$, the mass matrix has zeros. Diagonalizing this matrix we can find the massive and massless gauge bosons. For each symmetry generator that is broken by vacuum, the corresponding gauge field becomes massive, as is clear from (2.6.6). Thus, the Higgs phenomenon generalizes to arbitrary non-abelian groups.

Here we state one group-theoretical result due to Bludman and Klein [15], which throws light on the nature of the symmetry breaking in simple instances (like simple groups). Suppose a Higgs field has n real components and let us suppose v of them acquire v.e.v.s. Then the Bludman–Klein theorem states that

$$n - v = \text{dimension of the coset space} = N - M \tag{2.6.7}$$

if M is the number of generators of the subgroup that remains unbroken.

§2.7. Renormalizability and Triangle Anomalies

All the techniques for unified gauge theories have been laid out. The important reason for the attractiveness of spontaneously broken gauge theories for model building is the property of renormalizability. We saw in Chapter 1 that gauge invariance provides a natural explanation for universality of coupling strengths, as is observed in the case of weak interactions. The Higgs–Kibble mechanism provides a way to generate masses of the gauge bosons while at the same time maintaining the freedom to gauge transform the fields. This freedom would be lost if the masses for gauge fields were put in from outside. It was shown by 't Hooft [16] that this gauge freedom can be exploited to pass to a gauge where the propagator for massive gauge fields W can be written as

$$\Delta^W_{\mu\nu}(k, \xi) = \frac{-i}{(2\pi)^4}\left[\frac{\delta_{\mu\nu} - k_\mu k_\nu/k^2}{k^2 + M^2 - i\varepsilon} + \frac{k_\mu k_\nu}{k^2(k^2/\xi - M^2)}\right], \tag{2.7.1}$$

where ξ is the gauge parameter.

As a result

$$\Delta_{\mu\nu}(k) \xrightarrow[k\to\infty]{} \frac{1}{k^2}, \tag{2.7.2}$$

which leads to a renormalizable theory if we keep in the Lagrangian only terms with mass dimension less than or equal to four. Recall that for a normal massive vector boson

$$\Delta^{W}_{\mu\nu}(k) = \frac{-i}{(2\pi)^4}\left(\frac{\delta_{\mu\nu} + k_\mu k\nu/M^2}{k^2 + M^2 - i\varepsilon}\right). \tag{2.7.3}$$

We see that, $\Delta^{W}_{\mu\nu}(k) \xrightarrow[k\to\infty]{} 1$. Therefore, the theory is apparently nonrenormalizable by power counting arguments. The important point is that for $\xi \to \infty$,

$$\Delta^{W}_{\nu\mu}(k, \xi) \xrightarrow[\xi\to\infty]{} \Delta^{w}_{\mu\nu}(k). \tag{2.7.4}$$

Thus, the gauge freedom enables us to express the massive vector boson propagator in a canonical form (the U-gauge). Since, in the R_ξ-gauge, renormalizability can be explicitly checked, gauge invariance implies that the theory in the U-gauge must also be renormalizable, despite appearance, due to the cancellation of different terms [17].

Another point, important in the construction of renormalizable theories, is the absence of triangle anomalies [18]. The point is that, if there is an exact local chiral symmetry in a Lagrangian (at the tree level), the current is exactly conserved, i.e., $\partial_\mu J_\mu = 0$. However, at the one-loop level, appearance of triangle graphs lead to breakdown of current conservation, i.e.,

$$\partial_\mu J_\mu = \frac{g^2}{16\pi^2}\varepsilon^{\mu\nu\lambda\sigma} f_{\mu\nu} f_{\lambda\sigma}. \tag{2.7.5}$$

The breakdown of Ward identities (eqn. (2.7.5)) is equivalent to the introduction of dimension five terms into the effective Lagrangian, which therefore spoils renormalizability. Since the weak currents are chiral, in building weak interaction models using gauge theories, we must ensure that by appropriate choice of the fermion dector or gauge group G the current conservation is maintained, i.e., anomalies vanish. If θ_A's are the generators of the gauge group (in the space of chiral fermions) the condition for the disappearance of anomalies is that

$$T_r[\theta_A\{\theta_B, \theta_C\}]_L - T_r[\theta_A\{\theta_B, \theta_C\}]_R = 0. \tag{2.7.6}$$

This is an additional constraint on unified gauge models [19].

References

[1] For an excellent discussion of and references on symmetries and spontaneously broken symmetries, see
S. Weinberg, Brandeis Lectures, 1970;
M. A. B. Beg, Lectures Notes in Mexico, 1971;
G. Guralnik, C. R. Hagen, and T. W. B. Kibble, *Advances in High-Energy Physics* (edited by R. Cool and R. E. Marshak), Wiley, New York, 1969.
R. Gatto, *A Basic Course in Modern Weak Interaction Theory*, Bologna preprint (1979) (unpublished).

[2] Y. Chikashige, R. N. Mohapatra, and R. Peccei, *Phys. Lett.* **98B**, 265 (1981).

[3] For a survey of known limits on long-range forces, see
G. Feinberg and J. Sucher, *Phys. Rev.* **D20**, 1717 (1979).
[4] G. Gelmini, S. Nussinov, and T. Yanagida, *Nucl. Phys.* **B219**, 31 (1983);
H. Georgi, S. L. Glashow, and S. Nussinov, *Nucl. Phys.* **B193**, 297 (1981);
J. Moody and F. Wilczek, *Phys. Rev.* **D30**, 130 (1984).
[5] G. Gelmini and M. Roncadelli, *Phys. Lett.* **99B**, 411 (1981).
[6] R. Barbieri, R. N. Mohapatra, D. V. Nanopoulos, and D. Wyler, *Phys. Lett.* **107B**, 80 (1981).
[7] N. Ramsey and R. F. Code, *Phys. Rev.* **A4**, 1945 (1971).
[8] R. Barbieri and R. N. Mohapatra, *Z. Phys.* **C.11**, 175 (1981);
F. Wilczek, *Phys. Rev. Lett.* **49**, 1549 (1982);
D. Reiss, *Phys. Lett.* **115B**, 217 (1982).
[9] J. E. Kim, *Phys. Rev. Lett.* **43**, 103 (1979);
M. Dine, W. Fischler, and M. Srednicki, *Phys. Lett.* **101B**, 199 (1981);
D. Chang, R. N. Mohapatra, S. Nussinov, *Phys. Rev. Lett.* **55**, 2835 (1985).
[10] R. V. Eotvos, D. Pekar, and E. Fekele, *Ann. Phys.* **68**, 11 (1922).
[11] V. B. Braginsky and V. I. Panov, *Sov. Phys. JETP* **34**, 464 (1972);
For other related experiments, see
H. J. Paik, *Phys. Rev.* **D19**, 2320 (1979);
H. J. Paik, H. A. Chan, and M. Moody, *Proceedings of the Third Marcel Grossmann Meeting on General Relativity*, 1983, p. 839;
R. Spero, J. K. Hoskins, R. Newman, J. Pellam, and J. Schultz, *Phys. Rev. Lett.* **44**, 1645 (1980).
[12] D. R. Long, *Phys. Rev.* **D9**, 850 (1974);
Y. Fujii and K. Mima, *Phys Lett.* **79B**, 138 (1978);
Nature **260**, 417 (1976).
[13] D. Dicus, E. Kolb, V. Teplitz, and R. Wagoner, *Phys. Rev.* **D18**, 1829 (1978).
[14] M. Fukugita, S. Watamura, and M. Yoshimura, *Phys. Rev. Lett.* **18**, 1522 (1982).
[15] S. Bludman and A. Klein, *Phys. Rev.* **131**, 2363 (1962).
[16] G. 't Hooft, *Nucl. Phys.* **33B**, 173 (1971).
[17] For a detailed discussion of renormalizability of Yang–Mills theories, see
E. S. Abers and B. W. Lee, *Phys. Rep.* **9C**, 1 (1973);
G. 't Hooft and M. Veltman, *Nucl. Phys.* **B44**, 189 (1972);
H. Kluberg-Stein and J. B. Zuber, *Phys. Rev.* **D12**, 467, 482, 3159 (1975);
C. Becchi, A. Rouet, and R. Stora, *Commun. Math. Phys.* **42**, 127 (1975);
J. C. Taylor, *Nucl. Phys.* **B33**, 436 (1971);
J. Zino-Justin, Lecture Notes, Bonn, 1974.
[18] S. Adler, *Phys. Rev.* **177**, 2426 (1969);
J. Bell and R. Jackiw, *Nuovo Cimento*, **51A**, 47 (1969);
W. Bardeen, *Phys. Rev.* **184**, 1848 (1969).
[19] D. Gross and R. Jackiw, *Phys. Rev.* **D6**, 477 (1972);
C. Bouchiat, J. Illiopoulos, and Ph. Meyer, *Phys Lett.* **38B**, 519 (1972).

CHAPTER 3

The $SU(2)_L \times U(1)$ Model

§3.1. The $SU(2)_L \times U(1)$ Model of Glashow, Weinberg, and Salam

In this section we will apply the ideas of spontaneously broken gauge theories to construct the first successful model of electro-weak interaction of quarks and leptons. As we discussed in the Introduction the observed universality of the four-Fermi coupling of weak-decay processes suggests the existence of a hidden symmetry of weak interactions, and the symmetry manifests itself not through the existence of degenerate multiplets but through broken local symmetries. The $SU(2)_L \times U(1)$ model of Glashow, Weinberg, and Salam [1] provides a realization of this idea in the framework of a renormalizable field theory, and the recent discovery of $W^{\pm}$- and Z-bosons in the proton–antiproton collider experiments [2] has proved the correctness of these ideas and given a boost to the study of spontaneously broken non-abelian gauge theories as the way to probe further into the structure of quark–lepton interactions. This will be explored in the subsequent sections.

To discuss the $SU(2)_L \times U(1)$ model, we will start by giving (on p. 37) the assignment of quarks and leptons to representations of the gauge group. We will denote, by subscript a, the various generations; and by i, the $SU(3)_c$ index on quarks. (We will consider three generations of fermions.)

To implement the spontaneous breaking of the $SU(2)_L \times U(1)_Y$ symmetry, we have to include scalar bosons (the Higgs bosons) into the model. We choose the Higgs bosons ϕ to transform as the simplest nontrivial representation, i.e., doublets under the gauge group $SU(2)_L \times U(1)_Y \times SU(3)_C$

$$\phi \equiv \begin{pmatrix} \phi^+ \\ \phi^0 \end{pmatrix}, \qquad (2, +1, 1). \tag{3.1.1}$$

	$SU(2)_L \times U(1)_Y \times SU(3)_c$ representation	
$\psi_{a,L}^{(0)} \equiv \begin{pmatrix} \nu_a \\ E_a^- \end{pmatrix} L,$	$(2, -1, 1),$	
$E_{a,R}^{-(0)},$	$(1, -2, 1),$	
$Q_{a,iL} \equiv \begin{pmatrix} u_a^{(0)} \\ d_a^{(0)} \end{pmatrix} iL,$	$(2, \frac{1}{3}, 3),$	
$u_{a,iR}^{(0)},$	$(1, \frac{4}{3}, 3),$	
$d_{a,iR}^{(0)},$	$(1, -\frac{2}{3}, 3).$	(3.1.2)

(The superscript zero denotes the fact that the quarks are eigenstates of weak interactions but not of the mass matrices to be generated subsequent to spontaneous breaking.)

Except for the Higgs potential part and the Higgs–Yukawa couplings the rest of the Lagrangian is completely dictated by the requirements of gauge invariance and renormalizability, i.e., it is simply the gauge invariant kinetic term for the above fields $\psi_L^{(0)}$, $E_R^{(0)}$, $Q_L^{(0)}$, $U_R^{(0)}$, $d_R^{(0)}$, ϕ:

$$\mathscr{L} = \mathscr{L}_{\text{kin}}^{\text{matt}} + \mathscr{L}_{\text{kin}}^{\text{gauge}} - V(\phi) + \mathscr{L}_Y, \tag{3.1.3}$$

where

$$\begin{aligned} \mathscr{L}_{\text{kin}}^{\text{matt}} = & -\bar{Q}_L^{(0)}\gamma_\mu\left(\partial_\mu - \frac{ig}{2}\boldsymbol{\tau}\cdot\mathbf{W}_\mu - \frac{ig'}{6}B_\mu\right)Q_L^{(0)} \\ & -\bar{\psi}_L^{(0)}\gamma_\mu\left(\partial_\mu - \frac{ig}{2}\boldsymbol{\tau}\cdot\mathbf{W}_\mu + \frac{ig'}{2}B_\mu\right)\psi_L^{(0)} \\ & -\bar{E}_R^{(0)}\gamma_\mu(\partial_\mu + ig'B_\mu)E_R^{(0)} - \bar{U}_R^{(0)}\gamma_\mu\left(\partial_\mu - \frac{2i}{3}g'B_\mu\right)U_R^{(0)} \\ & -\bar{d}_R^{(0)}\gamma_\mu\left(\partial_\mu + \frac{i}{3}g'B_\mu\right)d_R^{(0)} - \left|\left(\partial_\mu\phi - \frac{ig}{2}\boldsymbol{\tau}\cdot\mathbf{W}_\mu\phi - \frac{ig'}{2}B_\mu\phi\right)\right|^2, \end{aligned} \tag{3.1.4}$$

where the color and generation indices are suppressed (and are summed over)

$$\mathscr{L}_{\text{kin}}^{\text{gauge}} = -\tfrac{1}{4}\mathbf{W}_{\mu\nu}\cdot\mathbf{W}_{\mu\nu} - \tfrac{1}{4}B_{\mu\nu}B_{\mu\nu}, \tag{3.1.5}$$

where

$$\mathbf{W}_{\mu\nu} = \partial_\mu\mathbf{W}_\nu - \partial_\nu\mathbf{W}_\mu + g\mathbf{W}_\mu \times \mathbf{W}_\nu.$$

Before discussing weak interactions, let us study the breakdown of $SU(2)_L \times U(1)_Y$ symmetry to $U(1)_{\text{em}}$, which is the only observed exact local symmetry in nature. For this purpose we choose the Higgs potential $V(\phi)$ as follows:

$$V(\phi) = -\mu^2\phi^+\phi + \lambda(\phi^+\phi)^2, \qquad \mu^2 > 0. \tag{3.1.6}$$

As noted earlier, the minimum of V corresponds to

$$\langle \phi \rangle = 1/\sqrt{2}\binom{0}{v}, \tag{3.1.7}$$

where

$$v = \frac{\mu}{\sqrt{\lambda}}. \tag{3.1.8}$$

It is easily checked that

$$\tfrac{1}{2}(\tau_3 + Y)\langle \phi \rangle = 0, \tag{3.1.9}$$

which, therefore, is the unbroken generator. We will call this "electric charge" and we get the formula

$$Q = I_{3W} + \frac{Y}{2}. \tag{3.1.10}$$

Y is a free parameter of the theory and we have adjusted it appropriately (anticipating eqn. (3.1.10)) so that the electric charges of the quarks and leptons come out correctly.

All other electro-weak generators are broken by vacuum. It therefore follows from our previous discussion (Chapter 2) that the gauge bosons corresponding to those generators, to be called $W^{\pm}$ and Z, pick up mass. We now calculate their masses.

Substituting eqn. (3.1.7) in the last term of eqn. (3.1.4) we isolate the following mass terms involving the gauge bosons:

$$\mathscr{L}_{\text{mass}} = -\tfrac{1}{4}g^2 v^2 W_\mu^+ W_\mu^- - \tfrac{1}{8}v^2(gW_{3\mu} - g'B_\mu)^2. \tag{3.1.11}$$

This implies that

$$m_W = \frac{gv}{2}, \tag{3.1.12}$$

and one linear combination of the two neutral gauge bosons, i.e.,

$$\frac{1}{\sqrt{g^2 + g'^2}}(gW_{3\mu} - g'B_\mu) \equiv Z_\mu \tag{3.1.13}$$

picks up the mass

$$m_Z = \tfrac{1}{2}(g^2 + g'^2)^{1/2} v. \tag{3.1.14}$$

The massless orthogonal combination

$$A_\mu = \frac{g'W_{3\mu} + gB_\mu}{\sqrt{g^2 + g'^2}} \tag{3.1.15}$$

associated with the unbroken generator Q is the electromagnetic potential (the photon field). The gauge coupling associated with A_μ is, of course, the electric charge which can now be expressed in terms of g and g', as follows.

Look at the gauge interactions involving W_3 and B, i.e., (keeping the $SU(2)_L \times U(1)_Y$ generators in place of currents and dropping Lorentz indices)

$$\mathscr{L}_{\text{int}}(W_3, B) \equiv -\frac{i}{2}(g\tau_3 W_3 + g'YB). \tag{3.1.16}$$

Using eqns. (3.1.13) and (3.1.15) we can rewrite $\mathscr{L}_{\text{int}}$ as follows (using $I_{3W} = \frac{1}{2}\tau_3$)

$$\mathscr{L}_{\text{int}} = -i\left[\left(\frac{g^2 I_{3W} - (g'^2/2)Y}{\sqrt{g^2 + g'^2}}\right)Z + \frac{gg'}{\sqrt{g^2 + g'^2}}QA\right]. \tag{3.1.17}$$

The second term implies that the magnitude of the electric charge of positron is

$$e = \frac{gg'}{\sqrt{g^2 + g'^2}}. \tag{3.1.18}$$

The first term predicts the structure of the neutral current interaction.

At this point it is convenient to introduce a reparametrization of g and g' in terms of an angle θ_W (known as the Weinberg angle) and the electric charge, e

$$\tan\theta_W \equiv \frac{g'}{g}. \tag{3.1.19}$$

Equation (3.1.18) then implies that

$$g = e \operatorname{cosec}\theta_W, \tag{3.1.20a}$$

$$g' = e \sec\theta_W. \tag{3.1.20b}$$

The W- and Z-masses can be rewritten as

$$m_W = \frac{ev}{2\sin\theta_W}, \tag{3.1.21a}$$

$$m_Z = \frac{ev}{2\sin\theta_W\cos\theta_W}. \tag{3.1.21b}$$

This leads to a relation between the W- and Z-boson masses

$$m_W = m_Z\cos\theta_W \tag{3.1.21c}$$

or if we define a parameter $\rho_W = m_W^2/(m_Z\cos\theta_W)^2$, the Glashow–Weinberg–Salam model predicts $\rho_W = 1$ at the tree level. This feature depends not just on the gauge structure of the model but also on the fact that symmetry breaking is implemented by the doublet Higgs bosons. Using eqn. (3.1.20a, b) in the first term of eqn. (3.1.17), i.e., the neutral current interaction involving the Z-boson can be written as

$$\mathscr{L}_{\text{N.C.}} = -i\frac{e}{\sin\theta_W\cos\theta_W}Z\cdot(I_{3W} - \sin^2\theta_W Q) \tag{3.1.22}$$

(where I_{3W} and Q symbolize the fermionic currents with respective $SU(2)_L \times U(1)$ transformation properties). Equation (3.1.22) predicts the form of the interaction of Z for any kind of matter multiplet, once their representation content is known. For arbitrary fermions we can write the $\mathscr{L}_{\text{N.C.}}$ as

$$\mathscr{L}_{\text{N.C.}} = -i\frac{e}{\sin\theta_W \cos\theta_W} Z_\mu \sum_i \bar{\psi}_i \gamma_\mu (I_{3W} - Q\sin^2\theta_W)\psi_i. \tag{3.1.23}$$

From this we can read off the neutral current interaction of any particle. This is an important prediction of the $SU(2)_L \times U(1)$ model and has been confirmed to a great degree of accuracy by neutral current experiments and provides the first curcial test of the ideas of the gauge theories. We will discuss this in somewhat greater detail in the next section. Here we consider the charged current aspects of the model.

We have seen that since the massive gauge bosons and their antiparticles are charged, this will generate the conventional charged current weak interaction which has the following form:

$$\mathscr{L}_{\text{wk}} = -\frac{ig}{2\sqrt{2}} W_\mu^+ [\bar{U}_a^{(0)}\gamma_\mu(1+\gamma_5)d_a^{(0)} + \bar{\nu}_a^{(0)}\gamma_\mu(1+\gamma_5)E^{-(0)}] + \text{h.c.} \tag{3.1.24}$$

From this we can conclude that the Fermi coupling can be expressed as

$$\frac{G_F}{\sqrt{2}} = \frac{g^2}{8m_W^2}. \tag{3.1.25}$$

Using eqn. (3.1.20a) we can write

$$m_W^2 = \frac{\pi\alpha}{G_F \sin^2\theta_W}\cdot\frac{1}{\sqrt{2}}. \tag{3.1.26}$$

Once the Weinberg angle is determined independently from neutral current interactions the mass of the charged W-boson can be predicted and vice versa. As we will see both the m_W and $\sin^2\theta_W$, determined from independent experiments, agree with eqn. (3.1.26) extremely well. It is important to note that this feature depends only on the gauge structure of the model and not on the detailed nature of symmetry breaking or on the fermionic structure of the model.

In eqn. (3.1.24) we have not displayed the mixing between the various generations. We will address this equation in a subsequent section. However, the universality of weak four-Fermi coupling (such as that between muon-decay and β-decay) already follows from eqn. (3.1.24).

§3.2. Neutral Current Interactions

In this section we will study the properties of the new kind of interactions induced by electro-weak gauge unification which do not involve the exchange of charged gauge bosons, the neutral current interactions. Its properties are dictated by the interaction Lagrangian in eqn. (3.1.23). The low-energy four-Fermi interaction induced by eqn. (3.1.23) is given by

$$\mathscr{L}_{\text{N.C.}} = -\frac{1}{2}\frac{e^2}{\sin^2\theta_W \cos^2\theta_W}\cdot\frac{1}{m_Z^2}J_\mu^{\text{N.C.}}J_\mu^{\text{N.C.}}, \tag{3.2.1}$$

where

$$J_\mu^{\text{N.C.}} = \sum_{i=\nu,e,u,d} \bar{\psi}_i\gamma_\mu(I_{3W} - \sin^2\theta_W\cdot Q)\psi_i. \tag{3.2.2}$$

If we define a parameter $\rho_W = (m_W/m_Z \cos\theta_Z)^2$, then we can write

$$H_{\text{N.C.}} = -\frac{4}{\sqrt{2}}G_F\rho_W J_\mu^{\text{N.C.}}J_\mu^{\text{N.C.}}. \tag{3.2.3}$$

From eqn. (3.2.3) we can write the various neutral current interactions between various particles in order to compare with experiments

$$H_{\text{N.C.}} = H_{\text{N.C.}}^{\nu} + H_{\text{N.C.}}^{e,H} + H_{\text{N.C.}}^{e\mu}, \tag{3.2.4}$$

$$H_{\text{N.C.}}^{\nu} = -\frac{G_F}{\sqrt{2}}\bar{\nu}\gamma_\mu(1+\gamma_5)\nu\sum_f[\varepsilon_L(f)\bar{f}\gamma_\mu(1+\gamma_5)f + \varepsilon_R(f)\bar{f}\gamma_\mu(1-\gamma_5)f], \tag{3.2.5}$$

where

$$\begin{aligned}
\varepsilon_L(u) &= \rho_W(\tfrac{1}{2} - \tfrac{2}{3}\sin^2\theta_W),\\
\varepsilon_R(u) &= \rho_W(-\tfrac{2}{3}\sin^2\theta_W),\\
\varepsilon_L(d) &= \rho_W(-\tfrac{1}{2} + \tfrac{1}{3}\sin^2\theta_W),\\
\varepsilon_R(d) &= \rho_W(+\tfrac{1}{3}\sin^2\theta_W),\\
\varepsilon_L(e) &= \rho_W(-\tfrac{1}{2} + \sin^2\theta_W),\\
\varepsilon_R(e) &= \rho_W(\sin^2\theta_W).
\end{aligned} \tag{3.2.6}$$

The parameters $\varepsilon_{L,R}$ can be extracted from the various neutrino neutral current experiments conducted at Fermilab and CERN and applying parton model for high-energy scattering of leptons off nucleons. A thorough analysis of neutral current data has been carried out by Kim *et al.* [3], who have extracted the following values for the various ε-parameters. The theoretical prediction of them using $\rho_W = 1$ and $\sin^2\theta_W = 0.233$ are noted on the right of Table 3.1. Here we have rised $g_{V,A}^f \equiv \varepsilon_L(f) \pm \varepsilon_R(f)$.

The agreement between theory and experiment is very good. The value of the ρ_W-parameter is actually better known than indicated by the neutrino data alone.

An important test of the standard model can be performed by improved $\nu_e e \to \nu_e e$, $\bar{\nu}_e e \to \bar{\nu}_e e$ scattering experiments. The reason is that, unlike the neutrino–hadron neutral current experiments, the $\overset{(-)}{\nu}_e e$ scattering experi-

Table 3.1

Neutral current parameters	Values extracted from experiments (ref. [4])	Predictions of the Glashow, Weinberg, and Salam model $\sin^2 \theta_W = 0.233$
$\varepsilon_L(u)$	0.339 ± 0.033	0.345
$\varepsilon_L(d)$	-0.424 ± 0.026	-0.423
$\varepsilon_R(u)$	-0.179 ± 0.019	-0.155
$\varepsilon_R(d)$	-0.016 ± 0.058	$+0.077$
$g_V(e)$	-0.043 ± 0.063	-0.034
$g_A(e)$	-0.545 ± 0.056	-0.5
$\rho_W = 1.001 \pm 0.21$		

ments involve interference between charged and neutral boson exchange diagrams in the standard model (see Fig. 3.1). The interference term in the standard model is proportional to $I_3(e_L) + \sin^2 \theta_W$ which is equal to $\approx -\frac{1}{2} + 0.233 \approx -0.267$ leading to destructive interference. In variants of standard model [5], the situation will, in general, be different; for instance, if there exist flavor-changing neutral current involving ν_e, then additional pieces which do not interfere with the charged current piece will appear. This experiment could therefore provide a sensitive test of (or reveal new physics beyond) the standard model. A main difficulty with this experiment is its low cross section compared to neutrino–hadron scattering.

Let us now look at the other pieces of the neutral current Hamiltonian.

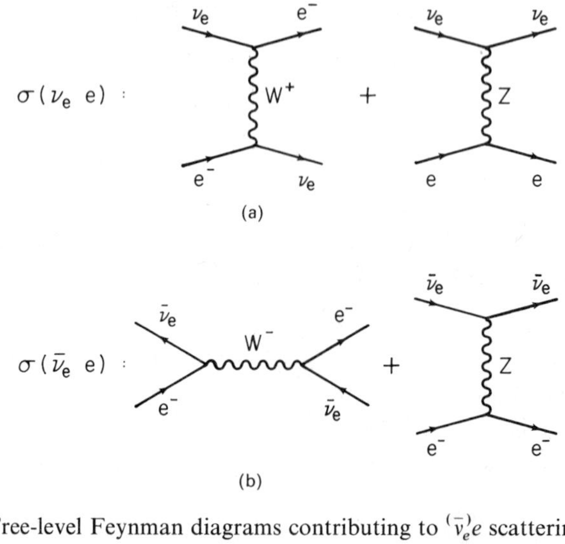

Figure 3.1. Tree-level Feynman diagrams contributing to $\overset{(-)}{\nu}_e e$ scattering in the standard model.

We parametrize $H^{eH}_{\text{N.C.}}$ as follows:

$$\mathscr{L}^{eH}_{\text{N.C.}} = -\frac{G_F}{\sqrt{2}}\sum_i (c_{1i}\bar{e}\gamma_\mu\gamma_5 e\bar{q}_i\gamma_\mu q_i + c_{2i}\bar{e}\gamma_\mu e\bar{q}_i\gamma_\mu\gamma_5 q_i), \tag{3.2.7}$$

$$C_{1u} = \frac{2}{\rho_W} g^e_A g^u_V,$$

$$C_{1d} = \frac{2}{\rho_W} g^e_A g^d_V,$$

$$C_{2u} = \frac{2}{\rho_W} g^e_V g^u_A,$$

$$C_{2d} = \frac{2}{\rho_W} g^e_V g^d_A. \tag{3.2.8}$$

Determination of g^f_V and g^f_A from neutrino neutral current data can be used to predict C_{1i} and C_{2i} which can then be used to predict parity violating effects in electronucleon interactions. Such effects can be measured by looking for parity violating effects in atomic physics, as well as looking for differences in left- and right-handed electron scattering off nucleons. By now, both of these kinds of experiments have been carried out and the results agree with the $SU(2)_L \times U(1)$ model to within the experimental accuracy.

A. Deep Inelastic Electron–Hadron Scattering

The SLAC experiment [6] on polarized electron-deep inelastic scattering $eD \to e + X$ measures the parity violating asymmetry

$$A = \frac{d\sigma(+) - d\sigma(-)}{d\sigma(+) + d\sigma(-)}. \tag{3.2.9}$$

Using the parton model approximations, and neglecting the effect of anti-quarks in nucleons for an isosinglet target such as deuteron, we can predict

$$\frac{A_D}{Q^2} = \frac{3G_F}{5\sqrt{2}\pi\alpha}[(C_{1u} - \tfrac{1}{2}C_{1d}) + F(y)(C_{2u} - \tfrac{1}{2}C_{2d})], \tag{3.2.10}$$

where

$$F(y) = \frac{1-(1-y)^2}{1+(1-y)^2},$$

where y is the dimensionless variable in parton models given by

$$y = \frac{E_i - E_f}{E_i}.$$

In the Glashow–Weinberg–Salam model

$$C_{1u} - \tfrac{1}{2}C_{1d} = -\tfrac{3}{4} + \tfrac{5}{3}\sin^2\theta_W,$$
$$C_{2u} - \tfrac{1}{2}C_{2d} = 3(\sin^2\theta_W - \tfrac{1}{4}), \tag{3.2.11}$$

and

$$\frac{3G_F}{5\sqrt{2}\pi\alpha} = 2.16 \times 10^{-4}\ \text{GeV}^{-2}. \tag{3.2.12}$$

SLAC data [6] yields

$$\frac{A_D}{Q^2} = [(-9.7 \pm 2.6) + (4.9 \pm 8.1)F(y)] \times 10^{-5},$$

which implies

$$C_{1u} - \tfrac{1}{2}C_{1d} = -0.45 \pm 0.12$$

and

$$C_{2u} - \tfrac{1}{2}C_{2d} = 0.23 \pm 0.38 \tag{3.2.13}$$

Putting $\sin^2\theta_W = 0.233$ yields $C_{1u} - \frac{1}{2}C_{1d} = -0.365$ and $C_{2u} - \frac{1}{2}C_{2d} = -0.051$, both values are in agreement with data.

B. Atomic Parity Violation

The dominant contribution to atomic parity violation in heavy nuclei comes from the C_{1i} term of eqn. (3.2.7). The reason for this is that in the non-relativistic limit $\bar{q}\gamma_\mu q \to q^+ q$ (which counts the number of quarks in a nucleus) whereas $\bar{q}\gamma_5\gamma_\mu q \xrightarrow[\text{N.R.}]{} q^+ \boldsymbol{\sigma} q$ (which measures the spin of the nucleus), which is a number of order 1 or 2 regardless of how heavy a nucleus is. For a typical nucleus, the effective weak parity violating electron nucleus interaction can be written as

$$H_{\text{wk}} = \frac{G_F}{\sqrt{2}} Q_w \boldsymbol{\sigma}_e \cdot \nabla\delta^3(\mathbf{r}), \tag{3.2.14}$$

where

$$Q_w(N, Z) = -2[C_{1u}(2Z + N) + C_{1d}(Z + 2N)], \tag{3.2.15}$$

where N and Z denote the number of protons and neutrons in a nucleus (note that the proton has two up-quarks and one down-quark, and vice versa for the neutron). For bismuth and thallium, on which experiments [7] have been carried out,

$$Q_w^{\text{Bi}}(126, 83) = -584C_{1u} - 670C_{1d} \approx -120.8,$$
$$Q_w^{\text{Tl}}(123, 81) = -570C_{1u} - 654C_{1d} \approx -118.$$

The results of these experiments are

$$Q_w^{\mathrm{Bi}} = -126 \pm 45$$

and

$$Q_w^{\mathrm{Tl}} = -155 \pm 63,$$

which are in agreement with the standard model.

C. Front–Back Asymmetry in $e^+e^- \to \mu^+\mu^-$

We now focus on the neutral weak interaction between electrons and muons. This interaction can be parametrized as follows:

$$\begin{aligned} H_{\mathrm{N.C.}}^{eh} = & -\frac{G_F}{\sqrt{2}}[h_{VV}(\bar{e}\gamma_\lambda e + \bar{\mu}\gamma_\lambda\mu)(\bar{e}\gamma_\lambda e + \bar{\mu}\gamma_\lambda\mu) \\ & - 2h_{VA}(\bar{e}\gamma_\lambda e + \bar{\mu}\gamma_\lambda\mu)(\bar{e}\gamma_\lambda\gamma_5 e + \bar{\mu}\gamma_\lambda\gamma_5\mu) \\ & + h_{AA}(\bar{e}\gamma_\lambda\gamma_5 e + \bar{\mu}\gamma_\lambda\gamma_5\mu) \times (\bar{e}\gamma_\lambda\gamma_5 e + \bar{\mu}\gamma_\lambda\gamma_5\mu)]. \end{aligned} \tag{3.2.16}$$

The forward–backward asymmetry measures the difference between the number of μ^-'s in the forward and backward hemisphere, with e^+e^- colliding direction as the polar axis, and is defined as follows:

$$A_{\mu\mu} = \frac{\int d\Omega[(d\sigma/d\Omega)(\theta) - (d\sigma/d\Omega)(\pi - \theta)]}{\int d\Omega[(d\sigma/d\Omega)(\theta) + (d\sigma/d\Omega)(\pi - \theta)]}. \tag{3.2.17}$$

The dominant contribution to the denominator is from electromagnetic interactions which cancel in the numerator leading to the result

$$A_{\mu\mu} = \frac{3}{16}\frac{1}{\sqrt{2}\sin^2\theta_W}\cdot\frac{S}{S - m_Z^2}. \tag{3.2.18}$$

It predicts $A_{\mu\mu} \simeq -9\%$ at the highest available e^+e^- energies right now, i.e., at the e^+e^- machine at DESY and this parameter has been measured by the various groups with the combined result [7a]:

$$A_{\mu\mu}^{\mathrm{expt}} = -10.4 \pm 1.4.$$

Again, there is good agreement between theory and experiment.

§3.3. Masses and Decay Properties of W- and Z-Bosons

In this section we will consider the predictions for W- and Z-boson masses in the standard model. This model has three free parameters (two gauge coupling constants, g and g', and the spontaneous symmetry breaking scale, v), which can be reexpressed in terms of: e, the electric charge of positron; $\sin\theta_W$, the weak mixing angle; and G_F ($\equiv G_\mu$), the Fermi coupling constant. We can therefore predict the W- and Z-boson masses. Let us discuss this step-by-step [8]. At the tree level, we have the following relation (eqn. (3.1.26)):

$$m_W^0 = \left[\frac{\pi\alpha^0}{\sqrt{2}G_\mu^0 \sin^2\theta_W}\right]. \tag{3.3.1}$$

We have denoted by a superscript zero the fact that no radiative corrections are applied. Let us first give physical argument that the radiative corrections are indeed important. The point is that radiative corrections will be defined after subtracting out the infinite contributions which introduce a subtraction point, μ, and will depend on it. In eqn. (3.3.1), the left-hand side must be defined at $\mu = m_W$; but conventional values of α^0 is at $\mu = 0$ since it is obtained from Compton scattering of real photons. Also, G_μ is obtained from μ-decay, where $\mu \simeq 0.1$ GeV. To get the tree level prediction for m_W we use:

$$\alpha^{0^{-1}} = 137\cdot035963(15), \tag{3.3.2}$$

$$G_\mu = (1.16634 \pm 0.00002) \times 10^{-5}\ \text{GeV}^{-2}. \tag{3.3.3}$$

Using this we get

$$m_W^0 = \frac{37.2810 \pm 0.0003}{\sin\theta_W^0}\,\text{GeV}. \tag{3.3.4}$$

The uncorrected tree level value of $\sin\theta_W^0$ can be obtained[3] from ν_μ-hadron scattering data:

$$\sin\theta_W^0 = 0.227. \tag{3.3.5}$$

This leads to

$$\text{m}_W^0 = 78.3\ \text{GeV} \qquad \text{(lowest order)},$$

$$m_Z^0 = 2\left(\frac{\pi\alpha^0}{\sqrt{2}G_\mu^0}\right)\frac{1}{\sin 2\theta_W^0} = 89.0\ \text{GeV} \qquad \text{(lowest order)}. \tag{3.3.6}$$

Following Marciano and Sirlin [8], if we denote the radiative corrections collectively by Δr, then we get

$$m_W = \frac{37.2810 \pm 0.0003\ \text{GeV}}{\sin\theta_W(1-\Delta r)^{1/2}} \tag{3.3.7a}$$

and

$$m_Z = \frac{74.562 \pm 0.0006\ \text{GeV}}{\sin 2\theta_W(1-\Delta r)^{1/2}}. \tag{3.3.7b}$$

The radiative corrections Δr have been discussed by a number of authors [8a]. The radiatively corrected $\sin\theta_W$ is obtained [9] to be

$$\sin^2\theta_W = 0.217 \pm 0.014. \tag{3.3.8}$$

The main sources for large contributions to Δr is the value of $\alpha(m_W)^{-1}$ coming from fermion vacuum polarization effects. One estimate [9] gives

$$\Delta r = 0.0696 \pm 0.0020$$

for

$$\sin^2\theta_W = 0.217,$$
$$m_H = m_W,$$
$$m_t = 36 \text{ GeV}. \tag{3.3.9}$$

Using this in eqn. (3.3.7) we predict

$$m_W = 83.0 \pm 2.7 \text{ GeV},$$
$$m_Z = 93.8 \pm 2.2 \text{ GeV}. \tag{3.3.10}$$

The W- and Z-bosons were discovered a year ago at the CERN $p\bar{p}$ colliding facility by the UA1 and UA2 experiments [10, 11] and the values for their masses are given as

$$M_W = 81.2 \pm 1.3 \text{ GeV},$$
$$M_Z = 92.8 \pm 1.5 \text{ GeV}. \tag{3.3.11}$$

These values are in excellent agreement with the predictions of the Glashow–Weinberg–Salam model, thus confirming the electro-weak symmetry $SU(2)_L \times U(1)_Y$ at low energies.

Let us now look at the decay properties of W and Z. The decay modes of W and Z are of two types: hadronic and leptonic.

$$W^+ \to u\bar{d},\ \bar{e}\nu_e,\ c\bar{s},\ \bar{\mu}\nu_\mu,\ t\bar{b},\ \bar{\tau}\nu_\tau, \tag{3.3.12}$$
$$Z \to u\bar{u},\ d\bar{d},\ c\bar{c},\ s\bar{s},\ t\bar{t},\ b\bar{b},\ \nu_e\bar{\nu}_e,\ \nu_\mu\bar{\nu}_\mu,\ \nu_\tau\bar{\nu}_\tau,\ e\bar{e},\ \mu\bar{\mu},\ \tau\bar{\tau}.$$

From the weak couplings given in previous sections we predict

$$\Gamma(W \to \text{hadrons}) = \frac{3G_\mu m_W^3}{2\sqrt{2}\pi}\left[1 - \frac{m_t^2}{2m_W^2} + \frac{1}{6}\frac{m_t^6}{m_W^6}\right] \times \left(1 + \frac{\alpha_s(m_W)}{\pi}\right)$$
$$\simeq 2.06 \text{ GeV} \quad \text{for } m_t = \frac{m_W}{2}, \tag{3.3.13}$$

where three colors have been taken into account, QCD corrections have been included and quark masses other than m_t have been ignored.

$$\Gamma(W \to l\bar{\nu}_l) = \frac{G_\mu m_W^3}{6\sqrt{2}\pi} \simeq 0.25 \text{ GeV}. \tag{3.3.14}$$

The total width of W is therefore about 2.8 GeV. Electro-weak correction to these results are small.

Coming to the Z-decay, its coupling can be inferred from eqn. (3.1.23) to be

$$Z = -i\frac{M_Z}{\sqrt{2}}\left(\frac{G_F}{\sqrt{2}}\right)^{1/2} Z_\mu \bar{f}\gamma_\mu(v_f - a_f\gamma_5)f, \tag{3.3.15}$$

where

$$|a_f| = 1,$$
$$|v_f| = (1 - 4Q\sin^2\theta_W). \tag{3.3.16}$$

Using this we can calculate

$$\Gamma(Z_0 \to l\bar{l}) = \frac{G_\mu M_Z^3}{12\sqrt{2}\pi}(1 - 4\sin^2\theta_W + 8\sin^2\theta_W) \simeq 0.092 \text{ GeV}, \tag{3.3.17}$$

$$\Gamma(Z_0 \to \nu_l\bar{\nu}_l) = \frac{G_\mu M_Z^3}{12\sqrt{2}\pi} \simeq 0.181 \text{ GeV}, \tag{3.318}$$

$$\Gamma(Z_0 \to \text{hadrons}) = \frac{G_\mu M_Z^3}{4\sqrt{2}\pi} f\left(\sin^2\theta_W, \frac{m_t^2}{m_t^2}\right)\left(1 + \frac{\alpha_s}{\pi}\right), \tag{3.3.19}$$

where

$$f\left(\sin^2\theta_W, \frac{m_t^2}{M_Z^2}\right) = 5 - \tfrac{28}{3}\sin^2\theta_W + \tfrac{80}{9}\sin^4\theta_W + \left(1 - \frac{4m_t^2}{M_Z^2}\right)^{1/2}$$
$$\times \left\{1 - \tfrac{8}{3}\sin^2\theta_W + \tfrac{32}{9}\sin^4\theta_W\right.$$
$$\left. - \frac{m_t^2}{M_Z^2}(1 + \tfrac{16}{3}\sin^2\theta_W - \tfrac{64}{9}\sin^4\theta_W)\right\},$$

$$\Gamma(Z_0 \to \text{hadrons}) \simeq 2.02 \text{ GeV}. \tag{3.3.20}$$

This gives a total width for Z_0 of about 2.85 GeV. We also see from here that each additional massless neutrino, with properties of similar neutrinos, will add 0.18 GeV to the width. Thus, measurement of Z-width will limit the number of neutrinos of type $m_\nu \ll M_Z$. Present results of the CERN experiment give $\Gamma_{\text{tot}}^Z < 6.5$ GeV which implies $N_\nu < 24$. More precise measurement of Γ_{tot}^Z is therefore of great theoretical interest. At this time, in the decay of the Z-boson, at CERN several unusual events involving $e^+e^-\gamma$ have been observed. The branching ratio for this decay in the standard model can be predicted to be [12]:

$$\frac{\Gamma(Z^0 \to e^+e^-\gamma)}{\Gamma(Z^0 \to e^+e^-) + \Gamma(Z^0 \to e^+e^-\gamma)} = \frac{\alpha}{\pi}\left[4\ln 2\varepsilon + 3)\ln\delta + \frac{\pi^2}{3} - \frac{7}{4}\right] \tag{3.3.21}$$

and

$$\frac{\Gamma(W \to e\nu\gamma)}{\Gamma(W \to e\nu) + \Gamma(W \to e\nu\gamma)} = \frac{\alpha}{2\pi}\left[(4 \ln \varepsilon + 3) \ln \delta + 2 \ln 2\varepsilon + \frac{\pi^2}{3} - \frac{1}{12}\right], \tag{3.3.22}$$

where ε is the lower limit on $E_\gamma/M_{W,Z}$ and 2δ denotes the lower limit on the opening angle of the photon relative to any charged fermion in the final state. These numbers imply a $Z \to e^+e^-\gamma$ branching ratio of about 2%, whereas the original observations indicated this branching ratio to be about 25%. This may be an indication of new physics. The above branching ratio has, however, gone down in subsequent data.

There is a very clear signature for pure $V - A$ (or $V + A$) structure of W-coupling to fermions in high-energy $p\bar{p}$ collision. There is an asymmetry in the $e^-(e^+)$ production with respect to the direction of collision that results from pure $V - A$ (or $V + A$) coupling. This can be seen as follows. We assume that W-production results from the collision of a quark (q), from proton and antiquark ($\bar{q}$), and from antiproton, via the diagram in Fig. 3.2.

If W-coupling to quarks is left-handed, then to antiquark it is right-handed. Consider Fig. 3.3(a). The u quark must be left-handed and the $\bar{d}$ must be right-handed giving rise to W^+ polarized towards the left. Since, by electric charge conservation $W^+ \to e^+\nu$, helicity conservation implies that for $V - A$ coupling e^+ must be right-handedly polarized and therefore emerge in the beam direction of the antiproton. By similarly analyzing Fig. 3.3(b) we conclude that e^- must emerge in the beam direction of the incident proton. This feature has been confirmed experimentally, thus proving the pure helicity structure of weak currents, which we can take as $V - A$ type.

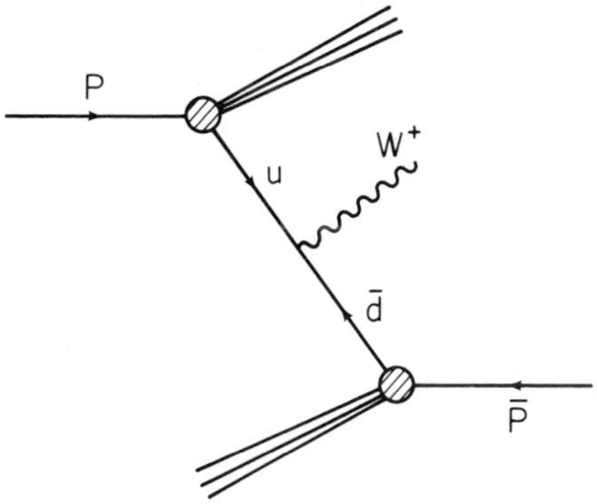

Figure 3.2. The Drell–Yan diagram for the production of W-bosons in $p\bar{p}$ collision.

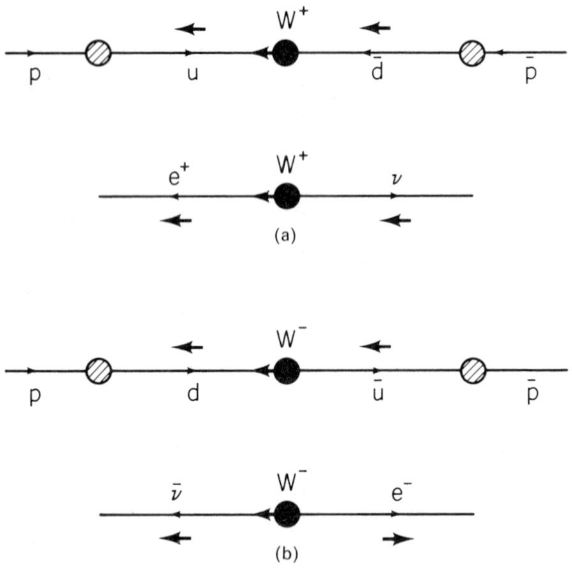

Figure 3.3. Asymmetry in $e^{\pm}$ distribution resulting from pure $V-A$ structure of weak currents.

§3.4. Fermion Masses and Mixing

So far we have not discussed the question of fermion masses as well as the higher generations of fermions, i.e., (c, s, μ, ν_μ), (t, b, τ, ν_τ). In this section we consider this question. Due to the chiral nature of weak interactions bare mass terms for fermions violate gauge invariance and therefore cannot be included in the Lagrangian if we want to preserve renormalizability. We, therefore, use the scalar Higgs doublet ϕ used to break gauge symmetries in Section 3.2 for this purpose. Recall that ϕ transforms under the $SU(2)_L \times U(1)$ gauge as follows:

$$\phi = \begin{pmatrix} \phi^+ \\ \phi^0 \end{pmatrix} (\tfrac{1}{2}, +1). \tag{3.4.1}$$

Its charge confugate field $\tilde{\phi}$ is denoted by

$$\tilde{\phi} = i\tau_2\phi^* = \begin{pmatrix} \phi_0^* \\ -\phi^- \end{pmatrix} (\tfrac{1}{2}, -1). \tag{3.4.2}$$

For the three generations of quarks and leptons denoted by $Q^0_{L,a}$, $U^0_{R,a}$, $d^0_{R,a}$, $\psi^0_{L,a}$, $E^0_{R,a}$, where a denotes the generation index $a = 1, 2, 3$, we can now write the gauge invariant Yukawa couplings as follows:

$$\mathscr{L}_Y = \sum_{a,b} (h^d_{ab}\bar{Q}^0_{La}\phi d^0_{Rb} + h^u_{ab}\bar{Q}^0_{La}\tilde{\phi}U^0_{Rb} + h^e_{ab}\bar{\psi}^0_{La}\phi E^0_{Rb}) + \text{h.c.} \tag{3.4.3}$$

Subsequent to spontaneous breakdown of gauge symmetry, $\mathscr{L}_Y$ leads to mass terms for fermions as follows:

$$\mathscr{L}_m = \bar{P}^0_{La} M^u_{ab} P^0_{Rb} + \bar{N}^0_{La} M^d_{ab} N^0_{Rb} + \bar{E}^0_{La} M^e_{ab} E^0_{Rb} + \text{h.c.,} \tag{3.4.4}$$

where

$$M^P = \frac{1}{\sqrt{2}} h^P v, \qquad p = u, d, e \tag{3.4.5}$$

denotes the mass matrices for up- and down-quarks and negatively charged leptons. Note that neutrinos are massless and will never acquire mass in higher orders since its chiral partner ν_R does not exist in the theory.

These mass matrices mix the weak eigenstates (denotes by superscript zero) of different generations and give rise to mixing angles such as the Cabibbo angle. Due to off-diagonal mixing terms in the mass matrix the quanta of the weak eigenstates are not eigenstates of the Hamiltonian (or mass). To get mass eigenstates we must diagonalize the mass matrices by means of bi-unitary transformations as given below

$$U_L^{(p)} M^{(p)} U_R^{(p)\dagger} = M^{(p)}_{\text{diag.}}, \tag{3.4.6}$$

where

$$M^{(u)}_{\text{diag.}} = (m_u, m_c, m_t, \ldots), \tag{3.4.7}$$

$$M^{(d)}_{\text{diag.}} = (m_d, m_s, m_b, \ldots), \tag{3.4.8}$$

$$M^{(e)}_{\text{diag.}} = (m_e, m_\mu, m_\tau, \ldots), \tag{3.4.9}$$

The mass eigenstates can be written as

$$\begin{aligned} P_{L,R} &= U^{(u)}_{L,R} P^0_{L,R}, \\ N_{L,R} &= U^{(d)}_{L,R} N^{(0)}_{L,R}, \\ E_{L,R} &= U^{(e)}_{L,R} E^{(0)}_{L,R}, \end{aligned} \tag{3.4.10}$$

where

$$\begin{aligned} P &= (u, c, t, \ldots), \\ N &= (d, s, b, \ldots), \\ E &= (e^-, \mu^-, \tau^-, \ldots). \end{aligned} \tag{3.4.11}$$

We will also call the mass eigenstate basis the flavor basis. If we now rewrite the gauge boson interactions given in (3.1.22) and (3.1.24) in the flavor basis we find that the neutral current interaction of the Z-boson remains diagonal, i.e., different flavors do not mix. This is an extremely desirable feature of the standard model since it implies that any neutral current process that changes flavor, like strangeness, can arise only at the one or higher loop level and must, therefore, be suppressed. This agrees with observations such as the suppression of $K_L^0 \to \mu\bar{\mu}$ decay, (compared to $K^+ \to \mu^+ \nu_\mu$ decay) $K_L - K_S$ mass difference, etc. We return to these questions in a subsequent section.

Turning to the interaction Lagrangian of the $W^\dagger$ with fermions, eqn. (3.1.24) can be written as

$$\mathscr{L}_{\text{gauge}} = -\frac{ig}{2\sqrt{2}} W_\mu^\dagger \bar{P} \gamma_\mu U_L^{(u)^\dagger} U_L^{(d)}(1 + \gamma_5)N + \text{h.c.} \tag{3.4.12}$$

We can write $U_{CKM} = U_L^{(u)^\dagger} U_L^{(d)}$ as the matrix that mixes different generations and is responsible for such phenomena as strangeness changing weak processes (e.g., $\Sigma \to p\pi^-$, $K \to \pi e\nu$, $K \to e\nu$, etc.). Since $U^{(p)}$'s unitary matrices, U_{CKM} is a unitary matrix; otherwise, its matrix elements are arbitrary and can only be fixed by experimental data. To give an example for two generations

$$U_{CKM} = \begin{pmatrix} \cos\theta_C & \sin\theta_C \\ -\sin\theta_C & \cos\theta_C \end{pmatrix}, \tag{3.4.13}$$

where θ_C is the Cabibbo angle. This matrix for two generations was first introduced by Glashow, Illiopoulos, and Maiani [13], and the absence of flavor-changing neutral current in lowest order is known as the Glashow–Illiopoulos–Maiani mechanism. Two other points about mixings are worth pointing out. The mixing matrices corresponding to U_R never appear in the final theory involving flavor eigenstates. Similarly, the left-handed mixing matrix for the charged leptons is also not observable since, in the charged current, we can redefine the neutrino states and absorb $U_L^{(l)}$, since neutrinos are massless all rotated bases are equivalent.

§3.5. Higher Order Induced Flavor-Changing Neutral Current Effects

As discussed in the previous section the lowest order neutral current couplings of Z^0 conserve all flavors such as strangeness, charm, etc. However, in higher orders these effects are induced by the charged currents. Since these effects are not present at the tree level and the theory is renormalizable, the magnitude of induced flavor-changing neutral current effects can be computed and used to test the standard model. These calculations have been carried out by Gaillard and Lee [14]. We briefly note the results for the two processes $K_L^0 \to \mu\bar{\mu}$ and $K_1 - K_2$ mass difference.

$K_L^0 \to \mu\bar{\mu}$ Decay

Experimentally the branching ratio for this decay is around 9×10^{-9} leading to an amplitude whose strength is $\sim G_F \times 10^{-4}$. There are two different sets of contributions to this process: (a) The 2γ intermediate state involving $K_L^0 \to$

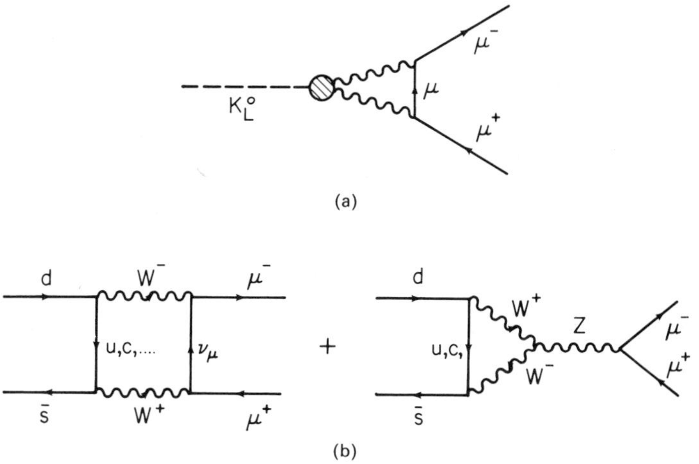

Figure 3.4. (a) Contribution of the two-photon intermediate state to $K_L^0 \to \mu\bar{\mu}$ decay. (b) Higher order weak corrections to $K_L^0 \to \mu\bar{\mu}$ decay.

$2\gamma \to \mu\bar{\mu}$. The absorptive part of this process involves the two photons or mass shell and can be predicted using known branching ratio for $K_L^0 \to 2\gamma$ decay [15] to be of the order $G_F\alpha^2$ which is of the same order as the experi mental observation (Fig. 3.4). (b) The pure weak interaction contribution involving the exchange of virtual $W^{\pm}$-bosons (Fig. 3.4(a)). The divergent parts of the diagram cancel between the up- and charm-quark intermediate states (the Glashow–Illiopoulos–Maiani cancellation) due to the unitary (or orthogonal) characters of the mixing matrices leading to a finite answer given by

$$A(K_L^0 \to \mu\bar{\mu}) \simeq \frac{G_F\alpha}{\pi} \frac{(m_c^2 - m_u^2)}{m_W^2 \sin^2\theta_W} \ln\left(\frac{m_W^2}{m_c^2}\right) \bar{d}\gamma_\mu\gamma_5 s\bar{\mu}\gamma_\mu\gamma_5\mu. \tag{3.5.1}$$

This explains the suppression of $K_L^0 \to \mu\bar{\mu}$ decay compared to $K \to \mu\nu$ decay and it also indicates that $m_c \ll m_W$.

$K_L - K_S$ Mass Difference

This arises from the higher order induced $\Delta S = 2$ matrix element in the standard model and is given by the contributions of the graphs shown in Fig. 3.5. The expression for the $\Delta S = 2$ matrix element in the two-generation model is given by

$$H_{\Delta S=2} = \frac{G_F^2}{4\pi^2} M_W^2 (\sin\theta_C \cos\theta_C)^2 \left(\frac{m_c^2}{M_W^2}\right) \bar{d}_L\gamma_\mu s_L \bar{d}_L\gamma_\mu s_L. \tag{3.5.2}$$

To obtain the value of the $K_1 - K_2$ mass difference, we have to evaluate the value of the $\langle K^0|\bar{d}_L\gamma_\mu s_L \bar{d}_L\gamma_\mu s|\bar{K}^0\rangle$ matrix element. It can be done using the

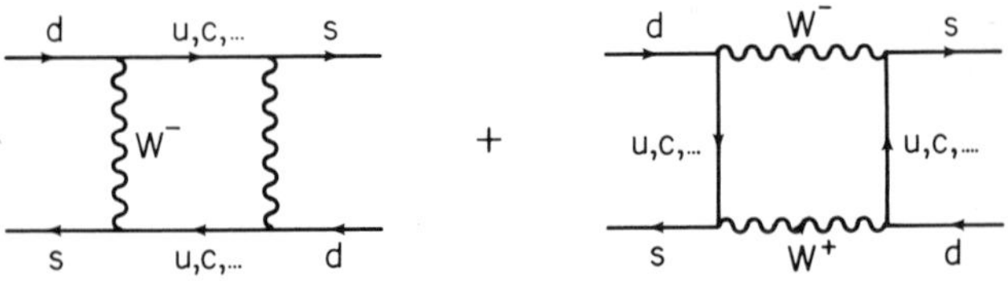

Figure 3.5. Higher order corrections leading to $\Delta S = 2$, $K^0 - \bar{K}^0$ mixing.

vacuum saturation approximation [16] to obtain

$$\Delta m_K \simeq \frac{G_F^2}{6\pi^2} M_W^2 f_K^2 M_K \left(\frac{m_c^2}{M_W^2}\right) \cos^2 \theta_C \sin^2 \theta_C \simeq 3.1 \times 10^{-15} \text{ GeV}. \qquad (3.5.3)$$

This can be compared with the experimental value of 3.5×10^{-15} GeV. This can be regarded as a successful prediction of the standard model, although many simplistic assumptions have been made such as keeping only the box graph which reflects only the short distance contribution and vacuum dominance of the matrix element. In any case, suppression of the $K_L - K_S$ mass difference is well understood.

$K_L^0 \to 2\gamma$ Decay

This decay can also be calculated in the standard model. The Feynman diagram is given in Fig. 3.6. In this case there is also the Glashow–Illiopoulos–Maiani cancellation but, unlike the previous cases, the suppression factor is

$$(m_c^2 - m_u^2)/(m_c^2 + m_u^2)$$

and we get

$$M(K_L^0 \to 2\gamma) \sim G_F \alpha \frac{\Delta m^2}{m_c^2}. \qquad (3.5.4)$$

Again, this is in accord with the experimental observations.

Thus, the standard $SU(2)_L \times U(1)$ model provides a good description of flavor-changing neutral current processes, which were not at all understood prior to the advent of gauge theories.

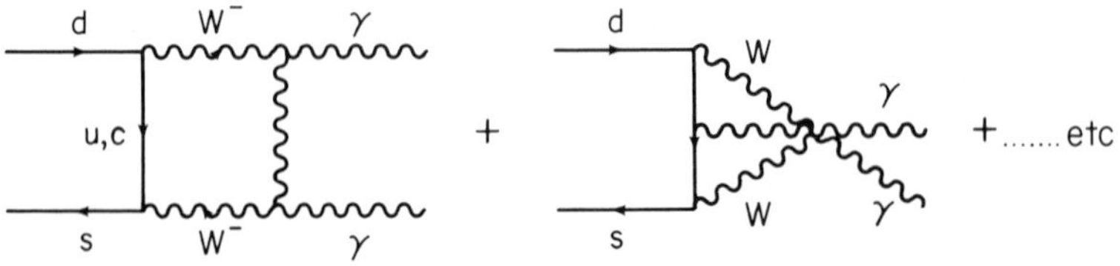

Figure 3.6. Higher order contributions leading to $K_L^0 \to 2\gamma$ decay.

§3.6. The Higgs Bosons

As we saw in Section 3.2, the gauge symmetry was spontaneously broken by means of an explicit scalar multiplet belonging to a doublet representation of the gauge group

$$\phi = \begin{pmatrix} \phi^+ \\ \phi^0 \end{pmatrix}. \tag{3.6.1}$$

We can write the complex field ϕ^0 in terms of real fields, i.e., $\phi^0 = (1/\sqrt{2})(\sigma + i\chi)$. Subsequent to spontaneous breakdown of the gauge symmetry, $\langle\sigma\rangle = v$, we can shift the σ-field. The $\phi^\pm$- and χ-fields, then get absorbed as the longitudinal components of the $W^\pm$- and Z-fields, respectively, leaving the shifted σ-field as a physical, real, scalar boson. We should therefore study the implications of the existence of this particle for physical processes and look for ways to detect it. Important for this purpose are the mass and couplings of the physical Higgs bosons, σ. To study its mass, we have to look at the Higgs potential

$$V = -\mu^2\phi^+\phi + \frac{\lambda}{2}(\phi^+\phi)^2. \tag{3.6.2}$$

On substituting $\phi^0 = (1/\sqrt{2})(v + \sigma + i\chi)$, and setting $\phi^\pm$ and χ to zero in the unitary gauge, we get the mass of the Higgs boson as

$$m_\sigma^2 = 2\lambda v^2. \tag{3.6.3}$$

This formula for the mass of the physical Higgs bosons is not valid for all λ; for $\lambda \simeq 10^{-4}$, the radiative contributions to the Higgs boson mass become important. In fact, it has been pointed out by Linde [17] and Weinberg [17] that radiative corrections provide a lower bound on m_σ. To illustrate the way to obtain the lower bound, we take the effective Higgs potential including the radiative corrections *à la* Coleman and E. Weinberg [18]

$$V(\sigma) = -\frac{\mu^2}{2}\sigma^2 + \frac{\lambda}{4}\sigma^4 + \frac{B}{4}\sigma^4 \ln\frac{\sigma^2}{V^2}, \tag{3.6.4}$$

where

$$B = \frac{3e^4(1 + \frac{1}{2}\sec^4\theta_W)}{64\pi^2\sin^4\theta_W} \simeq 6.4 \times 10^{-4}. \tag{3.6.5}$$

For $\mu^2 < 0$ we get

$$\frac{\partial V}{\partial\sigma} = 0 \tag{3.6.6}$$

leading to symmetry breaking, i.e., $\langle\sigma\rangle \neq 0$ only if $\lambda < -B/2$ in which case we get $\mu^2 = (\lambda + B/2)v^2$. There are then two minima of the potential: One at $\langle\sigma\rangle = 0$ and another at $\langle\sigma\rangle = v$. For

$$V|_{\langle\sigma\rangle=v} < V|_{\langle\sigma\rangle=0} \tag{3.6.7}$$

implies that

$$|\lambda| < B$$

Thus

$$\frac{B}{2} < |\lambda| < B. \tag{3.6.8}$$

Taking the second derivative of V with respect to σ, we find

$$m_\sigma^2 = (3B - 2|\lambda|)v^2 \geq Bv^2. \tag{3.6.9}$$

Using the value of B given above we find $m_\sigma \geq 6.6$ GeV.

Similarly, there is also an upper bound on the mass of the Higgs boson [19] which can be derived from many considerations. A naive way to see this is to look at the formula in eqn. (3.6.3) and observe that for the theory to remain perturbative we require $\lambda^2/4\pi \leq 1$. This would imply that

$$m_\sigma^2 \leq 8\pi v^2. \tag{3.6.10}$$

This implies $m_\sigma \leq 1$ TeV. Another way to obtain a bound is to look at the radiative corrections to the ρ_w-parameter. We find

$$\rho_w = 1 + \frac{\alpha}{4\pi} G\left(\frac{m_\sigma^2}{M_Z^2}, \cos^2\theta_W = c^2, \sin^2\theta_W = s^2 + \cdots\right), \tag{3.6.11}$$

where

$$G(\xi, c^2) = \frac{3\xi}{4s^2}\left[\frac{\ln(c^2/\xi)}{c^2 - \xi} + \frac{\ln \xi}{c^2(1-\xi)}\right]$$

$$\xrightarrow[\xi \to \text{large}]{} -\frac{3}{4}\left(\frac{\ln \xi}{c^2} + \frac{\ln c^2}{s^2}\right). \tag{3.6.12}$$

The experimental fact that $\Delta\rho_w \leq 0.01$ implies that $m_\sigma \leq 10^2 M_Z \simeq 9$ TeV. Thus we expect the physical Higgs boson mass to remain within 7 GeV to 1 TeV.

Let us now turn to its coupling to fermions. It is easy to see from eqn. (3.4.6) that

$$\mathscr{L}\sigma = \frac{\sigma}{v}(\bar{P}M^u P + \bar{N}M^d N + \bar{E}M^e E). \tag{3.6.13}$$

The heavier a fermion is, the stronger is its coupling to the Higgs boson σ. Therefore, in its decay, heavy fermions play an important role. Its decay modes to various final states can be calculated [20]

$$\Gamma(\sigma \to f\bar{f}) = \frac{(3)G_F}{4\pi\sqrt{2}} m_f^2 M_\sigma \left(1 - \frac{4m_f^2}{M_\sigma^2}\right)^{1/2}, \tag{3.6.14}$$

where factor 3 will appear only for quarks. For $M_\sigma > 2M_Z$

$$\Gamma(\sigma \to W^+W^-) = \frac{G_F M_\sigma^3}{8\pi\sqrt{2}}\left(1 - \frac{4M_W^2}{M_\sigma^2}\right)^{1/2} f\left(\frac{M_W}{M_\sigma}\right) \tag{3.6.15}$$

and

$$\Gamma(\sigma \to Z^0Z^0) = \frac{G_F M_\sigma^3}{16\sqrt{2}}\left(1 - \frac{4M_Z^2}{M_\sigma^2}\right)^{1/2} f\left(\frac{M_Z}{M_\sigma}\right), \tag{3.6.16}$$

where

$$f(x) = (1 - 4x^2 + 12x^4). \tag{3.6.17}$$

For $m_\sigma = 200$ GeV,

$$\Gamma_{t\bar{t}} : \Gamma_{W^+W^-} : \Gamma_{Z^0Z^0} = 1 : 3 \cdot 2 : 1. \tag{3.6.18}$$

Thus, coupling to the W^+W^- is the strongest. This would provide a clean signature of the Higgs boson by looking at two high transverse momentum leptons.

The production for heavy Higgs bosons in proton–proton and proton–antiproton collision proceeds through the gluon fusion mechanism given in Fig. 3.7. This process has been calculated [21]. This cross section is about 10^{-38} cm^2 and can be detected in the machines being planned for the 1990s.

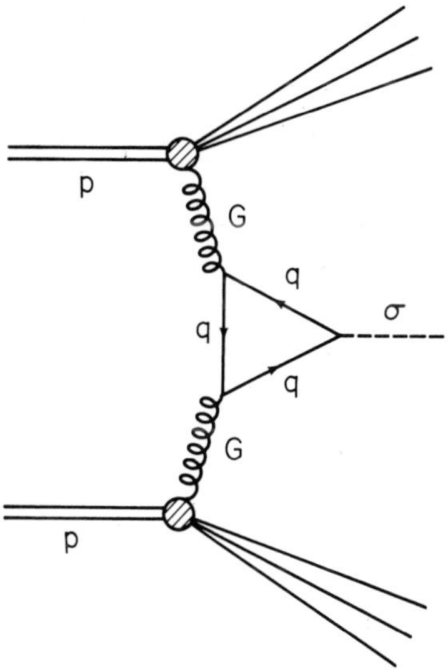

Figure 3.7. Gluon fusion diagram for the production of the Higgs boson in the standard model.

§3.7. $SU(2)_L \times U(1)$ Model with Two Higgs Doublets

Until the previous section we have discussed the minimal Higgs model. Often, from a theoretical point of view (such as the strong CP-problem to be discussed in Chapter 4) it is necessary to consider models with two Higgs doublets. In this section we present this model with a brief discussion of it. We keep all the features of the $SU(2)_L \times U(1)$ model except that we include two Higgs doublets ϕ_1 and ϕ_2 and let both of them participate in the spontaneous breakdown of the gauge symmetry

$$\phi_1 = \begin{pmatrix} \phi_1^{0*} \\ \phi_1^- \end{pmatrix} \quad \text{and} \quad \phi_2 = \begin{pmatrix} \phi_2^+ \\ \phi_2^0 \end{pmatrix}, \tag{3.7.1}$$

with

$$\langle \phi_1 \rangle = \frac{1}{\sqrt{2}} \begin{pmatrix} v_1 \\ 0 \end{pmatrix}, \qquad \langle \phi_2 \rangle = \frac{1}{\sqrt{2}} \begin{pmatrix} 0 \\ v_2 \end{pmatrix}. \tag{3.7.2}$$

The Yukawa coupling will now be given by

$$\begin{aligned} \mathscr{L}_Y = {} & \bar{Q}_L \phi_1 h_1 u_R + \bar{Q}_L \tilde{\phi}_1 h_1' d_R + \bar{Q}_L \phi_2 h_2 d_R + \bar{Q}_L \tilde{\phi}_2 h_2' d_R \\ & + \bar{L} \phi_2 h_3 E_R^- + \bar{L} \tilde{\phi}_1 h_3' \bar{E}_R + \text{h.c.} \end{aligned} \tag{3.7.3}$$

The W- and Z-boson masses now receive contributions from both vacuum expectation values, i.e.,

$$m_W = \frac{g}{2} \sqrt{v_1^2 + v_2^2}. \tag{3.7.4}$$

In this case, there are three neutral Higgs bosons, σ_1, σ_2, I, and one charged Higgs boson, $\phi^\pm$, which remain physical after spontaneous breakdown. The σ_1 and σ_2 are $\text{Re}\,\phi_1^0$ and $\text{Re}\,\phi_2^0$ and I is the imaginary part of a linear combination of ϕ_1 and ϕ_2. The states σ_1 and σ_2 will mix. If we denote their mass eigenstates by H_1 and H_2

$$\begin{aligned} H_1 &= \sigma_1 \cos\phi + \sigma_2 \sin\phi, \\ H_2 &= -\sigma_1 \sin\phi + \sigma_1 \cos\phi, \end{aligned} \tag{3.7.5}$$

their couplings to fermions is given by (denoting $v_1 = v\cos\alpha$ and $v_2 = v\sin\alpha$)

$$\begin{aligned} \mathscr{L}_Y = {} & \frac{1}{\sqrt{2}} \sigma_1 [\bar{P}_L f_1^{(u)} P_R + \bar{N}_L f_1^{(d)} N_R + \bar{E}_L f_1^{(e)} E_R] \\ & + \frac{1}{\sqrt{2}} \sigma_2 [\bar{P}_L f_2^{(u)} P_R + \bar{N}_L f_2^{(d)} N_R + \bar{E}_L f_2^{(e)} E_R] \\ & + \frac{i}{\sqrt{2}} I [\bar{P}_L f_3^{(u)} P_R + \bar{N}_L f_3^{(d)} N_R + \bar{E}_L f_3^{(e)} E_R] \\ & + H^- [\bar{d}_L f_4 u_R + \bar{d}_R f_5 u_L] + \text{h.c.}, \end{aligned} \tag{3.7.6}$$

where

$$
\begin{aligned}
\tilde{f}_1^{(u)} &= (h_1 \cos\phi + h_2' \sin\phi),\\
\tilde{f}_1^{(d)} &= (h_2 \sin\phi - h_1' \cos\phi),\\
\tilde{f}_1^{(e)} &= (h_3 \sin\phi - h_3' \cos\phi),\\
\tilde{f}_2^{(u)} &= (-h_1 \sin\phi + h_2' \cos\phi),\\
\tilde{f}_2^{(d)} &= (h_2 \cos\phi + h_1' \sin\phi),\\
\tilde{f}_2^{(e)} &= (h_3 \cos\phi + h_3' \sin\phi),\\
\tilde{f}_3^{(u)} &= (h_1 \sin\alpha + h_2' \cos\alpha),\\
\tilde{f}_3^{(d)} &= (h_2 \cos\alpha - h_1' \sin\alpha),\\
\tilde{f}_3^{(e)} &= (h_3 \cos\alpha - h_1' \sin\alpha),\\
f_4 &= U_L^{(d)}(h_1 \sin\alpha - h_2 \cos\alpha) U_R^{(u)^\dagger},\\
f_5 &= U_L^{(u)}(h_2 \cos\alpha + h_1 \sin\alpha) U_R^{(d)^\dagger},
\end{aligned}
\tag{3.7.7}
$$

where terms with and without a tilde are related as follows:

$$U_L^{(p)} \tilde{f}_i^{(p)} U_R^{(p)^\dagger} = f_i^{(p)}, \qquad p = u, d, e;\ i = 1, 2, 3. \tag{3.7.8}$$

The fermion mass matrices in this case are given by the formula

$$
\begin{aligned}
M_u &= \frac{v}{\sqrt{2}}(h_1 \cos\alpha + h_2' \sin\alpha),\\
M_d &= \frac{v}{\sqrt{2}}(h_2 \sin\alpha - h_1' \cos\alpha),\\
M_e &= \frac{v}{\sqrt{2}}(h_3 \sin\alpha - h_3' \cos\alpha).
\end{aligned}
\tag{3.7.9}
$$

Looking back at the Higgs couplings of σ_1 and σ_2 we find that, unlike the single Higgs doublet case, the Yukawa couplings of the physical quarks (mass eigenstates) are not flavor diagonal. This implies that at the tree level this model will lead to flavor-changing neutral currents. This difficulty can be avoided if we set the primed Yukawa couplings h_1', h_2', and h_3' equal to zero. It is then easy to convince oneself that all σ_1-couplings are flavor diagonal and parity conserving, whereas those of I are parity violating and flavor conserving. For the couplings in this case we get

$$
\begin{aligned}
\mathscr{L}_Y = 2^{1/4} G_F^{1/2} [&\sec\alpha \bar{P} M_u P(\cos\phi\sigma_1 - \sin\phi\sigma_2)\\
&+ \operatorname{cosec}\alpha \bar{N} M_d N(\sin\phi\sigma_1 + \cos\phi\sigma_2) + iI(\tan\alpha \bar{P}\gamma_5 M_u P\\
&+ \cot\alpha \bar{N}\gamma_5 M_d N) + H^+(\tan\alpha \bar{P}_R M_u U_{KM} N_L\\
&+ \cot\alpha \bar{P}_L U_{KM} M_d N_R)] + \text{h.c.}
\end{aligned}
\tag{3.7.10}
$$

We do not discuss this model any further here. The advantage of this model over the single Higgs doublet minimal model is that here couplings to quarks need not be suppressed due to their mass factor. The model also has a richer phenomenology [22].

§3.8. Beyond the Standard Model

The discovery of the $W^{\pm}$- and Z-bosons in the CERN collider experiments (with masses M_W and M_Z in striking agreement with the predictions of the standard model) has provided confirmation not only of the theoretical approach based on spontaneously broken gauge theories but also of the $SU(2)_L \times U(1)$ as the low-energy electro-weak (gauge?) symmetry. This model, however, introduces a number of free parameters into the theory to explain everything from the masses of the gauge bosons to the masses of leptons and quarks, as well their mixings. This is a highly unsatisfactory state of affairs and it is hoped that while the standard model provides a good starting point a lot of physics remains to be uncovered as we move into regimes beyond the energy scales explored in the standard model, i.e., $E > 100$ GeV or so. In the remaining chapters we present various theoretical schemes that extend our knowledge into this unexplored regime, each one attempting to cure, in one way or another, the shortcomings of the $SU(2)_L \times U(1)$ theory. It is, however, quite possible that the true reality may be a combination of some of these ideas or, even worse, outside their scope completely. It may, therefore, be useful to analyze ways to discover this new physics in a way that does not rely on any specific theoretical framework.

To this end, we must focus on aspects of physics that are unambiguously predicted in the standard model. The strategy would then be to look for deviations from these predictions. A number of papers [22, 23, 24, 25] have focused on this question and below we provide a short summary of these ideas. Before proceeding to this discussion we give a summary (in Table 3.2) of the experimental situation [26] for rare processes and the theoretical expectations for them based on the standard model.

Clearly, observation of any of the above rare processes would indicate the existence of new physics. The burden on the theorist, however, is the following: Once any indication of new physics appears, what can we say about the nature of new physics? To answer this question we note that from the observed magnitude of new physics a general idea can be had about the mass scale of the new physics (or new interaction). To say anything beyond that we must resort to a particular model. In this section we will give a rough description of how the first part of the answer can be given, focusing on a few specific examples.

(a) Neutrino Mass

The simplest (yet, extremely profound) example of new physics will be the discovery of nonzero neutrino mass. To see how and what we can learn about possible new physics from here, we remind ourselves that in the stan-

Table 3.2

Physical process	Expectation based on the standard $SU(2)_L \times U(1)$ model	Experimental situation
Higgs boson	One neutral with parity conserving fermion couplings	None observed
$\frac{B(Z \to e^+e^-\gamma)}{B(Z \to e^+e^-)}$	2–3%	few events observed at $sp\bar{p}s$
$\frac{B(W \to e\nu\gamma)}{B(W \to e\nu)}$	2–3%	none observed
Baryon nonconserving processes such as $p \to e^+\pi^0, \ldots, N \leftrightarrow \bar{N}$, etc.	0	$\tau_p > 10^{32}$ yr $\tau_{N-\bar{N}} > 10^6$ s or $> 6 \times 10^7$ s
Lepton number non-conserving processes (a) Neutrinoless double β-decay	0	$0^+ \to 0^+$ $\tau_{(\beta\beta)_{0\nu}} > 3.5 \times 10^{22}$ yr $0^+ \to 2^+$ $\tau_{(\beta\beta)_{0\nu}} > 1.2 \times 10^{22}$ yr
(b) $B(K^+ \to \pi^-e^+e^+)$	0	$< 10^{-8}$
(c) $B(\mu^- + N \to e^+ + N')$	0	$< 7 \times 10^{-11}$
Lepton flavor violation		
(a) $B(\mu \to e\gamma)$	0	$< 1.7 \times 10^{-10}$
$B(\mu \to e\gamma\gamma)$	0	$< 8.4 \times 10^{-9}$
(b) $B(\mu \to e + f)$ f = massless particle	0	$< 6 \times 10^{-6}$
(c) $B(\mu \to 3e)$	0	$< 1.9 \times 10^{-9}$
Other rare processes		
(a) $B(K_L^0 \to \mu\bar{e})$	0	$< 6 \times 10^{-6}$
(b) $B(K_L^0 \to e\bar{e})$	10^{-13}	$< 2 \times 10^{-7}$
(c) $B(K^+ \to \pi^+\mu^{\bar{e}})$	0	$< 7 \times 10^{-9}$
(d) $B(K^+ \to \pi^+\nu^{\bar{\nu}})$	$\geq 4 \times 10^{-10}$	$< 1.4 \times 10^{-7}$
Electric charge violation		
(a) Phonton mass, m_γ	0	$< 3 \times 10^{-27}$ eV
(b) $\tau(e^+ \to \nu + \gamma)$	0	$> 10^{22}$ yr
Neutrino masses		
m_{ν_e}	0	< 46 eV
m_{ν_μ}	0	< 0.25 MeV
m_{ν_τ}	0	< 70 MeV
Limits on parameters in μ-decay*		
ρ	3/4	0.752 ± 0.003
η	0	-0.12 ± 0.21
ξ	1	> 0.9959
δ	3/4	0.755 ± 0.008

* The parameters are defined in terms of the electron spectrum in μ-decay as follows:

$$dNe/x^2\,dx\,d(\cos\theta) = \text{const.}[(3 - 2x) + (4/3\rho - 1)(4x - 3) + 12(m_e/m_\mu)(1 - x)/x\eta$$
$$+ \{(2x - 1) + (4/3\delta - 1)(4x - 3)\}\xi P_\mu \cos\theta],$$

where $x = 2E_e/m_\mu$, θ is the angle between the electron momentum and polarization vector of the muon, and P_μ is the muon-polarization vector.

dard model $m_\nu = 0$. However, using the fields present in the standard model, i.e., the lepton doublet $\psi_L \equiv \binom{\nu_e}{e^-}_L$ and the Higgs doublet $\phi \equiv \binom{\phi^+}{\phi^-}$, we can write the following $SU(2)_L \times U(1)$ invariant operator:

$$L^{(1)} = \frac{\lambda}{M} \psi_L^T C^{-1} \tau_2 \tau \psi_L \cdot \phi^T \tau_2 \tau \phi. \tag{3.8.1}$$

The appearance $1/M$ is to make the operator have mass dimension 4 so it can be a legitimate Lagrangian. The $L^{(1)}$ must be $SU(2)_L \times U(1)_Y$ invariant since we are exploring physics at a higher mass scale $M \gg M_W$ where electro-weak symmetry is exact. After $SU(2)_L \times U(1)_Y$ symmetry is broken, $L^{(1)}$ can lead to an effective mass for neutrino as

$$m_{\nu_e} \simeq \frac{4m_W^2}{g^2 M} \lambda. \tag{3.8.2}$$

If m_{ν_e} is in the electron volt range and if λ is 10^{-6}–10^{-1}, the value of the new scale M is 10^6 GeV to 10^{11} GeV. Since λ is an unknown parameter, in the absence of a detailed theory, the precise value of M cannot be determined; however, the existence of a new scale will clearly be established.

The same kind of analysis can be carried out for most of the processes listed in Table 3.2. We give two more examples

(b) Proton Decay

The strategy here is to consider the lowest dimensional operator invariant under the $SU(2)_L \times U(1) \times SU(3)_c$ group that changes the baryon number. This question has been studied in detail in Ref. [23]. Here we simply present an illustrative example

$$\underset{(1)}{\mathscr{L}\Delta B \neq 0} = \frac{\varepsilon_{ijk}}{M^2} U_{R_i}^T c^{-1} d_{R_j} Q_{Lk}^T C^{-1} \tau_2 \psi_L. \tag{3.8.3}$$

Another operator of this type is

$$\underset{(2)}{\mathscr{L}\Delta B \neq 0} = \frac{\varepsilon_{ijk}}{M^2} Q_{L_i}^T \tau_2 C^{-1} Q_{L_j} U_{R_k}^T C^{-1} e_R. \tag{3.8.4}$$

Other operators of this type can be constructed, and it turns out that all these operators satisfy the selection rule $\Delta B = \Delta L$ and lead to decays of type

$$\begin{aligned} p &\to e^+ \pi^0 \\ &\to \bar{\nu} \pi^+. \end{aligned} \tag{3.8.5}$$

Observation of any of these processes would again indicate new physics with mass scale M. The present lower limits on $\tau_{p \to e^+ \pi^0}$ implies $M \geq 10^{15}$ GeV or so.

(c) Neutron–Antineutron Oscillation

The $\Delta B \neq 0$ operator that causes $N - \bar{N}$ transition must necessarily involve six quark fields and will therefore have dimension nine. A typical $SU(2)_L \times U(1) \times \mathrm{SU}(3)_c$ invariant operator of this type is

$$\underset{(3)}{\Delta B} = 2 = \frac{\varepsilon_{ijk}}{M^5} \varepsilon_{i'j'k'} U_{R_i}^T C^{-1} d_{R_{i'}} U_{R_j}^T C^{-1} d_{R_{j'}} d_{R_k}^T C^{-1} d_{R_{k'}}^R. \tag{3.8.6}$$

This involves a high power of the mass scale M and, therefore, is particularly interesting. The reason is that observation of $N - \bar{N}$ oscillation with mixing time of order ~ 1 yr would imply (see Chapter 6) $M \simeq 10^6$ GeV, indicating new physics in a nearby regime.

References

[1] S. L. Glashow, *Nucl. Phys.* **22**, 579 (1961);
A. Salam and J. C. Ward, *Phys. Lett.* **13**, 168 (1964);
S. Weinberg, *Phys. Rev. Lett.* **19**, 1264 (1967);
A. Salam, in *Elementary Particle Theory* (edited by N. Svartholm), Almquist and Forlag, Stockholm, 1968;
For an excellent review, see
E. S. Abers and B. W. Lee, *Phys. Rep.* **9C**, 1 (1973).

[2] UA1 Collaboration, G. Arnison *et al.*, *Phys. Lett.* **122B**, 103 (1983);
UA2 Collaboration, M. Banner *et al.*, *Phys. Lett.* **122B**, 476 (1983).

[3] J. E. Kim, P. Langacker, M. Levine, and H. H. Williams, *Rev. Mod. Phys.* **53**, 211 (1980).

[4] See, for instance,
M. Jonker *et al. Phys. Lett.* **99B**, 265 (1981).

[5] B. Kayser, E. Fishbach, S. P. Rosen, and H. Spivack, *Phys. Rev.* **D20**, 87 (1979).

[6] C. Y. Prescott *et al.*, *Phys. Lett.* **77B**, 347 (1978).

[7] P. Bucksbaum, E. Commins, and L. Hunter, *Phys. Rev. Lett.* **46**, 640 (1981);
L. M. Barkov, M. Zolotorev, and I. Khriplovich, *Sov. Phys. Usp.* **23**, 713 (1980);
M. A. Bouchiat *et al.*, *Phys. Lett.* **117B**, 358 (1982);
J. Hollister *et al.*, *Phys. Lett.* **46**, 643 (1981).

[7a] For a review, see
Albrecht Bohm, *Proceedings of the SLAC Summer Institute* (edited by M. Zipf *et al.*), Stanford, 1983.

[8] W. J. Marciano and A. Sirlin, *Phys. Rev.* **D29**, 945 (1984).

[8a] For an incomplete list of papers on radiative corrections in gauge theories, see
K. I. Aoki, Z. Hioki, R. Kawabe, M. Konuma, and T. Muta, *Prog. Theor. Phys.* Suppl. **73**, 1 (1982);
S. Sakakibara, *Proceedings of the Topical Conference on Radiative Corrections in* $\mathrm{SU}(2)_L \times \mathrm{U}(1)$ *Theories*, Trieste, 1983;
For other related work on radiative corrections in $\mathrm{SU}(2)_L \times \mathrm{U}(1)$ theories, see
W. J. Marciano and A. Sirlin, *Phys Rev.* **D22**, 2695 (1980);
A. Sirlin and W. J. Marciano, *Nucl. Phys.* **B189**, 442 (1981);
F. Antonelli, M. Consoli, and G. Corbo, *Phys. Lett.* **91B**, 90 (1980);
C. Llewellynsmith and J. Wheater, *Phys. Lett.* **105B**, 486 (1981);
J. Wheater and C. Llewellynsmith, *Nucl. Phys.* **B208**, 27 (1982);

S. Sakakibara, *Phys. Rev.* **D24**, 1149 (1981);
M. Veltman, *Phys. Lett.* **91B**, 95 (1980);
R. N. Mohapatra and G. Senjanovic, *Phys. Rev.* **D19**, 2165 (1979).

[9] A. Sirlin and W. Marciano, *Nucl. Phys.* **B189**, 442 (1981);
C. Llewellynsmith and J. Wheater, *Phys. Lett.* **105B**, 486 (1981).

[10] G. Arnison *et al.*, *Phys. Lett.* **126B**, 398 (1983);
P. Bagnaia *et al.*, *Phys. Lett.* **129B**, 130 (1983);
G. Arnison *et al.*, *Phys. Lett.* **129B**, 273 (1983).

[11] L. DiLella, Lecture Notes, CERN (1985), unpublished.

[12] W. Marciano, Talk Presented at Proton–Antiproton Collider Physics, Switzerland, 1984;
W. Marcian and Z. Parsa, *Proceedings of the 1982 DPE Summer Study*, 1982.

[13] S. L. Glashow, J. Illiopoulos, and L. Mariani, *Phys. Rev.* **D2**, 1285 (1970).

[14] M. K. Gaillard and B. W. Lee, *Phys. Rev.* **D10**, 897 (1974).

[15] B. R. Martin, E. de Rafael, and J. Smith, *Phys. Rev.* **D1**, (1970).

[16] R. N. Mohapatra, J. Subbarao, and R. E. Marshak, *Phys. Rev.* **171**, 1502 (1968).

[17] S. Weinberg, *Phys. Rev. Lett.* **36**, 294 (1976);
A. Linde, *JETP Lett.* **23**, 73 (1976).

[18] S. Coleman and E. Weinberg, *Phys. Rev.* **D7**, 1888 (1973).

[19] D. A. Dicus and V. S. Mathur, *Phys. Rev.* **D7**, [redacted] (1973);
B. W. Lee, C. Quigg, and H. Thacker, *Phys. Rev.* **D16**, 1519 (1977);
M. Veltman, *Acta Phys. Polon.* **B8**, 475 (1977).

[20] T. Rizzo, *Phys. Rev.* **D22**, 722 (1980).

[21] H. Georgi, S. L. Glashow, M. Machachek, and D. Nanopoulos, *Phys. Rev. Lett.* **40**, 692 (1978);
See also H. Gordon, W. Marciano, F. E. Paige, P. Grannis, S. Naculich, and H. H. Williams, *Proceedings of the 1982 DPF Summer Study on Elementary Particle Physics*, Snowmass, 1982, p. 161.

[22] L. Hall and M. Wise, *Nucl. Phys.* **B187**, 397 (1981);
M. Barnett, G. Senjanovic, and D. Wyler, ITP Santa Barbara preprint, (1984);
J. M. Frere, M. Gavela, and J. Varmaseren, *Phys. Lett.* **125**, 275 (1983).

[23] S. Weinberg, *Phys. Rev. Lett.* **43**, 1566 (1979);
F. Wilczek and A. Zee, *Phys. Rev. Lett.* **43**, 1571 (1979);
A. H. Weldon and A. Zee, *Nucl. Phys.* **B173**, 269 (1980);
R. N. Mohapatra, *Proceedings of the First Workshop on Grand Unification* (edited by P. Frampton, H. Georgi, and S. L. Glashow), Math Sci. Press, Brookline, MA, 1980.
For excellent recent reviews, see
J. D. Vergados, *Phys. Rep.* (1986) (to appear).
G. Costa and M. Zwirner, *Rev. Nuovo. Cim.* (1886) (to appear).

[24] C. Burges and H. Schnitzer, *Nucl. Phys.* **B228**, 464 (1983).

[25] M. K. Gaillard, *Proceedings of the Workshop on Intense Medium Energy Sources of Strangeness* (edited by T. Goldman, H. Haber, and H. F. Sadrozinski), (AIP), New York 1983, p. 54.

[26] The latest experimental situation in weak interaction has recently been summarized in
G. Barbiellini and C. Santoni, *Rev. Nuovo Cim.* (1986) (to appear).
Particle Data Group, *Rev. Mod. Phys.* **56**, S1 (1984).

CHAPTER 4

CP-Violation: Weak and Strong

§4.1. *CP*-Violation in Weak Interactions

The phenomenon of *CP*-violation was discovered in 1964 by Christenson, Cronin, Fitch, and Turlay in K^0-decays. To date Kaon systems have remained the only place where breakdown of *CP*-invariance has been observed. Since Lorentz invariant local field theories are *CPT*-invariant, breakdown of *CP*-symmetry implies breakdown of time-reversal invariance. This fact is used in experimental search for *CP*-violation by looking for kinematic effects odd under time-reversal symmetry. In the $K^0 - \bar{K}^0$ system, however, *CP*-violating interference effects appear which are experimentally measurable. The *CP*-violating phenomenology in K^0-decays has been extensively described in many places [1] and we do not repeat it here, except to note some salient points.

In the absence of weak interactions, $K^0 - \bar{K}^0$ are separate eigenstates of the Hamiltonian and they do not mix due to strangeness conservation. Once *CP*-conserving weak interactions are turned on, strangeness is no longer a good symmetry and this implies that the $K^0 - \bar{K}^0$ mix, and in the rest frame of the Kaons the mass matrix looks like the following:

$$M^{(+)}_{K^0-\bar{K}^0} = \begin{pmatrix} m_{K^0} & m_{K^0\bar{K}^0} \\ m_{K^0\bar{K}^0} & m_{\bar{K}^0} \end{pmatrix}, \tag{4.1.1}$$

with $m_{K^0} = m_{\bar{K}^0}$. The equality of off-diagonal elements follows from *CP*-invariance. The eigenstates of this matrix define the physical *K*-mesons K_1 and K_2

$$K_1^0 = \frac{K^0 - \bar{K}^0}{\sqrt{2}},$$

$$K_2^0 = \frac{K^0 + \bar{K}^0}{\sqrt{2}}. \tag{4.1.2}$$

If we assume $C|K^0\rangle = |\bar{K}^0\rangle$, $K_1^0(K_2^0)$ is CP-even (odd) and $m_{K_1^0} - m_{K_2^0} = 2m_{K^0\bar{K}^0}$. Also note that since K_1 and K_2 are not stable particles, $m_{K_{1,2}}$ are not real. If we now turn to the CP-violating interactions, then, in the $K_1 - K_2$ basis, the mass matrix is no longer diagonal and it looks as follows:

$$M_{K_1-K_2} = \begin{pmatrix} m_{K_1} & \delta m_{K_1^0-K_2^0} \\ -\delta m_{K_1^0-K_2^0} & m_{K_2} \end{pmatrix}. \tag{4.1.3}$$

CPT-invariance has been assumed in writing eqn. (4.1.3). Let us write

$$m_{K_1^0} = m_1 - \frac{i}{2}\gamma_1,$$

$$m_{K_2^0} = m_2 - \frac{i}{2}\gamma_2,$$

$$\delta m_{K_1^0-K_2^0} = i\left(m' - \frac{i}{2}\gamma'\right). \tag{4.1.4}$$

Then $M_{K_1^0-K_2^0}$ can be diagonalized, leading to the short- and long-lived eigenstates K_S and K_L, i.e.,

$$|K_S\rangle = \frac{|K_1^0\rangle + \rho|K_2^0\rangle}{1 + |\rho|^2},$$

$$|K_L\rangle = \frac{|K_2^0\rangle + \rho|K_1^0\rangle}{1 + |\rho|^2}, \tag{4.1.5}$$

where

$$\rho \simeq \frac{-i(m' - i\gamma'/2)}{[(m_1 - m_2) - (i/2)(\gamma_1 - \gamma_2)]}. \tag{4.1.6}$$

In terms of hadronic matrix elements

$$m' = \operatorname{Im} M_{K_1^0K_2^0}$$

and

$$\gamma' = \operatorname{Im} \Gamma_{K_1^0K_2^0}. \tag{4.1.7}$$

So far, we have calculated the CP-impurity in K_L and K_S systems. In order to apply this discussion to $K \to 2\pi$ decays, we have to realize that there are phases arising from the two-pion final states as follows:

$$\langle(2\pi); I|H_{\text{wk}}|K^0\rangle = A_I e^{i\delta_I}$$

and

$$\langle(2\pi); I|H_{\text{wk}}|\bar{K}^0\rangle = \bar{A}_I e^{i\delta_I}, \tag{4.1.8}$$

where $\langle(2\pi); I|$ denotes the definite isospin combination of the $\pi^+\pi^-$ and $\pi^0\pi^0$ states. The *CPT* theorem implies that $\bar{A}_I = A_I^*$, which can be used to calculate that

$$2^{-1/2}\langle(2\pi); I|H_{\rm wk}|K_1^0\rangle = i\, e^{i\delta_I}\,{\rm Im}\, A_I$$

and

$$2^{-1/2}\langle(2\pi); I|H_{\rm wk}|K_2^0\rangle = e^{i\delta_I}\,{\rm Re}\, A_I. \tag{4.1.9}$$

Let us now define

$$\eta_{ij} = \langle\pi_i\pi_j|H_{\rm wk}|K_L^0\rangle/\langle\pi_i\pi_j|H|K_S^0\rangle, \tag{4.1.10}$$

where $i, j = (+, -)$ or $(0, 0)$.

We can then write

$$\eta_{+-} = \varepsilon + \frac{\hat{\varepsilon}'}{1 + \hat{\omega}/\sqrt{2}},$$
$$\eta_{00} = \varepsilon - \frac{\hat{\varepsilon}'}{1 - \sqrt{2}\hat{\omega}}, \tag{4.1.11}$$

where

$$\varepsilon = \rho + i\frac{{\rm Im}\, A_0}{{\rm Re}\, A_0},$$
$$\varepsilon' = -\frac{i}{\sqrt{2}}\left\{\left(\frac{{\rm Im}\, A_0}{{\rm Re}\, A_0}\right) - \left(\frac{{\rm Im}\, A_2}{{\rm Re}\, A_2}\right)\right\}\hat{\omega},$$
$$\hat{\omega} = \left(\frac{{\rm Re}\, A_2}{{\rm Re}\, A_0}\right)\cdot e^{i(\delta_2-\delta_0)}. \tag{4.1.12}$$

From eqns. (4.1.6) and (4.1.12) we can calculate the phase of η_{+-} and η_{00} without reference to any models. Using $\Delta m = m_2 - m_1$ and $\gamma_1 \gg \gamma_2$ and $\gamma' \ll m'$, we find (using the fact that $\gamma'/\gamma_1 \approx {\rm Im}\, A_0/{\rm Re}\, A_0$, and $\gamma_1/2 \simeq \Delta m$)

$$\varepsilon \simeq \frac{i(m'/\Delta m) + i({\rm Im}\, A_0/{\rm Re}\, A_0)}{1 + (i/2)\gamma_1/\Delta m}$$
$$\approx \frac{m'/\Delta m + {\rm Im}\, A_0/{\rm Re}\, A_0}{\gamma_1/2\Delta m - i}. \tag{4.1.13}$$

Experimentally, $\gamma_1 \simeq 2\Delta m \simeq 0.70 \times 10^{-14}$ GeV implies that,

$$\varepsilon \simeq \frac{e^{i\pi/4}}{\sqrt{2}}\left(\frac{m'}{\Delta m} + \frac{{\rm Im}\, A_0}{{\rm Re}\, A_0}\right). \tag{4.1.14}$$

Furthermore, $\delta_2 - \delta_0 + \pi/2 \approx \pi/4$, which can be used to argue that $\varepsilon'/\varepsilon_1 \approx -\omega\xi/\sqrt{2}|\varepsilon + \xi|$ which is real (where we have denoted $\xi \equiv {\rm Im}\, A_0/{\rm Re}\, A_0$). Furthermore, since $\Delta I = 3/2 K \to 2\pi$, the amplitudes are known to be about twenty times smaller than the $\Delta I = 1/2$ amplitudes, and as we would expect, $\varepsilon' \ll \varepsilon$. This would imply that $|\eta_{+-}| \approx |\eta_{00}|$. Present experimental results [2] bear this out as we see below

$$
\begin{aligned}
|\eta_{+-}| &= (2.27 \pm 0.3) \times 10^{-3}, \\
\operatorname{Arg} \eta_{+-} &= 44.6 \pm 1.2^\circ, \\
(\eta_{00}/\eta_{+-})^2 &= 1.028 \pm 0.032 \pm 0.014 \quad \text{(Ref. [3])}.
\end{aligned}
\tag{4.1.15}
$$

In terms of ε', these results imply

$$
\frac{\varepsilon'}{\varepsilon} = (-4.5 \pm 5.3 \pm 2.4) \times 10^{-3}. \tag{4.1.16}
$$

Any model of CP-violation must explain the approximate equality of η_{+-} and η_{00} or the extreme smallness of ε'/ε. A phenomenological model, proposed by Wolfenstein [4], known as the superweak model predicts $\varepsilon' = 0$. The basic feature of this model is that all CP-violating interactions are $\Delta S = 2$ type and can, therefore, lead to $m' \neq 0$, but since Im A_0 and Im A_2 arise from $\Delta S = 1$ transitions, they must be zero. This phenomenological model can be embedded into gauge theories in many ways [5] and we will not pursue it further. In the following sections we discuss other, currently popular, models for CP-violation.

§4.2. *CP*-Violation in Gauge Models: Generalities

In order to introduce CP-violation into gauge models, we must make one or more parameters of the theory complex. Furthermore, this "complexity" must be complex enough so that by all possible phase redefinition of the fields in the theory we should not be able to remove it. This phase can reside in two sectors of the theory: (i) gauge interactions; and/or (ii) Higgs boson interactions of the fermions. Moreover, the CP-violation can be intrinsic to the parameters of the original Lagrangian prior to spontaneous symmetry breaking, or it could be of spontaneous origin in the sense that all parameters of the theory are real, but the vacuum expectation values of the Higgs fields are complex. Thus, it appears that, in general, there could be lots of arbitrariness in the discussion of CP-violation in gauge theories. However, a great deal of simplification can occur under certain circumstances. For instance, imagine a theory in which all parameters are real prior to spontaneous symmetry breaking and subsequent to spontaneous breaking only one Higgs field acquires complex vacuum expectation value. Then there will only be one phase parametrizing all CP-violating phenomena. Another circumstance was discovered by Kobayashi and Maskawa [6] in 1973. It was known earlier [7] that if we consider only left-handed currents participating in weak interactions, then for two generations, we cannot have a nontrivial CP-phase. In fact, it was suggested [7] that we may invoke right-handed currents to generate nontrivial CP-violating effects. Kobayashi and Maskawa pointed out that if we have only left-handed currents going to three generations of quarks (i.e., six flavors), only *one* nontrivial *CP*-phase appears in the theory for arbitrary values of the parameters of the Lagrangian.

To prove this statement, it is important to discuss the origin of the *CP*-violating phase in gauge models. First, we realize that, since the adjoint representation of the gauge group to which the gauge fields belong is a real representation, the gauge coupling must be real. We can therefore make the Yukawa couplings complex. Subsequent to spontaneous symmetry breaking, this gives rise to a complex mass matrix M for quarks. To identify the physical quark states which are mass eigenstates, the mass matrix must be diagonalized. In general, it requires a *bi*-unitary transformation to do this

$$U_L M U_R^\dagger = M_{\text{diag}}. \tag{4.2.1}$$

If we denote original quark fields, the weak eigenstates by $Q^0 \equiv (q_1^0, q_2^0, \ldots)$, then the mass eigenstates are denoted by

$$Q_{L,R} = U_{L,R} Q_R^0. \tag{4.2.2}$$

The $U_{L,R}$ matrices, in general, involve complex phases which are potential sources of *CP*-violation.

If we consider the standard model, then the gauge group is $\mathrm{SU}(2)_L \times \mathrm{U}(1)$ and the fermions transform as follows (see Chapter 3):

$$\text{doublet: } \begin{pmatrix} u_a^0 \\ d_a^0 \end{pmatrix}_L;$$

$$\text{singlets: } u_{aR}^0, d_{aR}^0, \qquad a = 1, \ldots, Ng \text{ for generations.}$$

Let us denote by $P^0 \equiv (u_1^0, u_2^2, \ldots)$ the up-quark vector, and by $N^0 \equiv (d_1^0, d_2^0, \ldots)$ the down-quark vector $(1, \ldots, Ng$ denote generations).

As discussed in Chapter 3, the gauge interactions for quarks can be written as follows:

$$\begin{aligned}\mathscr{L}_{\text{wk}} = & \frac{-ig}{\sqrt{2}} W_\mu^+ \bar{P}_L^0 \gamma_\mu N_L^0 \\ & + \frac{-ig}{\cos\theta_W} Z_\mu [\tfrac{1}{2}(\bar{P}_L^0 \gamma_\mu P_L^0 - \bar{N}_L^0 \gamma_\mu N_L^0) \\ & - \sin^2\theta_W (\tfrac{2}{3}\bar{P}^0 \gamma_\mu P^0 - \tfrac{1}{3}\bar{N}^0 \gamma_\mu N^0)].\end{aligned} \tag{4.2.3}$$

Using eqn. (4.2.2) to define physical quarks which are mass eigenstates, i.e., $P_{L,R} \equiv U_{L,R} P_{L,R}^0$ and $N_{L,R} \equiv V_{L,R} N_{L,R}^0$, we find the following expression for $\mathscr{L}_{\text{wk}}$:

$$\begin{aligned}\mathscr{L}_{\text{wk}} = & \frac{-ig}{\sqrt{2}} W_\mu^\dagger \bar{P}_L \gamma_\mu U_L V_L^\dagger N_L \\ & + \frac{-ig}{2\cos\theta_W} Z_\mu [\bar{P}_L \gamma_\mu P_L - \bar{N}_L \gamma_\mu N_L \\ & - \tfrac{4}{3}\sin^2\theta_W (\bar{P}\gamma_\mu P - \tfrac{1}{2}\bar{N}\gamma_\mu N)].\end{aligned} \tag{4.2.4}$$

$U_L V_L^\dagger$ is the charged current mixing matrix and will, in general, contain some complex phases. For N_g generations, $U_L V_L^\dagger \equiv U_{\text{KM}}$ is an $N_g \times N_g$ unitary matrix and contains N_g^2 parameters, out of which $N_g(N_g - 1)/2$ are real angles that parametrize the $N_g \times N_g$ orthogonal matrix. So, the starting number of complex phases is

$$n_p^0 = (N_g^2) - \frac{N_g(N_g - 1)}{2}$$

$$= \frac{N_g^2}{2} + \frac{N_g}{2} = \tfrac{1}{2}(N_g)(N_g + 1). \tag{4.2.5}$$

Now, the phases of both the N- and P-fields could be separately redefined and thereby $2N_g - 1$ phases could be removed. This, then, leaves us with n_p genuine CP-phases

$$n_p = \frac{N_g^2}{2} + \frac{N_g}{2} - 2N_g + 1$$

$$= \tfrac{1}{2}(N_g - 1)(N_g - 2). \tag{4.2.6}$$

We see that for $N_g = 1$ or 2, $n_p = 0$ and for three generations

$$n_p = 1. \tag{4.2.7}$$

Thus, in pure left-handed models, we needs three generations to get CP-violation. This is the Kobayashi–Maskawa model which we discuss in the next section.

We will show in a subsequent section that if we include right-handed charged currents the phase counting argument is different [7a], and we can obtain nontrivial phases even for two generations. We will show in Section 4.4 that this leads to a very interesting alternative model of CP-violation.

§4.3. The Kobayashi–Maskawa Model

It is the simplest extension of the standard one-generation model described in the previous chapter that can accommodate CP-violation. The Lagrangian describing weak interactions in this model is given in eqn. (4.2.4) where $U_{\text{KM}} = U_L V_L^\dagger$ contains both generation mixing as well as CP-violating effects. To study the implications of this model, we write down U_{KM} which is parametrized in terms of three real angles and a phase

$$U_{\text{KM}} = \begin{pmatrix} c_1 & -s_1 c_3 & -s_1 s_3 \\ +s_1 c_2 & c_1 c_2 c_3 - s_2 s_3 e^{i\delta} & c_1 c_2 c_3 + s_2 c_3 e^{i\delta} \\ +s_1 s_2 & c_2 s_2 c_3 + c_2 s_3 e^{i\delta} & c_1 s_2 s_3 - c_2 c_3 e^{i\delta} \end{pmatrix}, \tag{4.3.1}$$

where $c_i = \cos\theta_i$, $s_i = \sin\theta_i$, $i = 1, 2, 3$. We will restrict all the angles θ_i to the first quadrant, i.e., $0 \le \theta_i \le \pi/2$ and allow δ to vary between 0 to 2π. The detailed implications of this model have been studied in a number of recent papers and have been extensively reviewed by L. L. Chau [8].

To study the implications of this model for weak processes, we must know the values of θ_i and δ. Before doing this we can give an alternative representation for U_{KM} as follows:

$$U_{\mathrm{KM}} = \begin{pmatrix} V_{ud} & V_{us} & V_{ub} \\ V_{cd} & V_{cs} & V_{cb} \\ V_{td} & V_{ts} & V_{tb} \end{pmatrix}. \tag{4.3.2}$$

The various elements of U_{KM} have been determined from low-energy weak interaction data: the muon-decay rate defines the Fermi coupling constant G_F [9]:

$$G_F = (1.16638 \pm 0.00002) \times 10^{-5}\ \mathrm{GeV}^{-2}. \tag{4.3.3}$$

From $0^+ \to 0^+$ allowed Fermi transition and using eqn. (4.3.3) we can determine [10]

$$|V_{ud}| = 0.9738 \pm 0.025. \tag{4.3.4}$$

V_{us} can be determined [10] using semileptonic decays of hyperon and Ke_3 decays to be

$$|V_{us}| = 0.220 \pm 0.003. \tag{4.3.5}$$

Recent information [11, 12] on the lifetime and leptonic branching ratio can be used to determine V_{cb} and put an upper limit on V_{ub}

$$|V_{cb}| = 0.05 \pm 0.01$$

and

$$|V_{ub}| < 0.13|V_{cb}|. \tag{4.3.6}$$

These can be converted to give the following allowed ranges for the angles s_2 and s_3

$$0.02 < s_3 < 0.06,$$
$$0.05 < s_2 < 0.14. \tag{4.3.7}$$

To calculate ε we use eqn. (4.1.12) and calculate the imaginary part of the $\Delta S = 2K^0 - \bar{K}^0$ matrix element to find ρ, and the imaginary part of the $\Delta I = \frac{1}{2} K \to 2\pi$ decay amplitude to find $\mathrm{Im}\, A_0/\mathrm{Re}\, A_0$. The latter receives dominant contribution from the gluon mediated Penguin diagrams shown in Fig. 4.1. As far as the evaluation of ρ is concerned, reliable estimates can only be made of the short distance contribution coming from Fig. 4.2. To obtain an expression for ρ and the short distance contribution to $K_L - K_S$ mass difference, let us define a parameter λ_i, $i = e, t$

$$\lambda_i = V_{id}^* V_{is} \tag{4.3.8}$$

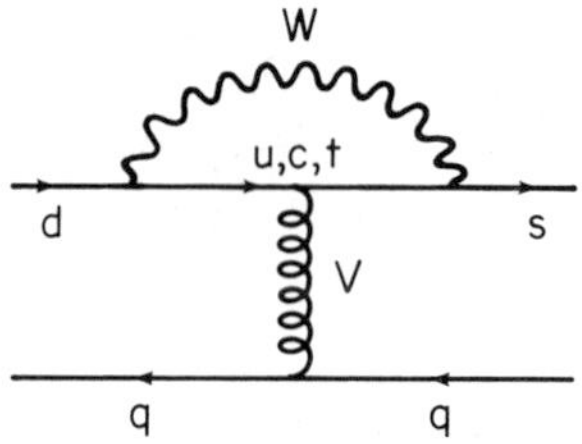

Figure 4.1. Penguin diagram which gives dominant contribution to the $\Delta I = \frac{1}{2} K \to 2\pi$ amplitudes.

and note that

$$m_{K_L} - m_{K_S} = 2 \text{ Re } M_{K^0-\bar{K}^0}. \tag{4.3.9}$$

The box graph of Fig. 4.2 makes the following contribution to $m_{K_L} - m_{K_S}$ [13]:

$$(m_{K_L} - m_{K_S})_{\text{box}} = \frac{G_F^2}{6\pi^2} M_W^2 f_K^2 m_K B F(x_i, \theta_j), \tag{4.3.10}$$

where

$$\begin{aligned} F(x_i, \theta_j) = {} & [(\text{Re } \lambda_c)^2 - (\text{Im } \lambda_c)^2]\eta_1 f(x_c) \\ & + [(\text{Re } \lambda_t)^2 - (\text{Im } \lambda_t)^2]\eta_2 f(x_t) \\ & + 2(\text{Re } \lambda_c \text{ Re } \lambda_t - \text{Im } \lambda_c \text{ Im } \lambda_t)\eta_3 f(x_t, x_c), \end{aligned} \tag{4.3.11}$$

where

$$\begin{aligned} f(x_i) &= x_i[\tfrac{1}{4} + \tfrac{9}{4}(1 - x_i)^{-1} - \tfrac{3}{2}(1 - x_i)^{-2}] + \frac{3}{2}\left[\frac{x_i}{x_i - 1}\right]^3 \ln x_i, \\ f(x_i, x_j) &= x_i x_j \left\{ [\tfrac{1}{4} + \tfrac{3}{2}(1 - x_j)^{-1} - \tfrac{3}{4}(1 - x_j)^{-2}] \frac{\ln x_j}{x_j - x_i} \right. \\ & \qquad \left. + (x_j \leftrightarrow x_i) - \tfrac{3}{4}[(1 - x_j)(1 - x_i)]^{-1} \right\}. \end{aligned} \tag{4.3.12}$$

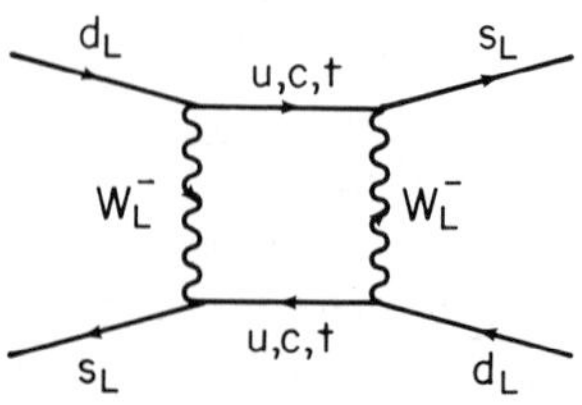

Figure 4.2. Short distance contribution to the $\Delta S = 2$ processes in the $SU(2)_L \times U(1)$ model.

$x_i = m_i^2/M_W^2$ and η_i are the various QCD corrections given [14] by $\eta_1 = 0.85$, $\eta_2 \simeq 0.6$, and $\eta_3 \simeq 0.39$. The parameter B denotes the matrix element $\langle K^0|\bar{d}\gamma_\mu(1+\gamma_5)s\,\bar{d}\gamma_\mu(1+\gamma_5)s|\bar{K}^0\rangle$. This is an unknown parameter whose value is expected [15] to be between 0.33 and 1, putting in the values of G_F, $f_K = 160$ MeV, $M_W = 82$ GeV, we find (using $m_c \simeq 1.4$ GeV)

$$\begin{aligned}(\Delta M)_{\text{box}} &\simeq 1.92 \times 10^{-10}\, BF(x_i, \theta_j)\ \text{GeV} \\ &\simeq 0.22 \times 10^{-14}\, B\ \text{GeV}.\end{aligned} \tag{4.3.13}$$

Thus, we need $B \simeq 1.5$ to understand the $K_L - K_S$ mass difference as coming entirely from short distance effects. It has, however, been argued that there may be significant long-distance effects in $K_L - K_S$ mass difference which are hard to estimate.

Let us now turn to the prediction of the Kobayashi–Maskawa model for ε and ε'. Again, we will ignore any possible long-distance effects. From eqn. (4.1.14) we see that we have to find m' which is the imaginary part of $M_{K^0\bar{K}^0}$ and $\xi \equiv \text{Im}\, A_0/\text{Re}\, A_0$: using the box graph in Fig. 4.2 we can write

$$|\varepsilon| = \frac{G_F^2}{12\sqrt{2}\pi^2}\frac{BM_W^2 f_K^2}{\Delta M} m_K D(x_i, \theta_j) + \xi \ldots, \tag{4.3.14}$$

where

$$\begin{aligned}D(x_i, \theta_j) = 2\, \text{Im}\, \lambda_t \{&-\text{Re}\, \lambda_c f(x_c)\eta_1 + \text{Re}\, \lambda_t f(x_t)\eta_2 \\ &+ (\text{Re}\, \lambda_c - \text{Re}\, \lambda_t) f(x_t, x_c)\eta_3\}.\end{aligned} \tag{4.3.15}$$

All the symbols in eqn. (4.3.15) are already explained. Using the Penguin diagram [16] of Fig. 4.1 we can write

$$\xi = Hs_2c_2s_3s_\delta, \tag{4.3.16}$$

where H is the matrix element of a four-quark $\Delta S = 1$ operator between the K and 2π states which has been estimated to be $H \approx -0.54$.

$$\begin{aligned}|\varepsilon| \simeq 0.4 s_2 s_3 s_\delta B \Bigg\{&-\eta_1 + \frac{\text{Re}\, \lambda_t}{\text{Re}\, \lambda_c}\eta_2 \frac{f(x_t)}{f(x_c)} \\ &+ \eta_3\left(1 - \frac{\text{Re}\, \lambda_t}{\text{Re}\, \lambda_c}\right) f(x_t, x_c)/f(x_c)\Bigg\} + \xi.\end{aligned} \tag{4.3.17}$$

Using eqn. (4.3.7) it is easy to check that for presently reported values [17] of m_t, i.e., 30 GeV $\leq m_t \leq$ 50 GeV, it is hard to get $\varepsilon \simeq 2 \times 10^{-3}$ for $B \simeq \frac{1}{3}$ although $B \simeq \frac{2}{3}$ can explain the value of ε if s_2 and s_3 are at their maximum value. This means that $b \to u$ semileptonic decay should be observable near the upper limit for the value of V_{ub} and $\sin \delta$ must be close to 1. Thus, all parameters must be stretched to their maximum allowed values to understand the CP-violating parameter ε.

Let us now study the predictions for ε' in this model starting from (4.1.12).

Since the Penguin diagrams are purely $\Delta I = \frac{1}{2}$ type, Im $A_2 = 0$ and we get

$$\varepsilon' = -\frac{\omega}{\sqrt{2}} \frac{\text{Im } A_0}{\text{Re } A_0}$$

or

$$\left|\frac{\varepsilon'}{\varepsilon}\right| = \frac{1}{\sqrt{2}} \left|\frac{\xi}{\varepsilon}\right| \left|\frac{A_2}{A_0}\right|$$

$$\simeq 15.6|\xi|. \tag{4.3.18}$$

Using eqn. (4.3.16) for ξ and the lower limits on the values of s_2 and s_3 then imply that

$$|\varepsilon'/\varepsilon| \geq 10^{-3}. \tag{4.3.19}$$

It can also be shown that within the framework of our approximations, ε'/ε is positive. This result is barely consistent with eqn. (4.1.16). So if ε'/ε keeps decreasing to zero or becomes negative, the Kobayashi–Maskawa model cannot fully explain observed *CP*-violation. In the next section, we will explore alternative models of *CP*-violation.

§4.4. Left–Right Symmetric Models of *CP*-Violation

Soon after the *CP*-violation models [5, 6] based on $SU(2)_L \times U(1)$ models were constructed, it was pointed out by Mohapatra and Pati [18] that extending the gauge group to $SU(2)_L \times SU(2)_R \times U(1)_{B-L}$ helps the introduction of *CP*-violation into gauge models even with two generations of quarks. The fact that with right-handed currents two generations are enough for *CP*-violation was actually noted in Ref. [7]. The $SU(2)_L \times SU(2)_R \times U(1)_{B-L}$ extends this idea and makes it much more appealing and phenomenologically acceptable. By that time (1974) even charm quarks were not discovered, let alone the third generation of fermions. As we will see below, an important aesthetic feature of the left–right symmetric model of *CP*-violation is that *CP*- and *P*-violation are linked to each other. In this section, we would like to discuss this model and its implications in brief.

The left–right symmetric electro-weak models will be discussed in detail in Chapter 6. Here we briefly introduce those aspects of the model needed for the discussion of *CP*-violation. The quarks and leptons will be assigned in a left–right symmetric manner under the gauge group. Denoting $Q = (u, d)$ and $\psi = (\nu, e^-)$ we assign (the numbers within parentheses stand for $SU(2)_L$, $SU(2)_R$, and $U(1)_{B-L}$ quantum numbers):

$$Q_L \equiv (\tfrac{1}{2}, 0, \tfrac{1}{3}), \qquad Q_R \equiv (0, \tfrac{1}{2}, \tfrac{1}{3}),$$
$$\psi_L \equiv (\tfrac{1}{2}, 0, -1), \qquad \psi_R \equiv (0, \tfrac{1}{2}, -1). \tag{4.4.1}$$

There are two sets of gauge bosons $\mathbf{W}_L$, $\mathbf{W}_R$, and B whose coupling to fermions is uniquely given by the requirement of gauge invariance. The details of gauge symmetry breaking will be discussed in Chapter 6. It is sufficient to note that we will choose the following pattern:

$$\mathrm{SU}(2)_L \times \mathrm{SU}(2)_R \times \mathrm{U}(1)_{B-L}$$
$$\downarrow m_{W_R}, m_{Z_R}$$
$$\mathrm{SU}(2)_L \times \mathrm{U}(1)_Y$$
$$\downarrow m_{W_L}$$
$$\mathrm{U}(1)_{\mathrm{em}}.$$

The first stage can be implemented by the choice of either left–right symmetric Higgs doublets (χ_L, χ_R) or triplets (Δ_L, Δ_R). The second stage is implemented by the mixed doublet $\phi(2, 2, 0)$ which acquires, in general, the following vacuum expectation value:

$$\langle \phi \rangle = \begin{pmatrix} \kappa & 0 \\ 0 & \kappa' e^{i\alpha} \end{pmatrix}. \tag{4.4.2}$$

This has the property of maintaining $\rho_W = 1$, a fact which is experimentally well established. An important point is that for both $\kappa, \kappa' \neq 0$ the W_L and W_R mix with a CP-violating phase $e^{i\alpha}$.

To discuss CP-violation we have to study the fermion masses. They arise from Yukawa coupling between Q and ϕ. We will consider the minimal set of Higgs bosons, i.e., one ϕ and its charge conjugate field $\tilde{\phi} \equiv \tau_2 \phi^* \tau_2$. Under $\mathrm{SU}(2)_L \times \mathrm{SU}(2)_R$ transformations we have

$$\phi \to U_L \phi U_R^+$$

and

$$\tilde{\phi} \to U_L \tilde{\phi} U_R^+. \tag{4.4.3}$$

The most general Yukawa coupling involving these fields, the quark fields Q, and leptonic fields ψ is

$$\begin{aligned} \mathscr{L}_Y = {} & h_{ij} \bar{Q}_{Li} \phi Q_{Rj} + \tilde{h}_{ij} Q_{Li} \tilde{\phi} Q_{Rj} + \text{h.c.} \\ & + h'_{ij} \bar{\psi}_{Li} \phi \psi_{Rj} + h''_{ij} \bar{\psi}_{Li} \tilde{\phi} \psi_{Ri} + \text{h.c.} \end{aligned} \tag{4.4.4}$$

We will consider the general case of complex Yukawa couplings. Under left–right symmetry transformation we have

$$\begin{aligned} \phi &\leftrightarrow \phi^+, \\ \tilde{\phi} &\leftrightarrow \tilde{\phi}^+. \end{aligned} \tag{4.4.5}$$

Thus, $\mathscr{L}_Y$ will be left–right symmetric provided [19]

$$h_{ij} = h_{ji}^*, \tag{4.4.6}$$

$$\tilde{h}_{ij} = \tilde{h}_{ji}^*, \tag{4.4.7}$$

and similarly for h' and h''.

The quark mass matrices are obtained from eqn. (4.4.4) on substituting vacuum expectation values for ϕ and $\tilde{\phi}$ in eqn. (4.4.2).

The first entry in eqn. (4.4.2) can be made real by a gauge transformation. It is worth emphasizing that for the most general Higgs potential and Yukawa couplings, eqn. (4.4.2) is the only allowed minimum. On the other hand, if in addition to parity invariance, Lagrangian is chosen to respect CP-invariance there exists a range of parameters for which $\alpha = 0$ naturally. The interesting point in this case is that the mass matrices are real and symmetric with equal left- and right-handed quark mixing angles.

If we require the weak Lagrangian to conserve both P and CP prior to spontaneous breakdown, two cases arise

(i) $\alpha = 0$. In this case, the mass matrices are real and symmetric leading to equal left- and right-fermion mixing matrices, i.e.,

$$U_L = U_R. \tag{4.4.8}$$

This has been called (in literature) manifest left–right symmetry [20].

(ii) $\alpha \neq 0$. This case turns out to be the most interesting since here the Yukawa couplings are real but CP-symmetry is spontaneously broken by vacuum leading to

$$M_u = M_u^{\mathrm{T}}, \qquad M_d = M_d^{\mathrm{T}}. \tag{4.4.9}$$

This implies [19]

$$U_R = U_L^* K, \tag{4.4.10}$$

where K is a diagonal unitary matrix. This case has an important bearing on CP-violation. It is important to emphasize here that with the straightforward definition of left–right transformation (eqn. (4.4.5)) of the ϕ-fields, we always have the real mixing angles (for both left- and right-hand sectors) equal. Thus, even though *a priori* the presence of right-handed charged currents could have brought in a whole new set of mixing angles, the symmetries of the theory in the simplest model do not permit this. The only new parameters in weak interactions therefore remain m_{W_R}, $W_L - W_R$ mixing angle ζ, and new complex phases in right-handed currents.

In the case of $\alpha \neq 0$ with arbitrary Yukawa couplings the mass matrices are no more Hermitian and, therefore, left- and right-mixing angles will be different, their difference being proportional to α.

The phase counting in left–right symmetric models is different due to the presence of right-handed charged currents. The situation is, of course, dependent on phase convention in one of which the number of significant phases (after arbitrary rotation of quark fields as in Section 4.3 in the left- and right-handed charged currents) is given by

$$N_L = \frac{(N-1)(N-2)}{2} \tag{4.4.11}$$

and

$$N_R = \tfrac{1}{2}N(N+1). \tag{4.4.12}$$

Note that this reduces to eqn. (4.2.6) in the absence of right-handed currents.

We find that for two generations, $N_L = 0$, $N_R = 3$; for three generations, $N_L = 1$, $N_R = 6$. Thus, for two generations it is possible to introduce *CP*-violation for two generations in the left–right symmetric models as claimed earlier.

It is now clear that all *CP*-violation resides in the W_R-sector, in the limit of $m_{W_R} \to \infty$, $\eta_{+-} \to 0$, which implies that

$$\eta_{+-} = \left(\frac{m_{W_L}}{n_{W_R}}\right)^2 \sin\delta. \tag{4.4.13}$$

To present the details we denote $P \equiv (u, c)$ and $N \equiv (d, s)$, and write the weak Lagrangian as

$$\mathscr{L}_{\text{wk}} = \frac{g}{2\sqrt{2}}[\bar{P}_L\gamma_\mu U_L N_L W_L^+ + \bar{P}_R\gamma_\mu U_R N_R W_R^+ + \text{h.c.}, \tag{4.4.14}$$

where

$$U_L = \begin{pmatrix} \cos\theta & \sin\theta \\ -\sin\theta & \cos\theta \end{pmatrix},$$

$$U_R = e^{i\tau_3\alpha_1}U_L e^{-i\tau_3\alpha_2}e^{i\beta} \qquad [\tau_3 \text{ is the diagonal Pauli matrix}]. \tag{4.4.15}$$

Ignoring Higgs- and $W_L - W_R$-mixing effects, we can write the $\Delta S = 1$ non-leptonic weak Hamiltonian in this model as ($\eta \equiv (M_{W_L}/m_{W_R})^2$)

$$H_{\text{wk}}^{\text{P.V.}} = \frac{G_F}{2\sqrt{2}} \sin 2\theta O_1^{(+)}(1 - + \eta e^{i\delta}) + \text{h.c.},$$

$$H_{\text{wk}}^{\text{P.C.}} = \frac{G_F}{2\sqrt{2}} \sin 2\theta O_2^{(+)}(1 + \eta e^{i\delta}) + \text{h.c.}, \tag{4.4.16}$$

where $\delta = -2\alpha_2$ and $O_i^{(+)}$ include both the long distance and the Penguin contributions to the effective nonleptonic weak Hamiltonian. This form of the weak Hamiltonian satisfies the relations

$$[I_3, H_{\text{wk}}^{(+)}] \simeq i(m_{W_L}/m_{W_R})^2 \sin\delta H_{\text{wk}}^{(-)} \tag{4.4.17}$$

for both parity-conserving and parity-violating parts. This was called the isoconjugate relation. It leads to

$$\frac{M(K_2^0 \to \pi_i\pi_j)}{M(K_1^0 \to \pi_i\pi_j)} = i\left(\frac{m_{W_L}}{m_{W_R}}\right)^2 \sin\delta \qquad (i, j = +, -, \text{ or } 00) \tag{4.4.18a}$$

and

$$\frac{M(K_1^0 \to \pi_i\pi_j\pi_k)}{M(K_2^0 \to \pi_i\pi_j\pi_k)} = -i\left(\frac{m_{W_L}}{m_{W_R}}\right)^2 \sin\delta \qquad (i, j, k = +, -, 0 \text{ or } 0, 0, 0). \tag{4.4.18b}$$

From these we obtain

$$\eta_{+-} = e^{i\pi/4}\left[\frac{\text{Im } M_{K^0\bar{K}^0}}{\text{Re } M_{K^0\bar{K}^0}} + \left(\frac{m_{W_L}}{m_{W_R}}\right)^2 \sin\delta\right] \tag{4.4.19}$$

and

$$\eta_{+-} - \eta_{+-0} = 2\left(\frac{n_{W_L}}{m_{W_R}}\right)^2 \sin\delta. \tag{4.4.20}$$

Thus, in the approximation that we ignore Higgs- and $W_L - W_R$-mixing effects we obtain $\varepsilon' = 0$ and $\eta_{+-} - \eta_{+-0} = 2\eta \sin\delta$, the former coinciding with the superweak result and the latter potentially different from superweak prediction. To discuss the second prediction, which could be tested in experiments involving intense Kaon beams, we have to study $M_{K^0\bar{K}^0}$. In addition to the $W_L - W_L$ contribution discussed earlier, there are $W_L - W_R$ and Higgs exchange graphs all of which are potentially large and can be written symbolically as

$$M_{K^0\bar{K}^0} = f_{LL}[1 + \eta e^{i\delta} A_{LR} + e^{i\delta} 2A_H] \ldots. \tag{4.4.21}$$

The absolute values of A_{LR} and A_H are much larger than 1, e.g., in the vacuum saturation approximation [21], $A_{LR} \simeq -430$ as we will discuss in Chapter 6. The complexion of the CP-violating effect will depend on whether A_{LR} and A_H cancel each other in $M_{K\bar{K}}$ (Case (a) or not (Case (b)).

Case (a). In this case, $\varepsilon \simeq \eta \sin\delta$ and $|\eta_{+-} - \eta_{+-0}| \simeq \eta_{+-}$, a prediction that can be tested by the next generation of Kaon beam experiments.

Case (b). In this case, A_{LR} dominates η_{+-} and $A_{LR} \gg 1$, and we obtain $\varepsilon \simeq \eta A_{LR} \sin\delta \approx \sin\delta$ and $|\eta_{+-} - \eta_{+-0}| \leq |\eta_{+-}|\,(10^{-2}$ to $10^{-3})$ which is like superweak prediction.

Note that, in either case, CP- and P-violation are related to each other, which is a unique feature of this model. In fact, if we are to take the model seriously as the only source of CP-violation in K-meson decays, W_R cannot be too heavy. This gives an upper bound on m_{W_R}

$$m_{W_R} \leq 35 \text{ TeV}. \tag{4.4.22}$$

Including QCD corrections [22], the upper limit goes up from 60 TeV to 80 TeV. Thus, the right-handed W-boson could be observable in machines with energies in the multi-TeV range.

Another noteworthy feature of this kind of model is the possible large CP-violation for semileptonic D-meson decays (for light W_R), a distinguishing feature from the Kobayashi–Maskawa and Higgs models.

In the discussion of CP-violation so far, we have ignored the $W_L - W_R$-mixing parameter. If we also ignore the Higgs-induced CP-violating effects by assuming heavy physical Higgs boson masses, we obtain that

$$\varepsilon' = 0 \quad \text{and} \quad d_n^e = 0, \tag{4.4.23}$$

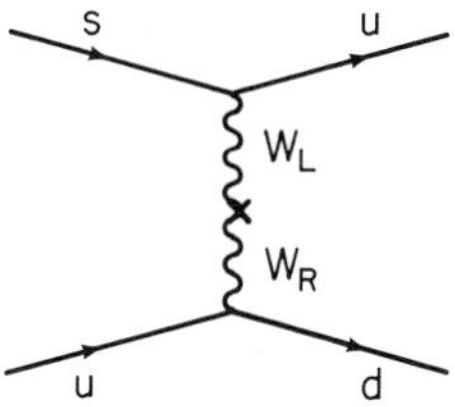

Figure 4.3. $W_L - W_R$ mixing contribution to ε'.

where d_n^e is the electric dipole moment of the neutron. Once the $W_L - W_R$ mixing effect is taken into account this leads to both ε' and d_n^e, by the graphs in Figs. 4.3 and 4.4, respectively [23].

$$\varepsilon' \approx \frac{1}{20}\left(\frac{\kappa'}{\kappa}\right)\left(\frac{m_{W_L}}{m_{W_R}}\right)^2 \sin\delta \tag{4.4.24}$$

and

$$d_n^e \approx 10^{-21} \sin\delta\left(\frac{\kappa'}{\kappa}\right)\left(\frac{m_{W_L}}{m_{W_R}}\right)^2 A, \tag{4.4.25}$$

where A is a model-dependent parameter which can be of order 1 to 1/10. Then we obtain the following relation between ε' and d_n^e, i.e.,

$$d_n^e \approx A\, 10^{-20}\, \varepsilon' \text{ e.c.m.} \tag{4.4.26}$$

Using the experimental bound that $\varepsilon'/\varepsilon \leq 2 \times 10^{-3}$ we get $d_n^e \leq 10^{-26}$–10^{-27} e.c.m. Thus, if ε' is measured at the level of 1% then, for this model to be valid, the electric dipole moment of the neutron should be $\approx 10^{-26}$–10^{-27} e.c.m.

Looking back for a moment at the prediction of $\varepsilon = (m_{W_L}/m_{W_R})^2\, 430 \sin\delta$, we realize that due to the box graph enhancement mentioned, which requires m_{W_R} in the TeV region, the smallness of *CP*-violation requires the phase δ to be extremely small, i.e., $\delta \leq 10^{-3}$. In fact, an amusing choice is $\eta = \delta = 1/430$ which reproduces the observed value of ε. It is then legitimate to question: How natural is it to choose δ so small? Actually, it turns out that with models

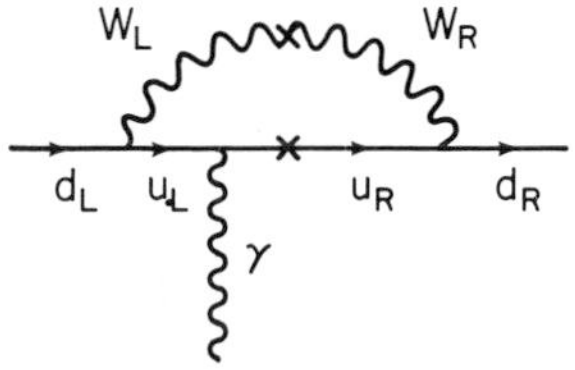

Figure 4.4. $W_L - W_R$ mixing contribution to d_n^e.

with soft *CP*-violation and gauge hierarchies we can show [24] that

$$\delta \approx \left(\frac{m_{W_L}}{m_{W_R}}\right)^2. \tag{4.4.27}$$

The basic argument behind eqn. (4.4.27) is the following. In the limit of $m_{W_R} \to \infty$, the effective low-energy theory is the standard model with only one Higgs doublet. In this model the minimum of the Higgs potential is *CP*-conserving. Thus, the soft *CP*-phase $\delta \to 0$ as $m_{W_R} \to \infty$. But, since it is nonvanishing when m_{W_R} is finite, we must have eqn. (4.4.27).

This model has recently been extended [25] to the six-quark model by the following observation: isoconjugate relations hold if the mixing matrices obey the form

$$U_L = 0$$

and

$$U_R = K_1 O K_2, \tag{4.4.28}$$

where K_1 and K_2 are diagonal unitary matrices and O is an orthogonal matrix. This kind of mixing matrices results if the matrices are such that MM^+ = real but M^+M not. The relation between *CP and P* remains valid in this model and so does the rest of the phenomenology. A natural implementation of this model, however, requires some *ad hoc* discrete symmetries in the starting Hamiltonian.

Another Way to Relate *CP*- and *P*-Violation

In a recent paper it has been noted by Chang [26] that left–right symmetric models provide another way to realize the connection between *P*- and *CP*-violation, if the theory is *CP*-conserving prior to spontaneous breakdown. The key to this observation is in eqn. (4.4.7) where we noted that if $\kappa' = 0$ (i.e., $W_L - W_R$ mixing vanishes) all *CP*-violation disappears. Since $\alpha \neq 0$ is the only source of *CP*-violation, all phases in the weak currents must be expressible in terms of α.

It has been shown that for the case of two generations the phases in eqn. (4.4.15) are

$$-4\beta = 2\alpha_2 = \alpha_1 = -\frac{m_c}{m_s}\left(\frac{\kappa'}{\kappa}\right)\sin\alpha. \tag{4.4.29}$$

For the parameters ε and ε' we obtain [26]

$$\varepsilon \simeq \frac{430}{2}\eta\left(\frac{\kappa'}{\kappa}\right)\frac{m_c}{m_s}\sin\alpha, \tag{4.4.30}$$

$$\varepsilon' \simeq \frac{\chi}{20\sqrt{2}}4\eta\left(\frac{\kappa'}{\kappa}\right)\sin\alpha, \tag{4.4.31}$$

where χ is a strong interaction parameter of order 10.

First point to note here is the connection between CP- and P-violation as in the isoconjugate model. Second, the phase α must also be very small to explain the observed magnitude. Finally, we obtain, $\varepsilon'/\varepsilon \approx \frac{1}{430} \approx 2 \times 10^{-3}$, which could be tested in the next round of experiments searching for ε'.

The model can be extended to three generations. Now, of course, there are seven phases all of which, in principle, are expressible in terms of α- and CP-violating phenomena which receive not only contributions from right-handed currents but also from the usual Kobayashi–Niskyima phase in the left-handed currents. The interesting point is that the box graph enhancement factor of 430 makes the right-handed current contributions dominate over the Kobayashi–Niskyima effect as the source of CP-violation, thus preserving the relation between CP- and P-violation.

Before closing, we point out that the difference between the isoconjugate model and the model of Ref. [26] is that ε'/ε can be arbitrarily small in isoconjugate models but not so in the other one. Thus, if the upper limit on ε'/ε is stretched to its maximum possible value of 10^{-4}, to model of Ref. [26] will be ruled out. However, in the framework of the isoconjugate model it will simply mean that the $W_L - W_R$ mixing of κ'/κ is very small.

Thus, right-handed currents provide a much more appealing framework for describing CP-violating interactions, and can be soon tested by improved experimental search for ε and neutron electric dipole moment.

§4.5. The Higgs Exchange Models

The idea, in these models, is to have CP-violation reside only in the Higgs sector while keeping the gauge sector CP-conserving as in the two-generation $SU(2)_L \times U(1)$ model. These models are of two types: (i) where the neutral Higgs couplings to quarks is flavor violating; and (ii) where it is flavor conserving. In the first case, $K_L - K_S$ mass difference implies that the masses of the Higgs bosons must be very heavy (typically $\alpha m_f^2/m_H^2 \leq 10^{-12}$ GeV, which implies $m_H \geq 10^2$ TeV or so). Such heavy masses are barely compatible with unitarity. Therefore, they can lead to superweak type models for CP-violation as discussed in Ref. [27]. In case (ii) CP-violation can arise through the exchange of charged Higgs bosons. Therefore, it can lead to milli-weak ($\sim G_F \times 10^{-3}$ as the strength of CP-violating interaction) type models of CP-violation. These have been discussed by Weinberg [28]. An important point has been made by Branco [29] that, for an arbitrary number of generations, the requirement of natural flavor conservation leads to CP-conserving gauge interactions in the $SU(2)_L \times U(1)$ model. In this model the only source of CP-violation will be from the Higgs exchanges. This gives a sort of uniqueness to the Higgs exchange models of CP-violation which otherwise would be complicated by the presence of phases in weak currents.

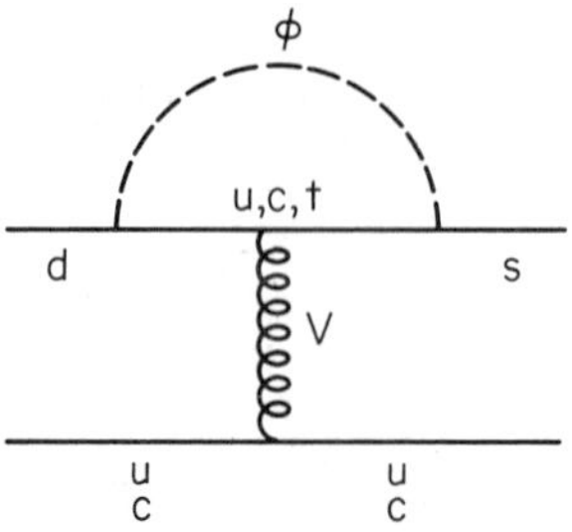

Figure 4.5. The Higgs–Penguin graph that contributes to ξ.

Higgs interactions with quarks in this class of models can be written as

$$\mathscr{L}_\phi = \frac{1}{\lambda_1^*}\bar{N}_R M^{\mathrm{d}} U_{\mathrm{KM}}^+ P_L \phi_1^- - \frac{1}{\lambda_2}\bar{P}_R M^{\mathrm{u}} U_{\mathrm{KM}} N_L \phi_2^+ + \text{h.c.}, \tag{4.5.1}$$

where $P = (u, c, t, \ldots)$; $N = (d, s, b, \ldots)$; U_{KM} is the Cabibbo–Kobayashi–Maskawa mixing matrix; $\phi_{1,2}$ are the physical charged Higgs bosons; and $M^{\mathrm{u,d}}$ are the diagonal mass matrices for the up- and down-quarks. *CP*-violation arises due to the complex propagator

$$\Delta_\phi = \frac{\langle 0|T(\phi_1^- \phi_2^+)|0\rangle_{\mathrm{F.T.}}}{\lambda_1^* \lambda_2}, \tag{4.5.2}$$

where F.T. stands for Fourier transform. It has been shown by Deshpande [30] and Sanda [31] that these models predict too large a value for ε'/ε and are in conflict with observations. To see this we note from eqn. (4.1.12) that

$$\frac{\varepsilon'}{\varepsilon} = -\frac{1}{20\sqrt{2}}\left(\frac{\xi}{\varepsilon + \xi}\right). \tag{4.5.3}$$

The dominant graph contributing to ξ and ε now involve Higgs bosons as shown in Figs. 4.5 and 4.6 and it has been found that $\varepsilon \ll \xi$. Therefore, $\varepsilon'/\varepsilon \simeq -\frac{1}{20}$ which is in conflict with experiments. Again, if long-distance effects [32], which strictly are not calculable, are assumed to be big the result could be weakened.

This model also tends to give a big contribution to the neutron electric

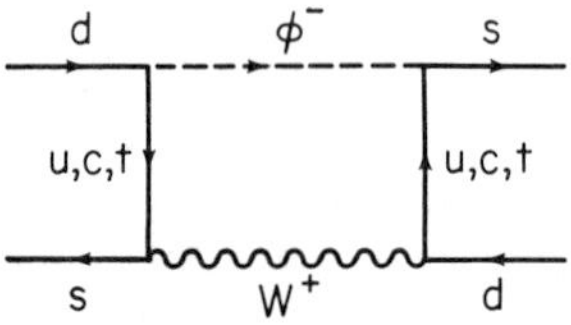

Figure 4.6. New contribution to $\Delta S = \mathrm{S}$ processes in the Higgs exchange models of *CP*-violation.

dipole moment [33], i.e., $d_n > 3 \times 10^{-26}$ e.c.m. and therefore could be in conflict with experiments [34].

§4.6. Strong CP-Violation and the θ-Problem

So far we have discussed CP-violation arising from weak interactions. It was observed in 1976 by Callan, Dashen, and Gross [35], and Jackiw and Rebbi [36] that quantum chromodynamics (QCD) leads to new CP-violating effects in strong interactions. This effect owes its origin to the fact that, in QCD, the axial baryon number current $A_\mu \equiv \sum_a \bar{q}_a \gamma_\mu \gamma_5 q_a$, where a is the flavor index and sum over color index is understood, has an anomaly, i.e.,

$$\partial_\mu A_\mu = \frac{g^2}{32\pi^2} \varepsilon_{\mu\nu\alpha\beta} G_{\mu\nu} G_{\alpha\beta} \tag{4.6.1}$$

this leads to degenerate multiple vacuum [37] structure (labeled by $|\nu\rangle$, $\nu = 0, 1, \ldots$) for QCD. This implies that the true vacuum must be in superposition of all $|\nu\rangle$ vacua with an arbitrary phase $e^{i\nu\theta}$; ν is the topological winding number characterizing the different vacuum solutions and is given by

$$\Delta Q_A = \frac{g^2}{32\pi^2} \int \varepsilon_{\mu\nu\alpha\beta} G_{\mu\nu} G_{\alpha\beta} = \nu, \tag{4.6.2}$$

Q_A is the axial charge.

This is equivalent to adding to the QCD Lagrangian a term

$$\mathscr{L}_\theta = \theta \frac{g^2}{32\pi^2} \varepsilon_{\mu\nu\alpha\beta} G_{\mu\nu} G_{\alpha\beta}, \tag{4.6.3}$$

where $G_{\mu\nu}$ is the gauge covariant antisymmetric gluon field. Note that $\mathscr{L}_\theta$ is P and CP odd. The new CP-violating Lagrangian will lead to observable consequences such as electric dipole moment of the neutron. To calculate this, we note the observation that θ can be expressed in terms of quark fields and for one generation, we can write [38]

$$\mathscr{L}_{CP} = i\theta \frac{m_u m_d}{m_u + m_d} (\bar{u}\gamma_5 u + \bar{d}\gamma_5 d). \tag{4.6.4}$$

From this, we can estimate that

$$d_n^e = (2.7 - 5.2) \times 10^{-16}\, \theta \text{ e.c.m.}$$

Using present limits on d_n^e we conclude that $\theta \leq 10^{-9}$. Thus, θ must either be zero or an extremely small number. The question now arises as to how to understand this small parameter in a natural manner. This is known as the strong CP-problem and, below, we discuss some of the proposed mechanisms to solve this problem. Before we discuss them it is important to point out that the additional term $\mathscr{L}_\theta$ in the Lagrangian is generated if we make a color and flavor singlet axial (or axial baryon number) transformation in the Lagrangian by an amount θ. This is because of Noether's theorem which says that if

$$q_j \to e^{i\theta\gamma_5} q_j, \tag{4.6.5}$$

then

$$\mathscr{L} \to \mathscr{L} + \delta\mathscr{L}, \tag{4.6.6}$$

where

$$\delta\mathscr{L} = i\theta\partial_\mu A_\mu. \tag{4.6.7}$$

Using eqn. (4.6.1) we see that strong CP-parameter θ is the parameter of chiral transformations. Therefore, if no other interactions are present and any of the quarks are massless then, by a chiral transformation, θ can be removed. This fact is reflected in eqn. (4.6.4) in that the physical effect of θ vanishes if either m_u or $m_d \to 0$.

Let us now briefly discuss whether the models of weak CP-violation described in this chapter lead to an acceptable θ. As a prototype example let us consider the Kobayashi–Maskawa model. In this model CP-violation is introduced through complex dimension-four Yukawa couplings. As a result, we would expect the quark mass matrices to have phases at the tree level which will receive infinite corrections in higher orders. Of course, not all the phases present in the mass matrix are related to the strong CP-problem. It is only the phase of the Det $M_u \cdot M_d$, which changes under axial baryon number transformation (4.6.5) and is therefore connected with θ. So, if in a theory $\text{Det}(M_u M_d)$ is finite to all orders, the θ will be finite. Unfortunately, in the Kobayashi–Maskawa model, $\text{Det}(M_u M_d)$ receives infinite contributions at the fourteenth order [38a]. Therefore, the Kobayashi–Maskawa model does not provide a natural framework for understanding the smallness of θ and extensions of this model are needed for this purpose.

Peccei–Quinn Mechanism

Perhaps the most popular solution to this problem is the one proposed by Peccei and Quinn [39], who proposed that the full Lagrangian of the world, i.e., $\mathscr{L} = \mathscr{L}_{\text{QCD}} + \mathscr{L}_{\text{wk}}$, be invariant under a chiral $U(1)_A$ symmetry (to be called $U(1)_{\text{PQ}}$ from now on) operating on the quark flavors and Higgs bosons. This symmetry can then be used to "rotate" θ away. A very nontrivial aspect of this interesting proposal is to show that quark masses, that arise subsequent to spontaneous breakdown, remain real as the $U(1)_{\text{PQ}}$ transformation is made to remove the θ. It was shown [39] that for the simple model considered there this indeed happens. Thus, after the strong CP-violation is removed by a chiral transformation, there remains no trace of CP-violation in the theory if there was no phase in weak interactions.

We now proceed to show how these ideas can be incorporated into the standard electro-weak model. We would like the weak interaction model to have realistic weak CP-violation. Let Q_{La} ($a = 1, 2, 3$ for generations) denote the left-handed quark doublets and u_{Ra} and d_{Ra} be right-handed singlet fields. We define the action of $\mathrm{U(1)_{PQ}}$ on them as follows:

$$Q_{La} \to e^{i\alpha} Q_{La},$$
$$(u)_{Ra}, (d)_{Ra} \to e^{-i\alpha}(u_{Ra}, d_{Ra}). \tag{4.6.8}$$

To generate fermion masses Yukawa couplings of quarks must also be $\mathrm{U(1)_{PQ}}$ invariant. If we have one doublet, called ϕ_d, such that it couples to Q_L and d_R, the coupling is $\bar{Q}_L \phi_d d_R$ where $\phi_d \equiv \begin{pmatrix} \phi_d^+ \\ \phi_d^0 \end{pmatrix}$; the local $\mathrm{U(1)}_Y$ invariance demands that $Y(\phi_d) = +1$ since $\mathrm{U}(Q_L) = \frac{1}{3}$ and $Y(d_R) = -\frac{2}{3}$. Equation (4.6.8) then requires that under $\mathrm{U(1)_{PQ}}$

$$\phi_d \xrightarrow{\mathrm{U}(1)_{PQ}} e^{2i\alpha} \phi_d. \tag{4.6.9}$$

Now $\mathrm{U(1)_{PQ}}$ forbids ϕ_d couplings to $\bar{Q}_L$ and u_R. Thus we must have another Higgs doublet ϕ_u with $Y(\phi_u) = -1$ or $\phi_u = \begin{pmatrix} \phi_u^0 \\ \phi_u^- \end{pmatrix}$ and

$$\phi_u \xrightarrow{\mathrm{U}(1)_{PQ}} e^{2i\alpha} \phi_u. \tag{4.6.10}$$

The full gauge invariant $\mathrm{U(1)_{PQ}}$ symmetric Yukawa coupling can now be written as

$$\mathscr{L}_Y = \sum_{a,b} (h_{ab}^u \bar{Q}_{L,a} \phi_u u_{R,b} + h_{ab}^d \bar{Q}_{La} \phi_d d_{R,b}) + \text{h.c.} \tag{4.6.11}$$

$h_{ab}^{u,d}$ are complex couplings which enable us to introduce CP-violation into the weak currents as in the Kobayashi–Maskawa model. The Higgs potential must be chosen in an $\mathrm{U(1)_{PQ}}$ invariant manner, yet such that the minimum of the potential will correspond to $\langle \phi_u^0 \rangle = v_u \neq 0$ and $\langle \phi_d^0 \rangle = v_d \neq 0$ leading to W, Z, and quark masses. In this theory the strong CP-violating parameter θ can be rotated to give zero.

The Axion

Subsequent to the work of Peccei and Quinn (who, in fact, proposed a model different from the one just described) it was shown by Weinberg [40] and Wilczek [40] that since $\mathrm{U(1)_{PQ}}$ symmetry is broken by the minimum of the potential, there must be a zero mass particle in the theory—called axion. They then pointed out that the axion actually picks up a very tiny mass from the instanton effects which breaks $\mathrm{U(1)_{PQ}}$ symmetry. To see how this happens we recall a result first obtained by 't Hooft which says that the instanton effects can be written in terms of an effective Lagrangian involving the quark fields as follows:

$$\mathscr{L}_{\text{eff}} \approx ce^{i\theta} \operatorname{Det} \|\bar{Q}_L Q_R\| + \text{h.c.} \tag{4.6.12}$$

$\bar{Q}_L Q_R$ in eqn. (4.6.12) symbolically denotes a matrix whose elements are quark bilinears and whose rows and columns denote the various flavors: for instance, for one flavor u, Det $\|\bar{Q}_L Q_R\| = \bar{u}_L u_R$; for two flavors u and d,

$$\operatorname{Det} \|\bar{Q}_L Q_R\| = \operatorname{Det} \begin{vmatrix} \bar{u}_L u_R & \bar{u}_L d_R \\ \bar{u}_R d_L & \bar{d}_L d_R \end{vmatrix}; \tag{4.6.13}$$

etc. Thus for N flavors, Det is a monomial in quark fields of degree $2N$. This, being a determinant, is not invariant under the axial baryon number transformation, and will therefore give mass to the axion which has been estimated by various authors [40, 41] to be the following:

$$\begin{aligned} m_a &\approx \frac{N m_\pi f_\pi}{2(m_u + m_d)^{1/2}} \left[\frac{m_u m_d m_s}{m_u m_d + m_d m_s + m_s m_u} \right]^{1/2} \frac{2^{1/4} G_F^{1/2}}{\sin 2\alpha} \\ &\approx 23 \text{ Kev} \times \frac{N}{\sin 2\alpha}, \end{aligned} \tag{4.6.14}$$

where $\tan \alpha = (v_d/v_u)$ and N is the number of quark flavors. For six flavors we expect $m_a \approx 138/\sin 2\alpha$ keV, i.e., less than 1 MeV. We now wish to study its coupling to matter, to look for processes, where it can be looked for. The axion is a mixture of the Higgs particles Im ϕ_u^0, Im ϕ_d^0, and the hadrons π^0 and η and can be written as

$$a = N\{\xi_{\pi^0}\pi^0 + \xi_\eta v^0 + \chi(\sin \alpha \operatorname{Im} \phi_u + \cos \alpha \operatorname{Im} \phi_d)\}. \tag{4.6.15}$$

The combination of the last two terms is obtained from orthogonality with the Higgs-Kibble particle which is fixed by spontaneous breaking of gauge symmetry. The admixture of π^0 comes entirely from the fact the m_u and m_d are nonzero and has been calculated [40] to be

$$\begin{aligned} \xi_\pi &= \xi \left[\frac{3m_d - m_u}{m_d + m_u} \tan \alpha - \frac{3m_u - m_d}{m_u + m_d} \cot \alpha \right], \\ \xi_\eta &= \xi \left[\sqrt{3} \tan \alpha + \frac{1}{\sqrt{3}} \cot \alpha \right], \\ \xi &= \frac{2^{1/4}}{4} G_F^{1/2} f_\pi, \end{aligned} \tag{4.6.16}$$

x is fixed by normalization. The axion couplings to heavy quarks and leptons are easily noted to be of γ_5-type since it is a pseudo-Goldstone boson (see Chapter 2) and is given by

$$\begin{aligned} \mathscr{L} = i 2^{1/4} G_F^{1/2} \mathscr{A} [&m_c \tan \alpha \bar{c} \gamma_5 c + m_b \cot \alpha \bar{b} \gamma_5 b \\ &+ (m_e \bar{e} \gamma_5 e + m_\mu \bar{\mu} \gamma_5 \mu)(\tan \alpha \text{ or } \cot \alpha)]. \end{aligned} \tag{4.6.17}$$

As far as the light quarks are concerned, we can find their coupling to nucleons from the π^0 and η admixture.

It is now possible to discuss the possible decay and production characteristics of the axion. If it has mass below $2m_e$, it will decay chiefly by processes $a^0 \to 2\gamma$. If $m_a > 2m_e$, we have a decay mode $a^0 \to e^+e^-$. The axions could be produced in decays of excited nuclei [42] and has indeed been looked for in such experiments [43] without success, although, another experiment by the Aachen group [43a] has seen effects which can be interpreted as an axionlike object. The present consensus appears to be that an axion with mass less than 1 MeV is inconsistent with experiments.

The apparent failure of the experimental searches [44] of the axion has prompted recent theoretical speculation that the $U(1)_{PQ}$ symmetry may be a symmetry at ultrahigh ($\sim 10^9$ GeV) energies [45] with breaking scale $V_{PQ} \gg n_W$. This hypothesis has the consequence that both the axion mass and its matter couplings become highly suppressed, i.e.,

$$m_a \approx \frac{f_\pi^2}{V_{PQ}} \tag{4.6.18}$$

and

$$g_{ffa} \approx \frac{f_\pi}{V_{PQ}}. \tag{4.6.19}$$

This particle is similar to the Majoron [46] discussed in Chapter 2 and therefore the astrophysical constraints from red giants imply [47] that $V_{PQ} \geq 10^9$ GeV. This implies that $m_a \approx 10^{-2}$ eV and $g_{ffa} \simeq 10^{-10}$. Because of this weak strength, this particle, like the Majoron, would escape experimental detection. Further experimental implications of this idea have been studied in a number of papers [48]. This idea is, however, criticized on the ground that to explain small θ we introduce a very high scale $V_{PQ} \leq 10^9$ GeV which requires additional fine tuning of parameters. However, if this concept is successfully accommodated into a grand unified theory, where the fine tuning is needed to study gauge hierarchies, then it appears less unnatural. Nevertheless, all these additional theoretical inputs, coupled with lack of the conventional light axion to show up, has made it more attractive to consider alternative solutions to the strong CP-problem which we briefly discuss in the next section.

§4.7. Other Solutions to the Strong CP-Problem

The basic idea of the previous section was to use a continuous symmetry that removes the θ-terms. Several alternative solutions use the idea of replacing the continuous symmetry by appropriate discrete symmetries so that it leads to $\theta = 0$ in the Lagrangian. Once these discrete symmetries are spontaneously broken we will not have any light axionlike particle. However, the number of possibilities for such theories is limited by the fact that since the discrete symmetries are broken a finite θ will arise as a result of quantum effects and only those theories will be acceptable which have $\theta_{\text{loop}} \leq 10^{-9}$. We discuss below four different possibilities which solve the strong CP-problem without a $U(1)_{PQ}$ symmetry.

The general strategy in all these models consists of the following:

(a) Choose $\theta = 0$ using some symmetry (or by appropriate choice of the gauge symmetry) and its particle content.
(b) Aftèr spontaneous symmetry breaking we obtain quark mass matrices $M_{u,d}$. On diagonalization these can also lead to nonvanishing θ_{QFD}. Thus, the theory must be such that subsequent to spontaneous breakdown

$$\theta_{\text{QFD}} \equiv \text{Arg Det}(M_u M_d) = 0. \tag{4.7.1}$$

Combination of conditions (a) and (b) guarantee that $\theta_{\text{tree}} = 0$ naturally. This implies that any contribution to θ arising in higher orders must be finite. To estimate these effects we calculate the radiative correction $\delta M_{u,d}$ to the mass matrices. The finite contribution to θ is then

$$\delta\theta_{\text{QFD}} = \text{Arg Det}(M_u + \delta M_u)(M_d + \delta M_d). \tag{4.7.2}$$

(i) The first series of such models [49] was constructed using the left–right symmetric models where parity is a good symmetry of the Lagrangian prior to spontaneous breakdown. Under parity $\theta \to -\theta$; thus, in left–right symmetric models, $\theta = 0$ naturally. Furthermore, if the model has manifest left–right symmetry subsequent to spontaneous breakdown (see Chapter 6), i.e.,

$$M_{u,d} = M_{u,d}^{+} \tag{4.7.3}$$

this satisfies eqn. (4.7.1) automatically. In Ref. [49] it has been shown that nonvanishing, finite contributions to θ arise at the two-loop level keeping $\theta \leq 10^{-9}$ without any unnatural adjustment of parameters.

Similar strategies have been employed outside the left–right symmetric models, by requiring additional discrete symmetries such as CP, which are then softly broken [50] to make the model realistic.

(ii) A second class of models, that do not use $U(1)_{PQ}$ symmetry, employs the observation that, in theories where CP-violation arises entirely out of vacuum expectation values, the value of the phases in the observable sector of quarks and leptons depends on whether there are high mass scales in the theory [51]. For instance, consider a theory with two mass scales μ and M with $\mu \ll M$. If, in the limit of $M \to \infty$, the effective low-energy theory is such that it is CP-conserving then any spontaneous CP-violating phase δ associated with the heavy scale M will manifest in the low-energy sector in such a way that

$$\delta \approx \left(\frac{\mu}{M}\right)^2. \tag{4.7.4}$$

Since θ_{tree} is zero in this theory because of CP-invariance any θ arising at the tree level will be $\approx (\mu/M)^2$. Chosing M appropriately we could make θ naturally small. How this can work in a realistic model has been demonstrated in Ref. [51] where $\mu = m_{W_L}$ and $M = M_{W_R}$. This would require $M_{W_R} \geq 10^6$ GeV. It would be interesting to implement this idea in other theories to test the generality of such models.

(iii) A third class of models recently proposed [52] considers models with softly broken Peccei–Quinn symmetry. If the soft breaking terms are properly chosen they will lead to a theory where, if $\theta_{\text{tree}} = 0$, θ_{loop} will be computable and finite.

A simple realization of this idea is to consider the $U(1)_{PQ}$ symmetric Lagrangian described in Section 4.6 and add the following soft breaking gauge invariant term to the Lagrangian such as

$$\mathscr{L}_B = -\mu^2 \phi_u \phi_d + \text{h.c.} \tag{4.7.5}$$

This class of models is devoid of a light axion and leads to nonvanishing θ at the three loop level. It has also been shown [52] that these theories arise naturally in $N = 1$ broken supergravity models.

(iv) Finally, a new class of models, that lead to finite and small θ without Peccei–Quinn symmetry, has recently been proposed [53]. Here we consider the CP-invariant Lagrangian (so that $\theta_{\text{QCD}} = 0$) with heavy fermions which belongs to real representations under the gauge group (i.e., $C + \bar{C}$ where C is a complex representation). If we denote the light fermions of the model by F, then to solve the strong CP-problem the following constraints must be imposed on the theory:

(a) no $\bar{C}F$- or CC-type mass or Yukawa interaction terms at the tree level;
(b) CP-violating phases only in coupling of F to $C + \bar{C}$ together.

The most general fermion mass matrix in such models is such that its determinant is real leading to vanishing θ_{QFD}.

§4.8. Summary

In summary we wish to note that CP-violation in gauge theories has two aspects to it: strong and observed weak CP-violation. The first is a theoretical problem that must be solved in any realistic model of weak CP-violation. It is worth pointing out that the Kobayashi–Maskawa model does not provide a solution to the strong CP-problem. If we calculate a higher loop, at the fourteenth order, an infinite contribution [38a] to θ appears. So θ must be renormalized and is therefore an arbitrary parameter. In fact, the third proposal [52], in Section 4.7, provides the simplest modification of the Kobayashi–Maskawa model which solves the strong CP-problem.

References

[1] R. E. Marshak, Riazuddin, and C. P. Ryan, *Theory of Weak Interactions in Particle Physics*, Wiley, New York, 1969;
L. Wolfenstein, *Theory and Phenomenology in Particle Physics* (edited by A. Zichichi), Academic Press, New York, 1969;
E. Paul, in *Elementary Particle Physics*, Springer Tracts in Modern Physics, vol. 79.
P. K. Kabir, *CP Puzzle*, Academic Press, New York, 1968.
[2] Particle Data Group, *Rev. Mod. Phys.* **56**, S1 (1984).
[3] B. Weinstein, Talk Given at XIth International Neutrino Conference, Dortmund, West Germany, June, 1984.
[4] L. Wolfenstein, *Phys. Rev. Lett.* **13**, 562 (1964).
[5] R. N. Mohapatra, J. C. Pati, and L. Wolfenstein, *Phys. Rev.* **D11**, 3319 (1975);
P. Sikivie, *Phys. Lett.* **65B**, 141 (1976);
A. Joshipura and I. Montray, *Nucl. Phys.* **B196**, 147 (1982);
S. M. Barr and P. Langacker, *Phys. Rev. Lett.* **42**, 1654 (1979).
[6] M. Kobayashi and T. Maskawa, *Prog. Theor. Phys.* **49**, 652 (1973).
[7] R. N. Mohapatra, *Phys. Rev.* **D6**, 2023 (1972).
[7a] R. N. Mohapatra and D. P. Sidhu, *Phys. Rev.* **D17**, 1876 (1978);
D. Chang, *Nucl. Phys.* **B214**, 435 (1983);
P. Herczeg, *Phys. Rev.* **D28**, 200 (1983).
[8] L. L. Chau, *Phys. Rep.* **95C**, 1 (1983).
[9] K. L. Giovanetti *et al.*, *Phys. Rev.* **D29**, 343 (1984).
[10] R. Shrock and L. L. Wang, *Phys. Rev. Lett.* **41**, 1692 (1978).
For recent survey, see
H. Leutwyder and M. Roos, CERN preprint, TH3830 (1984);
A. Buras, W. Slominski, and H. Steger, *Nucl. Phys.* **B238**, 529 (1984).
[11] E. Fernandez *et al.*, *Phys. Rev. Lett.* **51**, 1022 (1983);
N. S. Lockyer *et al.*, *Phys. Rev. Lett.* **51**, 1316 (1983);
The measured lifetimes of the b-quark are $(1.2 \pm 0.4 \pm 0.3) \times 10^{-12}$ s and $(1.6 \pm 0.4 \pm 0.3) \times 10^{-12}$ s, respectively.
[12] K. Klenknecht and B. Renk, *Z. Phys.* **C16**, 7 (1982); **C20**, 67 (1983).
[13] M. K. Gaillard and B. W. Lee, *Phys. Rev.* **D10**, 897 (1974).
[14] F. J. Gilman and M. Wise, *Phys. Rev.* **D27**, 1128 (1983).
[15] J. F. Donoghue, E. Golowich, and B. R. Holstein, *Phys. Lett.* **119B**, 412 (1982);
J. Trampetic, *Phys. Rev.* **D27**, 1565 (1983);
B. Guberina, B. Machet, and E. deRafael, *Phys. Lett.* **128B**, 269 (1983).
[16] F. J. Gilman and M. Wise, *Phys. Rev.* **D20**, 2392 (1979);
B. Gubering and R. D. Peccei, *Nucl. Phys.* **B163**, 289 (1980);
F. J. Gilman and J. Hagelin, *Phys. Lett.* **126B**, 111 (1983).
[17] Report of UA1 Collaboration at CERN, Leipzig (1984).
[18] R. N. Mohapatra and J. C. Pati, *Phys. Rev.* **D11**, 566 (1975).
[19] R. N. Mohapatra, F. E. Paige, and D. P. Sidhu, *Phys. Rev.* **D17**, 2642 (1978).
[20] M. A. B. Beg, R. V. Budny, R. N. Mohapatra, and A. Sirlin, *Phys. Rev. Lett.* **38**, 1252 (1977).
[21] G. Beall, M. Bender, and A. Soni, *Phys. Rev. Lett.* **48**, 848 (1982).
[22] I. I. Bigi and J. M. Frere, University of Michigan preprint (1983).
[23] G. Beal and A. Soni, *Phys. Rev. Lett.* **47**, 552 (1981);
G. Ecker, W. Grimus, and H. Neufeld, *Nucl. Phys.* **B229**, 421 (1983).
[24] A. Masiero, R. N. Mohapatra, and R. D. Peccei, *Nucl. Phys.* **B192**, 66 (1981).
[25] G. Branco, J. Frere, and J. Gerard, *Nucl. Phys.* **B221**, 317 (1983).
[26] D. Chang, *Nucl. Phys.* **B214**, 435 (1983).
R. N. Mohapatra, *Phys. Lett.* **159B**, 374 (1975).
[27] P. Sikivie, *Phys. Lett.* **65B**, 141 (1976).

[28] S. Weinberg, *Phys. Rev. Lett.* **37**, 657 (1976).
[29] G. Branco, *Phys. Rev. Lett.* **44**, 504 (1980).
[30] N. G. Deshpande, *Phys. Rev.* **D23**, 2654 (1981).
[31] A. I. Sanda, *Phys. Rev.* **D23**, 2647 (1981).
[32] D. Chang, *Phys. Rev.* **D25**, 1381 (1982).
[33] G. Beall and N. G. Deshpande, Oregon preprint (1983).
[34] W. Dress *et al.*, *Phys. Rev.* **D15**, 9 (1977);
I. Altarev *et al.*, *Phys. Lett.* **102B**, 13 (1981).
[35] C. Callan, R. Dashen, and D. Gross, *Phys Lett.* **63B**, 334 (1976).
[36] R. Jackiw and C. Rebbi, *Phys. Rev. Lett.* **37**, 172 (1976).
[37] G. 't Hooft, *Phys. Rev.* **D14**, 3432 (1976);
A. Belavin, A. M. Polyakov, A. Schwartz, and Yu S. Tyupkin, *Phys. Lett.* **59B**, 85 (1975).
[38] V. Baluni, *Phys. Rev.* **D19**, 2227 (1979);
R. Crewther, P. di Vecchia, G. Veneziano, and E. Witten, *Phys. Lett.* **88B**, 123 (1979).
[38a] J. Ellis and M. K. Gaillard, *Nucl. Phys.* **B150**, 141 (1979).
[39] R. D. Peccei and H. Quinn, *Phys. Rev. Lett.* **38**, 1440 (1977); *Phys. Rev.* **D16**, 1791 (1977).
[40] S. Weinberg, *Phys. Rev. Lett.* **40**, 223 (1978);
F. Wilczek, *Phys. Rev. Lett.* **40**, 279 (1978).
[41] W. Bardeen and S. H. H. Tye, *Phys. Lett.* **74B**, 229 (1978);
J. Kandaswamy, J. Schecter, and P. Salomonson, *Phys. Lett.* **74B**, 377 (1978).
[42] J. Barosso and N. Mukhopadhyaya, SIN preprint (1980).
[43] M. Zender, SIN preprint (1981).
[43a] H. Faissner *et al.* Aachen preprint (1983).
[44] For a recent experiment, see
L. W. Mo, J. D. Bjorken *et al* VPI preprint (1984).
[45] J. E. Kim, *Phys. Rev. Lett.* **43**, 103 (1979);
M. Shifman, A. Vainstein, and V. Zakharov, *Nucl. Phys.* **B166**, 493 (1980);
M. Dine, W. Fischler, and M. Srednicki, *Phys. Lett.* **104B**, 199 (1981).
[46] Y. Chikashige, R. N. Mohapatra, and R. D. Peccei, *Phys. Lett.* **98B**, 265 (1981).
[47] D. A. Dicus, E. Kolb, V. Teplitz, and R. V. Wagoner, *Phys. Rev.* **D18**, 1829 (1978);
M. Fukugita, S. Watamura, and M. Yoshimura, *Phys. Rev. Lett.* **48**, 1522 (1978).
[48] M. Wise, H. Georgi, and S. L. Glashow, *Phys. Rev. Lett.* **47**, 402 (1981);
R. Barbieri, R. N. Mohapatra, D. Nanopoulos, and D. Wyler, *Phys. Lett.* **107B**, 80 (1981);
J. Ellis, M. K. Gaillar, D. Nanopoulos, and S. Rudaz, *Phys. Lett.* **107B**, (1981);
A. Masiero and G. Segre, University of Pennsylvania preprint (1981).
[49] M. A. B. Bég and H. S. Tsao, *Phys. Rev. Lett.* **41**, 278 (1978);
R. N. Mohapatra and G. Senjanovic, *Phys. Lett.* **79B**, 283 (1978).
[50] H. Georgi, *Hadron J.* **1**, 155 (1978);
G. Segre and H. A. Weldon, *Phys. Rev. Lett.* **42**, 1191 (1979);
S. Barr and P. Langacker, *Phys. Rev. Lett.* **42**, 1654 (1979).
[51] A. Masiero, R. N. Mohapatra, and R. D. Peccei, *Nucl. Phys.* **B192**, 66 (1981).
[52] R. N. Mohapatra and G. Senjanovic, *Z. Phys.* **20**, 365 (1983);
R. N. Mohapatra, S. Ouvry, and G. Senjanovic, *Phys. Lett.* **126B**, 329 (1983);
D. Chang and R. N. Mohapatra, *Phys. Rev.* **D32**, 293 (1985).
[53] A. Nelson, *Phys. Lett.* **136B**, 387 (1984);
S. Barr, *Phys. Rev. Lett.* **53**, 329 (1984).

CHAPTER 5

Grand Unification and the SU(5) Model

§5.1. The Hypothesis of Grand Unification

We emphasized in the first chapter that the Yang–Mills theories provide a unique framework for describing interactions with universal couplings for different, apparently unrelated, processes such as μ-decay and β-decay, etc. This led to the successful electro-weak unification theories based on $SU(3)_c \times SU(2)_L \times U(1)_Y$ group (G_{321}). The 321-theory has three gauge couplings of different magnitudes and fermions assigned according to convenience, rather than any principle. Furthermore, the problem of quantization of electric charge remains unsolved in this model or in any unification group that has a U(1)-local symmetry. It is therefore logical to imagine a higher symmetry which unifies all three couplings and also simultaneously provides an explanation for the quantization of electric charge. The first attempt in this direction was made by Pati and Salam [1], who unified the quarks and leptons within the group $SU(2)_L \times SU(2)_R \times SU(4)_C$ by extending the color gauge group to include the leptons. They explained the quantization of electric charge, although they had three coupling constants, g_{2L}, g_{2R}, and g_c, since there was no natural left–right symmetry in their model. This shortcoming was soon removed [2] making this theory a two-coupling constant partial unification theory. This kind of theory will be discussed in Chapter 6.

A more ambitious approach was taken independently in the same year by Georgi and Glashow [3] who proposed the rank 4 simple group SU(5) as the grand unification group that not only explains the quantization of electric charge but also the unification of all coupling constants. In this chapter we will discuss the SU(5) model, its predictions, and its present experimental status. Before that, we discuss some general preliminaries for implementing the hypothesis of grand unification.

§5.2. SU(N) Grand Unification

Since $SU(3)_c \times SU(2)_L \times U(1)$ is a group of rank 4, the minimal unifying group is SU(5). For a general SU(N) ($N \geq 5$) grand unification (in order to ensure the vector nature of color gauge interactions, i.e., to have only 3 and 3* of color SU(3)), the simplest choice is to assign fermions only to antisymmetric representations. In other words, the irreducible representations will be chosen to be of type $\psi_{ijk\ldots}$ ($i, j, k, \ldots = 1, \ldots, N$ for SU(N)), antisymmetric in the interchange of any two indices. We remind the reader that the dimensionality of an mth rank antisymmetric tensor under SU(N) is

$$d(\psi_{i,\ldots,i_m}) = \frac{N!}{m!\,(N-m)!}. \tag{5.2.1}$$

Since we would like to have anomaly-free combinations, let us also give the anomalies $A_{m,N}$ associated with an antisymmetric representation of SU(N) with m boxes [3a]:

$$A_{m,N} = \frac{(N-3)!\,(N-2m)}{(N-m-1)!\,(m-1)!}. \tag{5.2.2}$$

Below we give some examples of sets of antisymmetric representations which are anomaly free and are therefore suitable candidates for grand unification ([N, m] means SU(N) with m antisymmetric indices)

SU(5): [5, 1] + [5, 3],

SU(6): 2[6, 1] + [6, 4],

SU(7): [7, 2] + [7, 4] + [7, 6],

SU(8): [8, 1] + [8, 2] + [8, 5],

SU(9): [9, 2] + [9, 5],

or [9, 1] + [9, 3] + [9, 5] + [9, 7],

SU(10): [10, 3] + [10, 6],

etc. We could of course keep going; but it turns out that the smallest anomaly-free combination beyond SU(10) makes the theory ultraviolet unstable [3b]. The formula for the β-function for SU(N) with fermions in [N, m] can be written as

$$\beta(g) = -\left[\tfrac{11}{3}N - \tfrac{1}{3}\sum_m \binom{N-2}{m-1} c_m\right]\frac{g^3}{16\pi^2} + O(g^5). \tag{5.2.3}$$

It is clear from the above that if we want to unify only one generation of fermions SU(5) is satisfactory, since the total number of components is 15 for the representation listed. (Some more about this later.) Furthermore, if the top quark is not found, a gauge group based on SU(6) may have a certain appeal.

§5.3. Sin$^2\,\theta_W$ in Grand Unified Theories (GUT)

Before proceeding to a detailed discussion of GUT models, we give a formula for the calculation of $\sin^2\theta_W$ in the exact symmetry limit in grand unified theories. In this subsection this formula will be derived in terms of the representation content of a particular kind of grand unification [4]. To do this, we note that, the grand unification group must contain diagonal generators (or normalized linear combinations of diagonal generators) which can be identified with T_{3_L} and Y, with the corresponding gauge bosons being called W_{3_L} and W_0. The corresponding couplings are not necessarily identical to the grand unifying coupling constant g_U. So let us call them g and g'. Next, if we calculate the fermion contributions to the leading divergences in the self-energies $\pi_{33}(q)$ and $\pi_{00}(q)$, then invariance under the grand unifying group implies that

$$\pi_{33}^{\text{div}}(q) = \pi_{00}^{\text{div}}(q). \tag{5.3.1}$$

But it is easy to see that the leading divergences of π_{33} and π_{00} are proportional, respectively, to $g^2\sum_i T_{3i}^2$ and $g'^2\sum_i Y_i^2$, where i goes over all the fermions in the theory.

Thus we obtain

$$g^2\sum_i T_{3i}^2 = g'^2\sum_i Y_i^2. \tag{5.3.2}$$

But $g'^2/g^2 = \tan^2\theta_W$, leading to

$$\cot^2\theta_W = \frac{\sum_i Y_i^2}{\sum_i T_{3i}^2}. \tag{5.3.3}$$

Using the fact that $\sum_i Q_i^2 = \sum_i(T_{3i}^2 + Y_i^2)$, the formula for $\sin^2\theta_W$ follows:

$$\sin^2\theta_W = \frac{\sum_i T_{3i}^2}{\sum_i Q_i^2}. \tag{5.3.4}$$

§5.4. SU(5)

In this section we will discuss the minimal grand unification based on SU(5) and its implications. As discussed in the previous chapter, the anomaly-free combination of SU(5) representations is $\{\bar{5}\} + \{10\}$ if both representations are chosen left-handed. Below we denote the $\{5\}$- and $\{10\}$-dimensional representations of fermions, where the $\{5\}$ is right-handed and the $\{10\}$ is chosen left-handed.

$$\begin{pmatrix} d_1 \\ d_2 \\ d_3 \\ e^+ \\ \nu^c \end{pmatrix}_R \begin{pmatrix} 0 & u_3^c & u_2^c & u_1 & d_1 \\ & 0 & u_1^c & u_2 & d_2 \\ & & 0 & u_3 & d_3 \\ & & & 0 & e^+ \\ & & & & 0 \end{pmatrix}_L. \tag{5.4.1}$$

There are $\{24\}$ gauge bosons associated with SU(5). Under $SU(3)_c \times SU(2)_L \times U(1)$ they decompose as follows:

$$\{24\} = \{8, 1, 0\} + \{1, 3, 0\} + \{3, 2, \tfrac{5}{3}\} + \{3^*, 2, -\tfrac{5}{3}\} + \{1, 1, 0\}. \tag{5.4.2}$$

To express this in matrix form we choose the following form for the SU(5) generators:

$$\lambda_i = \begin{pmatrix} & & & 0 & 0 \\ & \lambda_i & & 0 & 0 \\ & & & 0 & 0 \\ 0 & 0 & 0 & 0 & 0 \\ 0 & 0 & 0 & 0 & 0 \end{pmatrix}, \qquad i = 1, \ldots, 8,$$

$$\lambda_9 = \begin{pmatrix} & & & 1 & 0 \\ & 0 & & 0 & 0 \\ & & & 0 & 0 \\ 1 & 0 & 0 & & \\ 0 & 0 & 0 & & 0 \end{pmatrix}, \qquad \lambda_{10} = \begin{pmatrix} & & & -i & 0 \\ & 0 & & 0 & 0 \\ & & & 0 & 0 \\ i & 0 & 0 & & \\ 0 & 0 & 0 & & 0 \end{pmatrix},$$

$$\lambda_{11} = \begin{pmatrix} & & & 0 & 1 \\ & 0 & & 0 & 0 \\ & & & 0 & 0 \\ 0 & 0 & 0 & & \\ 1 & 0 & 0 & & 0 \end{pmatrix}, \qquad \lambda_{12} = \begin{pmatrix} & & & 0 & -i \\ & 0 & & 0 & 0 \\ & & & 0 & 0 \\ 0 & 0 & 0 & & \\ i & 0 & 0 & & 0 \end{pmatrix},$$

$$\lambda_{13} = \begin{pmatrix} & & & 0 & 0 \\ & 0 & & 1 & 0 \\ & & & 0 & 0 \\ 0 & 1 & 0 & & \\ 0 & 0 & 0 & & 0 \end{pmatrix}, \qquad \lambda_{14} = \begin{pmatrix} & & & 0 & 0 \\ & 0 & & -i & 0 \\ & & & 0 & 0 \\ 0 & i & 0 & & \\ 0 & 0 & 0 & & 0 \end{pmatrix},$$

$$\lambda_{15} = \begin{pmatrix} & & & 0 & 0 \\ & 0 & & 0 & 1 \\ & & & 0 & 0 \\ 0 & 0 & 0 & & \\ 0 & 1 & 0 & & 0 \end{pmatrix}, \qquad \lambda_{16} = \begin{pmatrix} & & & 0 & 0 \\ & 0 & & 0 & -i \\ & & & 0 & 0 \\ 0 & 0 & 0 & & \\ 0 & i & 0 & & 0 \end{pmatrix},$$

$$\lambda_{17} = \begin{pmatrix} & & & 0 & 0 \\ & 0 & & 0 & 0 \\ & & & 1 & 0 \\ 0 & 0 & 1 & & \\ 0 & 0 & 0 & & 0 \end{pmatrix}, \qquad \lambda_{18} = \begin{pmatrix} & & & 0 & 0 \\ & 0 & & 0 & 0 \\ & & & -i & 0 \\ 0 & 0 & i & & \\ 0 & 0 & 0 & & 0 \end{pmatrix},$$

$$\lambda_{19} = \begin{pmatrix} & & & 0 & 0 \\ & 0 & & 0 & 0 \\ & & & 0 & 1 \\ 0 & 0 & 0 & & \\ 0 & 0 & 1 & & 0 \end{pmatrix}, \qquad \lambda_{20} = \begin{pmatrix} & & & 0 & 0 \\ & 0 & & 0 & 0 \\ & & & 0 & -i \\ 0 & 0 & 0 & & \\ 0 & 0 & i & & 0 \end{pmatrix},$$

$$\lambda_{20+j} = \begin{pmatrix} & & & 0 & 0 \\ & 0 & & 0 & 0 \\ & & & 0 & 0 \\ 0 & 0 & 0 & & \\ 0 & 0 & 0 & \multicolumn{2}{c}{\tau_j} \end{pmatrix}, \qquad j = 1, 2, 3,$$

$$\lambda_{24} = \frac{2}{\sqrt{15}} \begin{pmatrix} 1 & & & & \\ & 1 & & & \\ & & 1 & & \\ & & & -3/2 & \\ & & & & -3/2 \end{pmatrix}. \tag{5.4.3}$$

This gives the following form for the gauge boson matrix:

$$\begin{pmatrix} \multicolumn{3}{c}{} & X_1^{-4/3} & Y_1^{-1/3} \\ \multicolumn{3}{c}{\frac{1}{\sqrt{2}}\boldsymbol{\lambda}\cdot\mathbf{V}_{\{8\}} + \frac{\sqrt{2}}{15}V_{24}} & X_2^{-4/3} & Y_2^{-1/3} \\ \multicolumn{3}{c}{} & X_1^{-4/3} & Y_3^{-1/3} \\ X_1^{4/3} & X_2^{4/3} & X_3^{4/3} & \multicolumn{2}{c}{} \\ Y_1^{1/3} & Y_2^{1/3} & Y_3^{1/3} & \multicolumn{2}{c}{\frac{1}{\sqrt{2}}\boldsymbol{\tau}\cdot\mathbf{W} - \frac{\sqrt{3}}{10}V_{24}} \end{pmatrix}. \tag{5.4.4}$$

One of the first things we can do now is to predict the value of the weak-electromagnetic mixing angle $\sin^2\theta_W$. For this purpose recall that θ_W is defined for the SU(2) × U(1) Glashow–Weinberg–Salam group as follows:

$$\tan\theta_W = \frac{g_{\mathrm{U}(1)}}{g_{\mathrm{SU}(2)}} \equiv \frac{g'}{g}. \tag{5.4.5}$$

Due to unification we can predict g'/g at the masses above which unification has occurred. To see this note that λ_{24} corresponds to Y. Let us take its eigenvalue for the e^+ and compare with the SU(2) × U(1) case: we have

$$\sqrt{\tfrac{3}{5}}\, g_{\mathrm{SU}(5)} = g'. \tag{5.4.6}$$

But

$$g_{\mathrm{SU}(5)} = g. \tag{5.4.7}$$

It therefore follows that $\tan\theta_W = \sqrt{\frac{3}{5}}$, leading to the famous prediction $\sin^2\theta_W = \frac{3}{8}$. Incidentally, the same result is also obtained if we use eqn. (5.3.4).

Next we would like to present the breakdown of the SU(5) group to $SU(3)_c \times U(1)_{em}$. Within the conventional Higgs picture this is achieved by introducing two Higgs multiplets: one belonging to the {24}-dimensional irreducible representation (denoted by Φ), and the other to the {5}-dimensional (denoted by H) representations. The stages of symmetry breakdown are given by

$$SU(5) \to SU(3)_c \times SU(2)_L \times U(1) \to SU(3)_c \times U(1)_{em}.$$

The first stage is achieved by $\langle \Phi \rangle \neq 0$ and the second stage by $\langle H \rangle \neq 0$ as follows: from the general group theory of spontaneous breakdown we can show that [5]

$$\langle \Phi \rangle = V \begin{pmatrix} 1 & & & & \\ & 1 & & & \\ & & 1 & & \\ & & & -3/2 & \\ & & & & -3/2 \end{pmatrix}, \tag{5.4.8}$$

i.e., the breaking is along the λ_{24} direction. This is responsible for the first stage of the symmetry breakdown. The second stage is caused by

$$\langle H \rangle = \begin{pmatrix} 0 \\ 0 \\ 0 \\ 0 \\ \rho/\sqrt{2} \end{pmatrix}. \tag{5.4.9}$$

In the presence of both H and Φ, the v.e.v. of Φ changes somewhat and looks like the following:

$$\langle \Phi \rangle = V \begin{pmatrix} 1 & & & & \\ & 1 & & & \\ & & 1 & & \\ & & & -3/2 - \varepsilon/2 & \\ & & & & -3/2 + \varepsilon/2 \end{pmatrix}. \tag{5.4.10}$$

It is then a straightforward matter to compute all the gauge boson masses

$$\text{Superheavy bosons } X, Y\text{:} \quad m_X^2 \approx m_Y^2 = \tfrac{25}{8} g^2 V^2, \tag{5.4.11}$$

$$W\text{-}, Z\text{-bosons:} \quad m_W^2 \simeq \frac{g^2 \rho^2}{4}(1 + \varepsilon), \qquad m_Z^2 \simeq \frac{g^2 \rho^2}{4 \cos^2 \theta_W}. \tag{5.4.12}$$

The reason why the X- and Y-bosons must be heavy can be seen by looking at the part of the gauge Lagrangian that involves X and Y. For one genera-

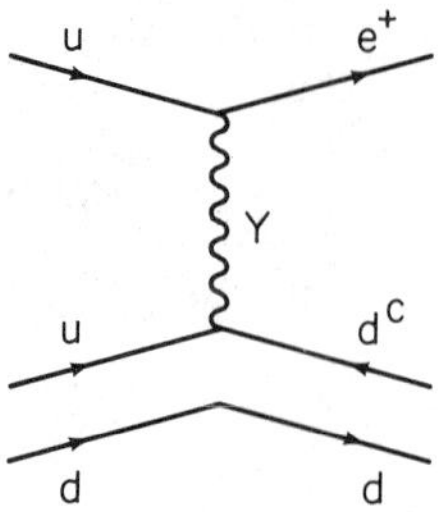

Figure 5.1

tion it is given by

$$\mathscr{L}_{X,Y} = \frac{ig_x}{\sqrt{2}} X_{\mu,i}(\varepsilon_{ijk}\overline{u}^c_{kL}\gamma_\mu u_{jL} + \bar{d}_i\gamma_\mu e^+)$$

$$+ \frac{ig_Y}{\sqrt{2}} Y_{\mu,i}(\varepsilon_{ijk}\overline{u}^c_{K_L}\gamma_\mu d_{hL} - \overline{u}_{i_L}\gamma_\mu e^+_L + \bar{d}_{i_R}\gamma_\mu \nu^c_R) + \text{h.c.} \tag{5.4.13}$$

We thus see that

$$\bar{X} \to uu \qquad (B = \tfrac{2}{3})$$
$$\to \bar{d}e^+ \qquad (B = -\tfrac{1}{3})$$

and

$$\bar{Y} \to ud \qquad (B = \tfrac{2}{3})$$
$$\to \bar{u}e^+ \qquad (B = -\tfrac{1}{3}).$$

Thus they can lead to proton decay of type $p \to e^+\pi^0$ via the diagrams shown in Fig. 5.1 with strength

$$M(p \to e^+\pi^0) \approx \frac{g^2_{X,Y}}{m^2_{X,Y}}$$

leading to a crude estimate for the lifetime of the proton to be about

$$\tau_p \approx \frac{m_X^4}{g^4 m_p^5}. \tag{5.4.14}$$

The lower limits [6] on the proton lifetime, existing up to 1979, then implied $m_{X,Y} \geq 10^{15}$ GeV.

Several dedicated searches for proton decay have been going on at various places around the world since that time [7–12] and we will discuss these results at the end of this chapter. On the basis of these experiments we can safely say that $\tau_p \geq 10^{31}$ yr with more stringent limits on particular decay modes such as $p \to e^+\pi^0$, etc. We now focus on some theoretical aspects of the SU(5) model, especially the problem of understanding such large masses previously nonexistent in elementary particle physics.

One aspect of the SU(5) model is that there exist two totally different mass scales. The theoretical question then arises as to how, in a theory with divergences (albeit of renormalizable type), such large mass ratios can be maintained. We will aruge that, in order to maintain this disparity between mass scales, certain parameters of the theory must be adjusted to one part in 10^{13}. This is the so-called "gauge hierarchy" problem. There is, of course, a related problem which can be called "supposed gauge hierarchy impasse," which we will also mention below.

Gauge Hierarchy Problems

To study this problem we write the potential involving the $\Phi\{24\}$ and $H\{5\}$ Higgs multiplets

$$\mathscr{V} = -\tfrac{1}{2}\mu^2 \operatorname{Tr}\phi^2 + \tfrac{1}{4}a(\operatorname{Tr}\phi^2)^2 + \tfrac{1}{2}b\operatorname{Tr}\phi^4 - \tfrac{1}{2}v^2H^+H + \frac{\lambda}{4}(H^+H)^2$$
$$+ \alpha H^+H\operatorname{Tr}\phi^2 + \beta H^+\phi^2 H. \tag{5.4.15}$$

Substituting the expected minima given in eqns. (5.4.9) and (5.4.10) we obtain

$$\mathscr{V}_{\min} = -\tfrac{1}{2}\mu^2(\tfrac{15}{2}V^2 + \tfrac{1}{2}\varepsilon^2V^2) + \frac{a}{4}(\tfrac{15}{2}V^2 + \tfrac{1}{2}\varepsilon^2V^2)^2$$
$$+ \frac{b}{2}(\tfrac{105}{8}V^4 + \tfrac{27}{4}V^4\varepsilon^2 + \cdots) - \tfrac{1}{4}v^2\rho^2$$
$$+ \frac{\lambda}{16}\rho^4 + \frac{\alpha}{2}\rho^2(\tfrac{15}{2}V^2 + \tfrac{1}{2}\varepsilon^2V^2) + \beta\frac{\rho^2V^2}{8}(3-\varepsilon)^2. \tag{5.4.16}$$

The conditions for minimum give the following equations:

$$\varepsilon \simeq \frac{3\beta}{20b}(\rho/V)^2, \tag{5.4.17a}$$

$$\mu^2 = \tfrac{15}{2}av^2 + \tfrac{7}{2}bv^2 + \alpha\rho^2 + \tfrac{3}{10}\beta\rho^2 + O(\varepsilon\rho^2), \tag{5.4.17b}$$

$$v^2 = \tfrac{1}{2}\lambda\rho^2 + 15\alpha v^2 + \tfrac{9}{2}\beta v^2 - 3\varepsilon\beta v^2 + O(\varepsilon\rho^2). \tag{5.4.17c}$$

From (5.4.17c) we see that if $v^2 \approx \rho^2$, then both α, β must be of order ρ^2/V^2 10^{-26} so that ρ can be much smaller than V; the other possibility is that α, β are of order 1 and v^2 is of order V^2, however, there must then be a cancellation between v^2 and $15\alpha V^2 + \tfrac{9}{2}\beta V^2$ (to an incredible degree of accuracy) to have a quantity of order $10^{-26}\ V^2$. Both of these are technically allowed (though extremely unnatural) and may indeed be a symptom of something wrong with such simple-minded grand unification.

Let us now discuss the second point, i.e., the "gauge hierarchy impasse [13]." If we chose α and β at the tree level to an accuracy of 10^{-26}, once we include radiative corrections, α and β "may acquire" a lower bound of

$\geq g^4/32\pi^2$ and therefore may make such a choice technically unrealizable. This, if true, would indeed have been a fatal blow to SU(5) and other such simple grand unification schemes. It has, however, been noted (by explicit computation of the one-loop potential) that no such impasse occurs [14], and indeed (at least up to the one-loop level) it is possible to adjust parameters arbitrarily to obtain any desired gauge hierarchy. From now on we will assume the existence of a gauge hierarchy which is still an open problem, and which has been one of the strongest motivations for considering supersymmetric models.

§5.5. Grand Unification Mass Scale and Sin² θ_W at Low Energies

The philosophy of grand unification is that all interactions are described by a single coupling constant. However, we know quite well that at low energies (i.e., $E \approx 1$ GeV), strong interaction couplings are much bigger than the weak and electromagnetic couplings. The next problem in the discussion of grand unification is how to reconcile these two ideas. The solution was first suggested by Georgi, Quinn, and Weinberg [15]. Their observation was that coupling constants are scale-dependent quantities whose rate of variation with scale is governed by Callan–Symanzik equations. Therefore these coupling constants must change with energy and, if the hypothesis of grand unification is to hold, they must be equal to the SU(5) coupling at some mass M_U. We will call M_U the grand unification scale. For $\mu < M_U$, SU(5) breaks down to $SU(3)_c \times SU(2) \times U(1)$ and the separate couplings behave differently. To study this behavior we need to know how the gauge coupling constants change with energy. This is given by a powerful theorem first proved by Appelquist and Carazone [16].

Decoupling Theorem

If a gauge invariant (under group G) Lagrangian field theory contains particles with two very different masses m and M ($m \ll M$) and is described by a renormalizable Lagrangian, the behavior of the light particles in the theory for $E \ll M$ can be described completely by a renormalizable Lagrangian involving only the light particles. The effect of the heavy particles [17] is simply to rescale the coupling constants and renormalization parameters of the theory.

In the context of the SU(5) model this means that if, at mass scale $M \approx V$, the SU(5) symmetry breaks down to $SU(3)_c \times SU(2)_L \times U(1)$, then the behavior of the three gauge couplings g_3, g_2, and g_1 corresponding to these symmetries must evolve according to the β-functions corresponding to SU(3),

SU(2), and U(1), respectively, with no memory whatsoever of SU(5). We can therefore write

$$\frac{dg_i}{dt} = \beta_i(g_i) \equiv +b_{0i}\frac{g_i^3}{16\pi^2}, \qquad i = 3, 2, 1, \tag{5.5.1}$$

where

$$t = \ln \mu.$$

We know that

$$\beta_3(g_3) = -\frac{g_3^3}{16\pi^2}\left(11 - \tfrac{2}{3}N_f - \frac{N_H}{6}\right), \tag{5.5.2}$$

where N_f is the number of four-component color triplet fermions and N_H is the number of the color triplet Higgs bosons. Similarly,

$$\beta_2(g_2) = -\frac{g_2^3}{16\pi^2}\left(\tfrac{22}{3} - \tfrac{2}{3}N_f - \frac{N_H}{6}\right), \tag{5.5.3}$$

$$\beta_1(g_1) = +\tfrac{2}{3}N_f \cdot \frac{g_1^3}{16\pi^2} + \text{Higgs boson contributions.} \tag{5.5.4}$$

Let us first omit the contribution of the Higgs bosons and define $\alpha_i = g_i^2/4\pi$. Then, $d\alpha_i/dt = (1/2\pi)\cdot g_i(dg_i/dt)$. Equation (5.5.1) can then be written as

$$\begin{aligned} \frac{d\alpha_3}{dt} &= -\frac{\alpha_3^2}{2\pi}(11 - \tfrac{2}{3}N_f), \\ \frac{d\alpha_2}{dt} &= -\frac{\alpha_2^2}{2\pi}(\tfrac{22}{3} - \tfrac{2}{3}N_f), \\ \frac{d\alpha_1}{dt} &= \frac{\alpha_1^2}{2\pi}\cdot\tfrac{2}{3}N_f. \end{aligned} \tag{5.5.5}$$

The boundary condition on these differential equations is

$$\alpha_1(M_X) = \alpha_2(M_X) = \alpha_3(M_X) = \alpha_U. \tag{5.5.6}$$

Using (5.5.5) and (5.5.6) we will find $\alpha_i(\mu)$ for $\mu \approx m_W$. To identify low-energy parameters we note that $\alpha_3(m_W) = \alpha_s(m_W)$, $\alpha_2(m_W) = \alpha_w(m_W)$, and $\tfrac{3}{5}\alpha_1(m_W) = g'^2(m_W)/4\pi$, with $\sin^2\theta_W = g'^2(m_W)/\{g'^2(m_W) + g^2(m_W)\}$. The solutions for these equations can be written as

$$\frac{1}{\alpha_i(\mu)} = \frac{1}{\alpha_i(M_X)} + \frac{b_{0_i}}{2\pi}\ln\frac{M_X}{\mu}. \tag{5.5.7}$$

One equation that we obtain from this is

$$\frac{1}{\alpha_2(\mu)} - \frac{1}{\alpha_1(M_X)} = -\frac{1}{2\pi}\cdot\frac{22}{3}\ln\frac{M_X}{\mu}, \tag{5.5.8a}$$

which leads to a connection between $\sin^2\theta_W$ and the unification scale M_X, i.e.,

$$\sin^2\theta_W(m_W) = \frac{3}{8} - \frac{55\alpha(m_W)}{24\pi}\ln\frac{M_X}{m_W}. \tag{5.5.8b}$$

If we include one Higgs doublet this changes the above expression as follows:

$$\sin^2\theta_W(m_W) = \tfrac{3}{8} - \alpha(m_W)\cdot\frac{109}{48\pi}\ln\frac{M_X}{m_W}. \tag{5.5.9a}$$

Another equation that can be obtained from eqn. (5.5.7) is (call $\alpha_3(m_W) \equiv \alpha_s(m_W)$)

$$\frac{\alpha(m_W)}{\alpha_s(m_W)} = \frac{3}{8} - \frac{67}{16\pi}\alpha(m_W)\ln\frac{M_U}{m_W}. \tag{5.5.9b}$$

The experimentally measured value of α is at $\mu = 0$ since it involves physical photons. Therefore we must evaluate $\alpha(m_W)$. To do this we note that α satisfies a renormalization group equation as follows:

$$\frac{\partial e^2}{\partial t} = \frac{1}{6\pi^2}\sum_f Q_f^2 e^4. \tag{5.5.10}$$

This implies that

$$\frac{1}{\alpha(m_W)} \simeq \frac{1}{\alpha(0)} - \frac{2}{3\pi}\sum_f Q_f^2 \ln\frac{m_W}{m_f} + \frac{1}{6\pi} + \ldots, \tag{5.5.11}$$

where m_f is the fermion mass which includes all quarks and leptons. Putting in $m_t \simeq 40$ GeV we find [18] $\alpha^{-1}(m_W) \simeq 128$. From eqn. (5.5.9) we can evaluate M_X by using a value of $\alpha_s(m_W)$.

The fine structure constant for QCD (α_s depends on the value of the QCD scale parameter $\Lambda_{\overline{MS}}$ which roughly denotes the scale at which the chromodynamic interactions become strong), is given by the following formula:

$$\alpha_s(Q^2) = \frac{12\pi}{25\ln(Q^2/\Lambda_{\overline{MS}}^2)} + \text{high order terms}. \tag{5.5.12}$$

A plausible value [19] for $\Lambda_{\overline{MS}}$ is

$$\Lambda_{\overline{MS}} = 160^{+100}_{-80}\ \text{MeV}. \tag{5.5.13}$$

This value is derived from decays of the $B(\gamma \to \text{hadrons} + \text{photon})/B(\gamma \to \text{hadrons})$ for which the next order in α_s corrections is small. For a general $\Lambda_{\overline{MS}}$ the predicted value [20] of M_X is

$$M_X = 2.1 \times 10^{14} \times (1.5)^{\pm 1}\left[\frac{\Lambda_{\overline{MS}}}{0.16\ \text{GeV}}\right]. \tag{5.5.14}$$

This leads to a prediction for $\sin^2\theta_W(m_W)$ as follows:

$$\sin^2\theta_W(M_W) = 0.214 \pm 0.003 \pm 0.006 \ln\left[\frac{0.16\ \text{GeV}}{\Lambda_{\overline{MS}}}\right]. \tag{5.5.15}$$

This prediction for $\sin^2\theta_W$ is in excellent agreement with the values obtained from the study of neutral currents [21] and the recent discovery [22] of W- and Z-masses

$$\sin^2\theta_W = 0.22 \pm 0.02. \tag{5.5.16}$$

§5.6. Detailed Predictions of the SU(5) Model for Proton Decay and Uncertainties in the Predictions

In this section we study the detailed prediction of the SU(5) model for the lifetime of the proton and the various uncertainties involved in this estimate. The attractive feature of the SU(5) model is that being a two-scale (m_W and M_X) model, the constraint of grand unification helps to predict the unification scale M_X and thereby the lifetime of the proton. In the previous section we have evaluated M_X which gives the strength of the baryon violating interaction of the form $qqql$, i.e., g_U^2/M_X^2. The value of α_U is predicted to be about $\frac{1}{50}$; however, to translate this into a lifetime for proton decay, several intermediate steps are needed which we will outline after giving the baryon numbers violating the Hamiltonian and some of their properties.

Using eq. (5.4.1) we can write the $\Delta B \neq 0$ Hamiltonian obtained from the gauge boson exchange for one generation of fermions as follows:

$$\begin{aligned} H_{\Delta B\neq 0} = {} & \frac{g_X^2}{2M_X^2}[\varepsilon_{ijk}\bar{u}^c_{kL}\gamma_\mu u_{jL} + \bar{d}_i\gamma_\mu e^+][\varepsilon_{ilm}\bar{u}_{lL}\gamma_\mu u^c_{mL} + \bar{e}^+\gamma_\mu d_i] \\ & + \frac{g_Y^2}{2M_Y^2}[\varepsilon_{ijk}\bar{u}^c_{kL}\gamma_\mu d_{jL} - \bar{u}_{iL}\gamma_\mu e^+_L + \bar{d}_{iR}\gamma_\mu \nu^c_R] \\ & \times [\varepsilon_{ilm}\bar{d}_{lL}\gamma_\mu u^c_{mL} - \bar{e}^+_L\gamma_\mu u_{i_L} + \bar{\nu}^c_R\gamma_\mu d_{i_R}] \\ & + \text{Higgs contributions.} \end{aligned} \tag{5.6.1}$$

If the Higgs boson masses are of order 10^{13} GeV or so, the Higgs contribution to proton decay can be neglected since the associated Yukawa couplings are $\approx G_F^{+1/2} m_f$. Looking at this Hamiltonian it is clear that it conserves the quantum number $B - L$ where L is the lepton number. This is also true for the Higgs bosons contributions as will be discussed in the next section. This can be understood by realizing that even though the gauge bosons X, Y have no definite baryon number (thus leading to proton decay), they have a definite $B - L$ quantum number 2/3. From this we can conclude that only the following decay modes of proton and bound neutron are allowed by the SU(5) model.

$$
\begin{aligned}
p &\to e^+\pi^0 \\
&\to e^+\rho^0 \\
&\to e^+\omega^0 \\
&\to e^+\eta \\
&\to \bar{\nu}\rho^+ \\
&\to \bar{\nu}\pi^+ \\
&\to \mu^+K^0 \\
&\to \bar{\nu}_\mu K^+ \qquad \text{(through flavor mixing effects)}, \\
n &\to \bar{\nu}\omega \\
&\to \bar{\nu}\rho^0 \\
&\to \bar{\nu}\pi^0 \\
&\to e^+\rho^- \\
&\to e^+\pi^0 \\
&\to \bar{\nu}_\mu K^0 \qquad \text{(through flavor mixing effects)}.
\end{aligned}
$$

To estimate the total decay width for the proton and branching ratios for the various decay modes we have to go through several steps.

(a) The effective $\Delta B \neq 0$ Hamiltonian is defined at the grand unification scale M_X, and therefore has to be extrapolated down to the $\mu \simeq 1$ GeV scale by calculating weak as well as strong QCD corrections to the vertex [23]. The QCD correction is given by

$$A_3 \approx \left[\frac{\alpha_3(1\ \text{GeV})}{\alpha_U}\right]^{2/11-2/3f}. \tag{5.6.2}$$

The electro-weak corrections are

$$A_{21} \approx \left[\frac{\alpha_2(100\ \text{GeV})}{\alpha_U}\right]^{27/(86-8f)} \times \begin{cases} \left[\dfrac{\alpha_1(100\ \text{GeV})}{\alpha_U}\right]^{-69/(6+40f)}, \\ \left[\dfrac{\alpha_1(100\ \text{GeV})}{\alpha_U}\right]^{-33/(6+40f)}. \end{cases} \tag{5.6.3}$$

The top factor in eqn. (5.6.3) applies to the operators involving e_L^- and the bottom factor to that involving ν_L and $e_{\bar{R}}$; f is the number of flavors. These factors have the effect of enhancing the matrix element by a factor of 3.5 to 4, decreasing the proton lifetime by roughly a factor 15.

(b) The next problem is to go from the quark–lepton form of the Hamiltonian to calculate the proton decay matrix elements. We may use [23], [24]

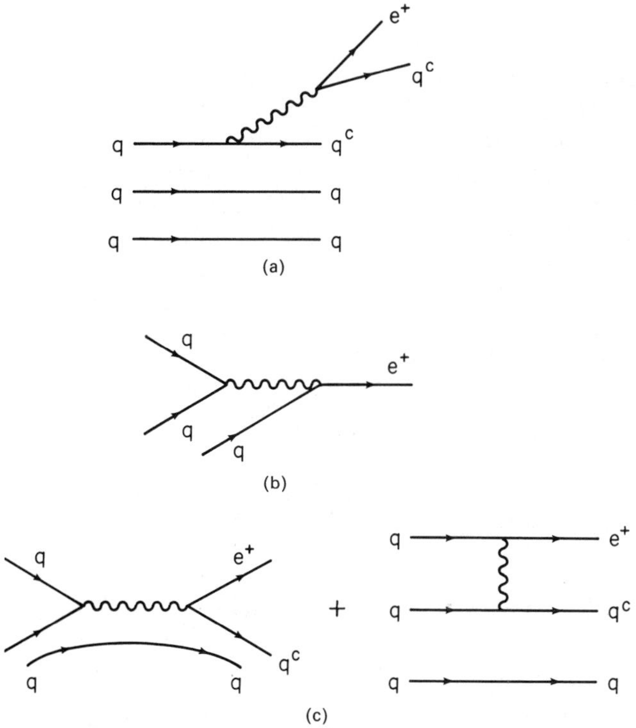

Figure 5.2. Different classes of contributions to $p \to e^+\pi^0$ decay amplitude.

the naive nonrelativistic SU(6) wave functions (for the proton and the various mesons) to evaluate the proton decay matrix element or ultrarelativistic bag model [25]. In terms of quark diagrams we can draw the three types shown in Fig. 5.2. Figure 5.2(a) is not expected to be significant due to phase space suppression. Recent estimates [26] of proton decay have therefore focused on the diagrams in Figs. 5.2(b) (c). All these calculations lead to the value of τ_p as follows:

$$\tau_p = 1.6 \times 10^{30} \text{ yr to } 2.5 \times 10^{28} \text{ yr.} \tag{5.6.4}$$

In most models $p \to e^+\pi^0$ is predicted to occur most often with a branching ratio of 40–60%, and the next most frequent being $p \to e^+\omega$, $e^+\rho^0$, and $\bar{\nu}_e\pi^+$ with branching ratios of 5–20%, 1–10%, and 16–24%, respectively. This gives a maximum value for the partial lifetime of $p \to e^+\pi^0$ decay of about 3×10^{30} yr. The present experimental value for this number comes from the IMB experiments which is

$$\tau_{p \to e^+\pi^0} \geq 2 \times 10^{32} \text{ yr.} \tag{5.6.5}$$

This result rules out the minimum SU(5) model as the grand unification symmetry.

(c) *Effect of Flavor Mixing on Proton Decay*. This was first studied in Ref.

[27] and subsequently in Ref. [28]. The basic question, first raised by Jarlskog [29], is that in the presence of flavor mixing effects in SU(5) models with several generations, it is *a priori* possible that the mixing matrix that appears in the baryon number conserving weak charged currents, the Cabibbo–Kobayashi–Maskawa matrix, need not be the same as the one in the baryon number violating interactions. This would have profound implications on the predictions for proton decay in the SU(5) model. For instance, if the two mixing matrices are unrelated, we could "rotate" away proton decay by adjusting the mixing angles in the $\Delta B \neq 0$ sector. It was, however, shown in Ref. [27] that in the minimal SU(5) model this is not possible, i.e., both mixing matrices in the gauge interactions of $\Delta B = 0$ and $\Delta B \neq 0$ types are the same. To prove this we note that

$$\begin{aligned}\mathscr{L}_I = {} & \frac{i}{\sqrt{2}} g W_\mu^+ [\bar{Q}_{1L}\gamma_\mu Q_{2L} + \overline{M}{}^c_{1L}\gamma_\mu M^c_{2L}] \\ & + \frac{i}{\sqrt{2}} g_X X_{\mu i}[\varepsilon_{ijk}\bar{Q}^c_{1kL}\gamma_\mu Q_{1jL} + \bar{Q}_{2i}\gamma_\mu M_1) \\ & + \frac{i}{\sqrt{2}} g_Y Y_{\mu i}[\varepsilon_{ijk}\bar{Q}^c_{1kL}\gamma_\mu Q_{2jL} - \bar{Q}_{1iL}\gamma_\mu M_{1L} + \bar{Q}_{2iR}\gamma_\mu M_{2R}] + \text{h.c.},\end{aligned} \tag{5.6.6}$$

where Q and M are column vectors for unmixed like-charge fermions of different generations (and $\psi^c = C\bar{\psi}^T$)

$$\begin{aligned} Q_{1i} &= (p_i^1, p_i^2, \ldots), \\ Q_{2i} &= (n_i^1, n_i^2, \ldots), \\ M_1 &= (\varepsilon^{+^1}, \varepsilon^{+^2}, \ldots), \\ M_2 &= (L^{0^1}, L^{0^2}, \ldots) \end{aligned} \tag{5.6.7}$$

(superscripts stand for generations and subscripts for color).

The fermion mass term in the Lagrangian, subsequent to spontaneous breakdown, can be written as

$$\mathscr{L}_m = \bar{Q}_{1L} M^{(+)} Q_{1R} + \bar{Q}_{2L} M^{(-)} Q_{2R} + \overline{M}_{1L} M^{(-)} M_{1R}. \tag{5.6.8}$$

The above mass matrices will, in general, be complex and can be diagonalized by the following bi-unitary transformations:

$$U_L M^{(+)} U_R^+ = D^{(+)}, \tag{5.6.9}$$

$$V_L M^{(-)} V_R^+ = D^{(-)}. \tag{5.6.10}$$

The physical fermions are then given by

$$\begin{aligned} P_L &= U_L Q_{1L}, \\ N_L &= V_L Q_{2L}, \\ E_L^+ &= V_L M_{1L}, \end{aligned} \tag{5.6.11}$$

and similarly for right-handed helicity components where

$$P = (u, c, t, \ldots),$$

$$N = (d, s, b, \ldots),$$

$$E^+ = (e^+, \mu^+, \tau^+, \ldots),$$

and

$$D^{(+)} = \text{diag}(m_u, m_c, m_t, \ldots),$$

$$D^{(-)} = \text{diag}(m_d, m_s, m_b, \ldots). \tag{5.6.12}$$

In terms of the physical fermion fields, eqn. (5.6.6) can be rewritten as

$$\begin{aligned}\mathscr{L}_I &= \frac{ig}{\sqrt{2}} W_\mu^+ [\bar{P}_L \gamma_\mu U_L V_L^+ N_L + \bar{E}_R^+ \gamma_\mu E_R^0] \\ &\quad + \frac{ig_X}{\sqrt{2}} X_{\mu i}[\varepsilon_{ijk} \bar{P}_{kL}^c \gamma_\mu U_R^* U_L^+ P_{jL} + \bar{N}_i \gamma_\mu E^+] \\ &\quad + \frac{ig_Y}{\sqrt{2}} Y_{\mu i}[\varepsilon_{ijk} \bar{P}_{kL}^c \gamma_\mu U_R^* V_L^+ N_{jL} - \bar{P}_{iL} \gamma_\mu U_L V_L^+ E_L^+ + \bar{N}_{iR} \gamma_\mu E_R^0] + \text{h.c.},\end{aligned} \tag{5.6.13}$$

where $E_R^0 = V_R M_{2R}$. Note that in this model, since all neutrinos are massless, E_R^0 is also an eigenstate of the Hamiltonian. From eqn. (5.6.13) we find that charged current weak interactions are described by the matrix $U_L V_L^+$ which, for two families, can be made completely real by suitable redefinitions of the phases of the fermion fields. For three families it can be written in the form first suggested by Kobayashi and Maskawa as discussed in Chapter 4. The interaction of X- and Y-bosons, however, contain mixing matrices like $U_R^* U_L^+$ or $U_R^* V_L$, etc., which will involve mixing angles very different from those in $U_L V_L^+$. However, we now show that the SU(5) model possesses an intrinsic symmetry for the mass matrix of the $I_{3w} = +\frac{1}{2}$ quarks which enables us to simplify eqn. (5.6.13) further. To see this note that the Yukawa coupling leading to $M^{(+)}$ is

$$\sum_{a,b} h_{ab} \varepsilon^{pqrst} \chi_{pq}^{Ta} H_r C^{-1} \chi_{st}^b + \text{h.c.}, \tag{5.6.14}$$

where C is the Dirac charge-conjugation matrix. An examination of eqn. (5.6.14) reveals that the mass matrix $M^{(+)}$ resulting from it has the following property

$$M_{ab}^{(+)} = M_{ba}^{(+)} = \frac{\rho}{\sqrt{2}}(h_{ab} + h_{ba}), \tag{5.6.15}$$

where $\rho/\sqrt{2} = \langle H_5 \rangle$. It then follows that

$$U_L = K U_R^*, \tag{5.6.16}$$

where K is a diagonal unitary matrix. Using eqn. (5.6.16) we find that

$$U_R^* U_L^+ = K^{-1}$$

and

$$U_R^* V_L^+ = K^{-1} U_L V_L^+. \tag{5.6.17}$$

Thus, up to a diagonal unitary matrix K, all gauge boson interactions in eqn. (5.6.13) are expressible in terms of a single mixing matrix $U_L V_L^+$. From this it follows that, since the mixing angles between the light-mass quarks (u, d, ...) and the heavier quarks (c, s, ...), etc. are small, the proton decay Hamiltonian is not substantially altered in the presence of more generations of fermions.

We now proceed to show that for the simplest Higgs system with one $\{\underline{5}\}$ and one $\{\underline{24}\}$ Higgs multiplet even the Higgs–boson–fermion interactions become very simple. As is well known, only the $\{\underline{5}\}$-multiplet H interacts with fermions prior to spontaneous breakdown

$$\mathscr{L}_H = \sum_{a,b} f_{ab} \bar{\chi}^a_{pqL} H_p \psi_{qR} + \sum_{a,b} h_{ab} \varepsilon^{pqrst} \chi^T_{pq} H_r \chi^b_{st} + \text{h.c.} \tag{5.6.18}$$

We write

$$H = \begin{cases} H_i \\ H_4, \\ H_5 \end{cases}$$

where H_i are the heavy, colored factionally charged Higgs mesons that couple to baryon-number-nonconserving currents. We will concern ourselves with their interactions with the physical fermions. From straightforward algebra it follows that their interaction with physical fermions can be written as

$$\mathscr{L}_H = -H_i\{\varepsilon_{ijk} \bar{P}^c_{kL} U_R^* V_R^+ N_{jR} + \bar{P}_{iL} U_L V_R^+ E_R^+ \tag{5.6.19}$$

$$+ \varepsilon_{ijk} P_{jL}^T C^{-1} U_L^* H V_L^+ N_{kL} + \varepsilon_{ijk} N_{jL}^T C^{-1} V_L^* H U_L^+ P_{kL}$$

$$+ P_{iL}^{C^T} C^{-1} U_R H (V_L^+ E_L^+)\} + \text{h.c.}, \tag{5.6.20}$$

where $F_{ab} = f_{ab}$; $H_{ab} = h_{ab}$. But note that $F = \rho^{-1}\sqrt{2}M^{(-)}$ and $H = \rho^{-1}\sqrt{2}M^{(+)}$. Therefore, using eqns. (5.6.10) and (5.6.11) we find

$$\begin{aligned} U_R^* F V_R^+ &= K^{-1} U_L V_L^+ D^{(-)} \rho^{-1}\sqrt{2}, \\ U_L F V_R^+ &= U_L V_L^+ D^{(-)} \rho^{-1}\sqrt{2}, \\ U_L^* H V_L^+ &= D^+ K^{-1} U_L V_L^+ \rho^{-1}\sqrt{2}, \\ V_L^* H U_L^+ &= (U_L V_L^+)^T K^{-1} D^{(+)} \rho^{-1}\sqrt{2}, \\ U_R H V_L^+ &= D^+ U_L V_L^+ \rho^{-1}\sqrt{2}. \end{aligned} \tag{5.6.21}$$

Thus, all Higgs baryon-number-nonconserving interactions also become expressible in terms of the same mixing matrix $U_L V_L^+$ which describes the charged current weak interactions. Therefore, the Higgs contribution to the proton decay can also be safely estimated using the model with a single mixing matrix. It is now obvious that if there are more than one $\{\underline{5}\}$ Higgs multiplets coupling to fermions, eqn. (5.6.21) does not hold and, therefore, the nature of the baryon-number-nonconserving Higgs boson interactions becomes more complicated.

We would now like to comment on the impact of adding Higgs fields belonging to the $\{\underline{45}\}$ representation H^p_{qr} of SU(5). This will generate new kinds of Yukawa couplings involving $\{\underline{5}^*\}$ and $\{\underline{10}\}$ as well as those involving only the $\underline{10}$ representations of fermions. These interactions will be given by

$$\mathscr{L}'_Y = \sum_{a,b} f'_{ab}\bar{\chi}^a_{pq,L}\psi_{r,R}H^r_{pq} + \sum_{a,b} h'_{ab}\varepsilon^{pqrst}\chi^{T_a}_{pq,L}C^{-1}\chi^b_{lr,L}H^l_{st}. \tag{5.6.22}$$

The H fields acquire vacuum expectation values as follows: $\langle H^4_{45}\rangle = -3\langle H^i_{i5}\rangle$ (no sum over i). By substituting $\langle H\rangle$ into eqn. (5.6.15) it becomes clear that the contribution of $\mathscr{L}_{Y'}$ to the up-quark mass matrix $M^{(+)}$ does not satisfy eqn. (5.6.15). This enables us to draw the following conclusions:

(a) If the $\{\underline{45}\}$ Higgs multiplet contributes only to the down-quark mass matrix $M^{(-)}$, the conclusions about the dominant structure of the proton decay Hamiltonian remain valid.
(b) On the other hand, if H^p_{qr} contributes to $M^{(+)}$, then the situation will be very different and the conclusions will not hold in general.

Another source of uncertainty in the proton lifetime has to do with superheavy Higgs boson masses [29a]. The problem is that, due to uncertainties in $\lambda\phi^4$ type couplings, the masses of the Higgs bosons could easily be uncertain by a factor of 100. This would imply that the evolution of the coupling constants will change leading to changes in M_X. These uncertainties have been estimated in Ref. [29a] and could change M_X by a factor of 2 to 3.

§5.7. Some Other Aspects of the SU(5) Model

In this section we focus our attention on some other aspects of the SU(5) model: this will include predictions for fermion masses, selection rules for baryon nonconservations, and possible extensions of the SU(5) model.

(i) Fermion Masses in the SU(5) Model

In the minimal SU(5) model with only the Higgs multiplets in the $\{24\}$- and $\{5\}$-dimensional (denoted by H) representations, the down-quark and lepton masses will arise from the following Yukawa couplings (the up-quark mass has been discussed in eqn. (5.6.14))

$$\mathscr{L}_Y = h\bar{\psi}_{pR}H^+_q\chi_{pq} + \text{h.c.} \tag{5.7.1}$$

Substituting $\langle H_5\rangle = \rho/\sqrt{2}$, we find equal mass for electron and down-quarks

$$m_e = m_d = h\rho/\sqrt{2}. \tag{5.7.2}$$

In the presence of all generations of quarks and leptons the generalized mass matrix looks like

$$M_{ab}^{(l^-)} = M_{ab}^{(d)} = h_{ab}\rho/\sqrt{2}. \tag{5.7.3}$$

Diagonalization of this mass matrix therefore says that all its eigenvalues are equal, i.e.,

$$m_e = m_d, \qquad m_\mu = m_s, \qquad m_\tau = m_b. \tag{5.7.4}$$

Note, however, that these equations are valid only at mass scales where SU(5) is a good symmetry. But since observed masses correspond to energies of order 1 GeV these relations must be extrapolated. The extrapolation formula can be written as follows:

$$\ln \frac{m_{f_1}(\mu)}{m_{f_2}(\mu)} = \ln \frac{m_{f_1}(M)}{m_{f_2}(M)} + \int \sum_i [\gamma_{f_1}^i - \gamma_{f_2}^i] \frac{d\mu'}{\mu'}, \tag{5.7.5}$$

where the notation is self-explanatory except for i which stands for subgroup G_i. However, if we look at certain mass ratios such as

$$\frac{m_e}{m_\mu} = \frac{m_d}{m_s}, \tag{5.7.6}$$

without doing any calculation it is easy to argue that they will remain the same. The reason is that, since the SU(2) and U(1) quantum numbers of particles in each ratio are equal, their contributions will cancel. In fact, before calculating any fermion masses, it is quite clear that the above relation is grossly violated by experiments and therefore requires drastic changes in the simple structure of the model.

One way to correct [30], [31] for this is to include a $\{45\}$-dimensional H_{qs}^p Higgs boson in the theory. In the presence of this we will have additional Yukawa couplings of the form

$$\mathscr{L}_Y = h' \bar{\psi}_{p,R} H_{sq}^{p^+} \chi_{sq,L} + \text{h.c.} \tag{5.7.7}$$

The property of the $\{45\}$-dimensional representation of the Higgs meson is that

$$H_{sq}^p = -H_{qs}^p, \qquad \sum_{p=1}^{5} H_{ps}^p = 0. \tag{5.7.8}$$

From this it follows that the only allowed nonzero vacuum expectation values of H are

$$\langle H_{i5}^i \rangle = -\tfrac{1}{3}\langle H_{45}^4 \rangle = \Lambda. \tag{5.7.9}$$

Thus, using (5.7.9) will yield lepton–quark mass relations of type $m_q = 3m_l$. Now, using this, let us see how we can fix the lepton–quark mass relations: consider only the $e - \mu$ sector of the mass matrix. Then, by means of suitable discrete symmetries, we can have the following type of Yukawa coupling:

$$\mathscr{L}_Y = h_1(\bar{\psi}_{p,R}^{(1)} H_q^+ \chi_{pq,L}^{(2)} + \bar{\psi}_{p,R}^{(2)} H_q^+ \chi_{pq,L}^{(1)}) + h_2 \bar{\psi}_{p,R}^{(2)} H_{gr}^{p^+} \chi_{gr,L}^{(2)} + \text{h.c.} \tag{5.7.10}$$

This leads to mass matrices of the form

$$\begin{array}{c} \\ e \\ \mu \end{array}\!\!\begin{array}{c} \begin{array}{cc} e & \mu \end{array} \\ \begin{pmatrix} 0 & a \\ a & 3b \end{pmatrix} \end{array}, \qquad \begin{array}{c} \\ d \\ s \end{array}\!\!\begin{array}{c} \begin{array}{cc} d & s \end{array} \\ \begin{pmatrix} 0 & a \\ a & b \end{pmatrix} \end{array}. \tag{5.7.11}$$

This implies

$$\frac{m_e}{m_\mu} = \frac{1}{9}\frac{m_d}{m_s}. \tag{5.7.12}$$

This relation is, of course, quite well satisfied by experiment. But we pay the price of including several Higgs multiplets in the theory.

(ii) $B - L$ Conservation in Proton Decay

Next we want to display a very important property of the minimal SU(5) model that contains only {24}- and {5}-dimensional Higgs multiplets. We will show that even after symmetry breaking there exists an exact global symmetry in the theory which can be identified with the $B - L$ quantum number [32].

To see this, note that only {5}-dimensional Higgs {H} couples to fermions with couplings of the form $\bar{\psi} H^\dagger \chi$, $\chi\chi H$ (ψ and χ defined in eqn. (4.8)). These couplings are therefore invariant under a global $\tilde{U}(1)$ transformation with "charges" $\tilde{Q}$

$$\begin{aligned} \tilde{Q}(\psi) &= +1, \\ \tilde{Q}(H) &= -\tfrac{2}{3}, \\ \tilde{Q}(\chi) &= +\tfrac{1}{3}, \qquad \tilde{Q}(\phi) = 0. \end{aligned} \tag{5.7.13}$$

Of course, the gauge interactions and the Higgs potential clearly respect this symmetry. The second stage of SU(5) symmetry breaking is implemented by $\langle H_5 \rangle \neq 0$ (the first stage being implemented by $\langle \phi \rangle \neq 0$ clearly leaves the symmetry intact). Therefore, it breaks $\tilde{\mathrm{U}}(1)$; however, it keeps a linear combination with ω still invariant

$$\omega = -\left(\frac{2}{\sqrt{15}}\lambda_{24} - \frac{3}{5}\tilde{Q}\right) \tag{5.7.14}$$

since $\omega(H_5) = 0$. Looking at ω it is apparent that

$$\omega = B - L. \tag{5.7.15}$$

Thus all baryon and lepton nonconserving processes in SU(5) must respect the $B - L$ quantum number. This is, of course, very important and can be used to distinguish the minimal SU(5) model from other models such as the minimal $\mathrm{SU(2)}_L \times \mathrm{SU(2)}_R \times \mathrm{U(1)}_{B-L}$ model, where the baryon nonconserving processes respect the $\Delta B = 2$ selection rule. A typical allowed $\Delta B \neq 0$ process in SU(5) is $p \to e^+\pi^0$, $n \to e^+\pi^-$, etc., whereas in the latter case it is the

$N \leftrightarrow \bar{N}$ oscillation and $N + P \to \pi$'s, etc. Note that in this connection the process $n \to e^- \pi^+$, which obeys the $\Delta(B + L) = 0$ selection rule, is indeed a rare process not allowed in any simple unified model. Another consequence of $B - L$ conservation is that the neutrino cannot acquire Majorana mass due to higher orders and must therefore remain massless to all orders.

We wish to note two further points at this stage.

(i) It has been pointed out by Weinberg [33], and Wilczek and Zee [33] that exact $B - L$ conservation follows for all four-fermion operators (operators of dimension 6) that break the baryon number but which respect $SU(3)_c \times SU(2)_L \times U(1)$ symmetry and is therefore independent of any particular grand unification scheme. Thus, violation of $B - L$ conservation at the level of $\tau_{\Delta B \neq 0} \simeq 10^{30}$ yr must therefore mean that operators of dimension higher than six are involved, and must require the existence of mass scale below 10^{15} GeV to be observable in low-energy processes.

(ii) Another important point worth noting is that if we extend the SU(5) model by including an extra $\{15\}$-dimensional Higgs representation, then that leads to $B - L$ violation. To see this, denote the $\{15\}$-dimensional Higgs meson by S_{pq}: this allows additional couplings into the theory

$$\mathscr{L}_Y = h_{15} \psi_{pR}^T C^{-1} \psi_{qR} S_{pq}^+ + \lambda_{15} M_{15} H_p^+ H_q^+ S_{pq} + \text{h.c.} \qquad (5.7.16)$$

It is clear that this interaction violates the new $\tilde{U}(1)$ symmetry present in the minimal SU(5) model. To study its effects observe that (due to λ_{15} term, we obtain, regardless of the sign of its mass term)

$$\langle S_{55} \rangle = \kappa \neq 0. \qquad (5.7.17)$$

This leads to the Majorana mass term for the neutrino: $h_{15}\kappa$. *A priori*, there exists a rather weak restriction of the magnitude of κ arising from the fact that it contributes to the violation of the weak $\Delta I_W = 1/2$ rule (or $\rho_W = 1$) and must therefore satisfy the constraint

$$(g\kappa/m_{W_L})^2 \ll 1. \qquad (5.7.18)$$

This still, however, leave κ big enough so that to understand the smallness of neutrino mass h_{15} may have to be tuned to a small value.

(iii) $N - \bar{N}$ Oscillation

In this model $\Delta B = 2$ transitions, such as $N - \bar{N}$ oscillation, proceed via the Feynman graph shown in Fig. 5.3 and have strength $(h_{15} h_5^2 / m_{15}^2 m_5^4) \lambda M$. But $\{15\}$ also contributes a piece to the proton decay in addition to the usual $\{5\}$ and X, Y mediated graphs (Fig. 5.4). The sum total of this contribution is (symbolically)

$$\frac{h_5^2}{m_5^2} + \frac{h_5 h_{15} \lambda_{15} M_{15} \langle H_5 \rangle}{m_5^2 m_{15}^2}. \qquad (5.7.19)$$

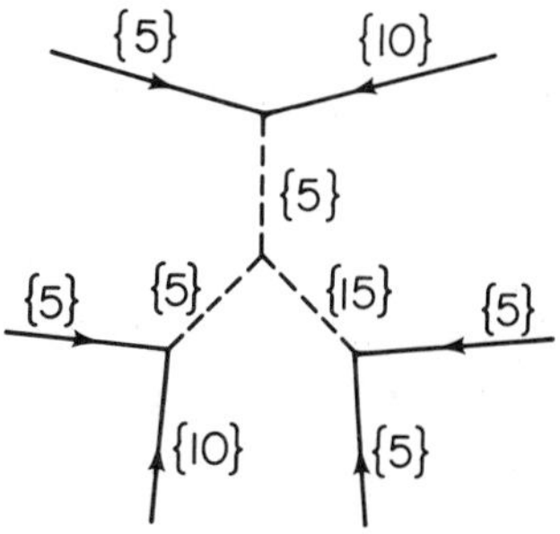

Figure 5.3

Observed stability of proton with $\tau_p > 10^{30}$ yr implies

$$h_5^2/m_5^2 < 10^{-30}\ \text{GeV}^{-2} \quad \text{and} \quad \frac{h_{15}}{h_5}\frac{\lambda_{15}M_{15}\langle H_5\rangle}{m_{15}^2} < 1. \tag{5.7.20}$$

Equation (5.7.19) implies that

$$\begin{aligned} A_{N\leftrightarrow\bar{N}} &\approx \left(\frac{h_{15}\lambda_{15}M_{15}}{m_{15}^2}\right)\left(\frac{h_{15}^2}{m_5^4}\right) < \left(\frac{h_5^4}{m_5^4}\right)\frac{1}{h_5\langle H_5\rangle} \\ &\leq 10^{-60}\ \text{GeV}^{-4}\frac{1}{h_5\langle H_5\rangle} \leq 10^{-58}\ \text{GeV}^{-5} \end{aligned} \tag{5.7.21}$$

(assuming $h_5\langle H_5\rangle$ which is a typical fermion mass ≤ 10 MeV). Such a small $N - \bar{N}$ transition amplitude would lead to $\tau_{N-\bar{N}} \geq 10^{28}$ yr which is insignificant. Thus, unless the model is made considerably complicated, SU(5) will not allow a significant $N - \bar{N}$ oscillation.

To summarize this chapter, a beautiful illustration of the hypothesis of grand unification is the SU(5) model discussed in this chapter. Even though experimental results on proton decay have ruled out this model in its simplest form, the basic concept of grand unification may be right, and indeed, has

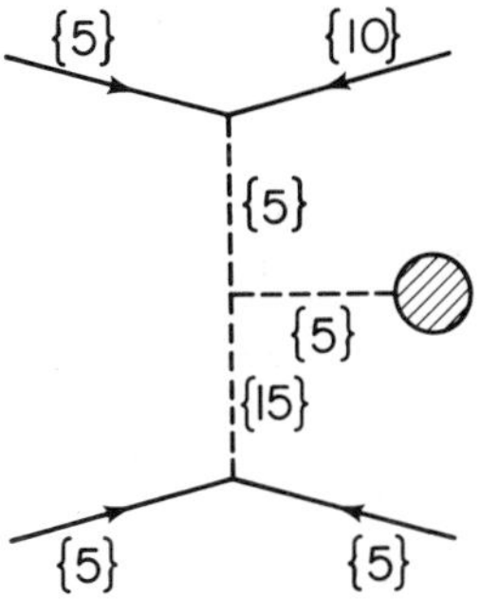

Figure 5.4

generated a great deal of activity and excitement. Part of this excitement will be recorded in some of the subsequent chapters.

References

[1] J. C. Pati and A. Salam, *Phys. Rev.* **D10**, 275 (1974).

[2] R. N. Mohapatra and J. C. Pati, *Phys. Rev.* **D11**, 2558 (1975); G. Senjanovic and R. N. Mohapatra, *Phys. Rev.* **D12**, 1502 (1975).

[3] H. Georgi and S. L. Glashow, *Phys. Rev. Lett.* **32**, 438 (1974).

[3a] S. Okubo, *Phys. Rev.* **D16**, 3528 (1977); J. Banks and H. Georgi, *Phys. Rev.* **D14**, 1158 (1976).

[3b] J. Chakrabarti, M. Popovic, and R. N. Mohapatra, *Phys. Rev.* **D21**, 3212 (1980).

[4] J. D. Bjorken, SLAC Summer Institute Lectures, 1976.

[5] L. F. Li, *Phys. Rev.* **D9**, 1723 (1974).

[6] J. Learned, F. Reines, and A. Soni, *Phys. Rev. Lett.* **43**, 907 (1979).

[7] M. Krishnaswamy, M. G. K. Menon, N. Mandal, V. Narasimhan, B. V. Sreekantan, S. Ito, and S. Miyake, *Phys. Lett.* **106B**, 339 (1981); **115B**, 349 (1892).

[8] R. Bionta *et al.*, *Phys. Rev. Lett.* **51**, 27 (1983); B. Cortez *et al.*, *Phys. Rev. Lett.* **52**, 1092 (1984).

[9] M. Koshiba, *Proceedings of the Leipzig Conference*, 1894.

[10] E. Fiorini, *Proceedings of the Leipzig Conference*, 1984.

[11] D. Cline, Washington APS Meeting, 1984.

[12] M. L. Cherry *et al.*, *Phys. Rev. Lett.* **47**, 1507 (1981).

[13] E. Gildener, *Phys. Rev.* **D14**, 1667 (1976).

[14] R. N. Mohapatra and G. Senjanovic, *Hadron. J.* **1**, 903 (1978); I. Bars, *Orbis Sciential*, 1979; K. T. Mahanthappa and D. Unger, *Phys. Lett.* **78B**, 604 (1978); T. N. Sherry, *Phys. Lett.* **88B**, 76 (1979); M. A. Namazie and W. A. Sayed, ICTP/78–79/9; S. Weinberg, *Phys. Lett.* **82B**, 387 (1979).

[15] H. Georgi, H. Quinn, and S. Weinberg, *Phys. Rev. Lett.* **33**, 451 (1974).

[16] T. Appelquist and J. Carrazone, *Phys. Rev.* **D11**, 2856 (1975).

[17] For further discussion of heavy particle effects at low energies, see D. Toussaint, *Phys. Rev.* **D18**, 1626 (1978); G. Senjanovic and A. Sokorac, *Phys. Rev.* **D18**, 2708 (1978); J. Kubo and S. Sakakibara, Dortmund preprint, 1980; Y. Kazama and Y. P. Yao, Michigan preprint, 1980; S. Weinberg, *Phys. Lett.* **91B**, 51 (1980); L. Hall, *Nucl. Phys.* **B178**, 75 (1981); N. P. Chang, A. Das, and J. P. Mercader, *Phys. Lett.* **39B**, 137 (1980).

[18] W. Marciano, Brookhaven preprint 34728, 1984; *Phys. Rev.* **D20**, 274 (1979).

[19] P. B. MacKenzie and G. P. Lapage, *Phys. Rev. Lett.* **47**, 1244 (1981); A. Buras, *1981 International Symposium on Lepton and Photon Interactions*, 1981, p. 636.

[20] For excelent reviews see: P. Langacker, *Phys. Rep.* **72**, 185 (1981); J. Ellis, in *21st Scottish University Summer School* (edited by K. C. Bowler and D. G. Sutherland), Edinburgh, 1981, p. 201;

W. Marciano, in *Fourth Workshop on Grand Unification* (edited by H. A. Weldon *et al.*), Birkhauser, Boston, 1983, p. 13.

[21] J. E. Kim, P. Langacker, M. Levine, and H. Williams, *Rev. Mod. Phys.* **53**, 211 (1980).

[22] G. Arnison *et al.*, *Phys. Lett.* **126B**, 398 (1983);
P. Bagnaia *et al.*, *Phys. Lett.* **129B**, 130 (1983).

[23] A. Buras, J. Ellis, M. K. Gaillard, and D. V. Nanopoulos, *Nucl. Phys.* **B135**, 66 (1978).

[24] T. Goldman and D. A. Ross, *Nucl. Phys.* **B171**, 273 (1980);
C. Jarlskog and F. J. Yndurain, *Nucl. Phys.* **B149**, 29 (1979);
M. Machachek, *Nucl. Phys.* **B159**, 37 (1979);
M. B. Gavela, A. LeYouanc, L. Oliver, O. Pene, and J. C. Raynal, Orsay preprint, PTHE 80/6, 1980.

[25] A. Din, G. Girardi, and P. Sorba, *Phys. Lett.* **91B**, 77 (1980);
J. Donoghue, *Phys. Lett.* **92B**, 99 (1980);
E. Golowich, University of Massachusetts preprint, 1980.

[26] For a recent review and references see
P. Langacker, *Proceedings of the Fermilab Conference on Inner Space/Outer Space Connection*, 1984.

[27] R. N. Mohapatra, *Phys. Rev. Lett.* **43**, 893 (1979).

[28] J. Ellis, M. K. Gaillard, and D. V. Nanopoulos, *Phys. Lett.* **80B**, 360 (1979).

[29] C. Jarlskog, *Phys. Lett.* **82B**, 401 (1979).

[29a] G. Cook, K. T. Mahanthapa, and M. Sher, *Phys. Lett.* **90B**, 298 (1980).

[30] P. Frampton, S. Nandi, and J. Scanio, *Phys. Lett.* **85B**, 255 (1979).

[31] H. Georgi and C. Jarlskog, *Phys. Lett.* **86B**, 297 (1979).

[32] M. T. Vaughn, Northeastern University preprint, 1979.

[33] S. Weinberg, *Phys. Rev. Lett.* **43**, 1566 (1979);
F. Wilczek and A. Zee, **43**, 1571 (1979).

CHAPTER 6

Left–Right Symmetric Models of Weak Interactions

§6.1. Why Left–Right Symmetry?

While the standard electro-weak model, based on the spontaneously broken local symmetry $SU(3)_c \times SU(2)_L \times U(1)_Y$, has been extremely successful in the description of low-energy weak phenomena, it leaves a lot of questions unanswered. One of the unsolved problems is understanding the origin of parity violation in low-energy physics. An interesting approach is to assume that the interaction Lagrangian (or dynamics) is intrinsically left–right symmetric, the asymmetry observed in nature (i.e., β-decay and μ-decay, etc.) arising from the vacuum being noninvariant under parity symmetry. Within the framework of gauge theories this idea has found its realization in the $SU(2)_L \times SU(2)_R \times U(1)_{B-L}$ models [1] constructed in 1973–1974. An important feature of this model is that, at low energies, it reproduces all the features of the $SU(2)_L \times U(1)$ model, and as we move up in energies new effects associated with parity invariance of the Lagrangian (such as a second natural Z-boson, right-handed charged currents, right-handed neutrino) are supposed to appear.

There exist several other considerations having to do with weak interaction that find their place naturally in a left–right symmetric model rather than the standard model. Foremost among them is the neutrino mass. We do not know whether the neutrino has mass or not. Laboratory experiments involving tritium decay [2], which indicate endpoint behavior of the decay spectrum characteristic of a nonvanishing neutrino mass, are controversial. There exist astrophysical considerations [3] having to do with missing mass of the universe, galactic clusters, and galaxy formation, etc., which are easily understood if the neutrino has a nonvanishing mass in the electron volt range. If, however, $m_\nu \neq 0$ and is in the electron volt range, the most natural framework to understand it is the left–right symmetric models as will be shown below.

Second, if weak interaction symmetries are to arise out of a more fundamental substructure of quarks and leptons, and if the forces at the substructure level are assumed to be similar to those operating in nuclear physics, i.e., QCD, then $SU(2)_L \times SU(2)_R \times U(1)_{B-L}$ arises as the natural weak interaction symmetry rather than $SU(2)_L \times U(1)$.

Another deficiency of the standard model is the lack of any physical meaning of the U(1) generator, which in the left–right symmetric models becomes the $B - L$ quantum number [4]. All of the weak interaction symmetry generators then have a physical meaning. As if suggesting a deeper symmetry structure in the $SU(2)_L \times U(1)_Y$ model the only anomaly-free quantum number left ungauged is $B - L$, and once $B - L$ is included as a gauge generator the weak gauge group becomes $SU(2)_L \times SU(2)_R \times U(1)_{B-L}$, and electric charge is given by [4]

$$Q = I_{3L} + I_{3R} + \frac{B - L}{2}. \tag{6.1.1}$$

Our final comment has to do with the status of CP-violation in gauge theories. It is interesting that in the standard $SU(2)_L \times U(1)$ model, three generations are required to have nontrivial CP-violation and all CP-violations are parametrized by only one phase, δ_{KM}, the Kobayashi–Maskawa phase. But the model provides no hint as to why observed CP-violation has milliweak strength. As noted in Chapter 4 the left–right models provide a more appealing alternative [5], where the smallness of CP-violation is related to the suppression of $V + A$ currents, i.e.,

$$\eta_{+-} \simeq \left(\frac{m_{W_L}}{m_{W_R}}\right)^2 \sin\delta. \tag{6.1.2}$$

If both parity and CP-violations owe their origin to spontaneous breakdown of gauge symmetries, as is our belief, eqn. (6.1.2) can be proved [6] for three generations and becomes valid regardless of the contribution of the Higgs sectors.

Having summarized our motivations for studying left–right models, we give the outline and scope of the present chapter. In Section 6.2 we introduce the model, discuss the bounds on the right-handed gauge boson masses, and possible ways of detecting the $W_R^{\pm}$ and Z_R in high-energy experiments. In Section 6.3 we discuss neutrino masses, lepton numbers violation, and other implications for low-energy experiments. In Section 6.4 we discuss neutrino mass and neutrinoless double β-decay. In Section 6.5 we discuss the selection rules for baryon number violating processes and higher unification constraints on the scale of right-handed currents. In Section 6.6 we discuss some recent work where parity and $SU(2)_R$ scales are decoupled, and their impact on physics.

§6.2. The Model, Symmetry Breaking, and Gauge Boson Masses

The left–right symmetric models of weak interactions are based on the gauge group $SU(2)_L \times SU(2)_R \times U(1)_{B-L}$ with the following assignment of fermions to the gauge group: denoting $Q \equiv \binom{u}{d}$ and $\psi \equiv \binom{\nu}{e}$ we have

$$Q_L\colon\ (\tfrac{1}{2}, 0, \tfrac{1}{3}), \qquad Q_R\colon\ (0, \tfrac{1}{2}, \tfrac{1}{3}),$$
$$\psi_L\colon\ (\tfrac{1}{2}, 0, -1), \qquad \psi_R\colon\ (0, \tfrac{1}{2}, -1), \tag{6.2.1}$$

where the U(1) generator corresponds to the $B - L$ quantum numbers of the multiplet. The electric charge formula is given by [4]

$$Q = I_{3L} + I_{3R} + \frac{B - L}{2}. \tag{6.2.2}$$

Due to the existence of the discrete parity symmetry the model has only two gauge coupling constants prior to symmetry breaking, i.e., $g_2 \equiv g_{2L} = g_{2R}$ and g' and, as before, we can define

$$\sin\theta_W = e/g_{2L} \tag{6.2.3}$$

and we will parametrize the neutral current Hamiltonian in terms of θ_W.

Since we assume the weak interaction symmetry to be $SU(2)_L \times SU(2)_R \times U(1)_{B-L} \times P$, we can break it down to the $SU(2)_L \times U(1)$ model in two stages [7]: $SU(2)_L \times SU(2)_R \times U(1)_{B-L} \times P \underset{M_P}{\rightarrow} SU(2)_L \times SU(2)_R \times U(1)_{B-L} \underset{M_{W_R}}{\longrightarrow} SU(2)_L \times U(1)_Y \underset{M_{W_L}}{\longrightarrow} U(1)_{em}$.

At the first stage only the parity symmetry is broken and weak gauge symmetry is unbroken–leaving the $\mathbf{W}_L$ and $\mathbf{W}_R$ massless, their masses arise at the subsequent stages of symmetry breaking. The parity breaking at the first stage would manifest as different gauge couplings $g_{2L} \neq g_{2R}$ at $\mu \geq M_{W_R}$. Often the choice of Higgs multiplets breaks both parity and $SU(2)_R$ at the same scale. In the bulk of this chapter we will assume this: i.e., $M_P = M_{W_R}$, which happens with the minimal choice of the Higgs multiplets discussed below.

We use the Higgs multiplets [8, 9] dictated by intuitive dynamical requirement that they be bilinears in the basic fermionic multiplets. Then the unique minimal set required to break the symmetry down to the $U(1)_{em}$ is

$$\Delta_L(1, 0, +2) + \Delta_R(0, 1, +2)$$

and

$$\phi(\tfrac{1}{2}, \tfrac{1}{2}, 0) \tag{6.2.4}$$

under left–right symmetry $\Delta_L \leftrightarrow \Delta_R$ and $\phi \leftrightarrow \phi^+$.

It can be show [8] that, for a range of parameters, an exactly left–right symmetric potential would lead to parity violating minima that will break the gauge symmetry

$$\langle \Delta_R \rangle = \begin{pmatrix} 0 & 0 \\ v_R & 0 \end{pmatrix},$$

$$\langle \Delta_L \rangle = \begin{pmatrix} 0 & 0 \\ v_L & 0 \end{pmatrix},$$

$$\langle \phi \rangle = \begin{pmatrix} \kappa & 0 \\ 0 & \kappa' e^{i\alpha} \end{pmatrix}, \tag{6.2.5}$$

where

$$v_L \simeq \gamma \frac{\kappa^2}{v_R}. \tag{6.2.6}$$

At the first stage of breaking $W_R^{\pm}$ picks up the mass

$$m_{W_R} = g v_R$$

and the combination

$$Z_R = \{\sqrt{\cos 2\theta_W}(\mathrm{Sec}\,\theta_W) W_{3R} - \tan \theta_W B\}$$

acquires the following mass

$$m_{Z_R}^2 = m_{W_R}^2 \frac{2\,\mathrm{Cos}^2\,\theta_W}{\mathrm{Cos}\,2\theta_W}. \tag{6.2.7}$$

This is the analog of the formula (3.1.21c) for the left–right symmetric models. For the case of symmetry breaking by doublet Higgs mesons with $B - L = 1$, the relation becomes

$$\frac{m_{Z_R}^2}{m_{W_R}^2} = \frac{\mathrm{Cos}^2\,\theta_W}{\mathrm{Cos}\,2\theta_W}. \tag{6.2.8}$$

In general, if a right-handed Higgs multiplet with weak right-handed isospin I and $B - L$ quantum number y were to break the symmetry, the above mass relation would read

$$\frac{m_{Z_R}^2}{m_{W_R}^2} = \frac{y^2}{(I + y/2)(I - y/2 + 1)} \cdot \frac{\mathrm{Cos}^2\,\theta_W}{\mathrm{Cos}\,2\theta_W}. \tag{6.2.9}$$

Thus, we see that, measuring the W_R and Z_R mass would throw light on the Higgs sector of the theory. This is important because, for the general case of the Higgs boson with quantum number (I, y), the highest charge of the physical Higgs bosons would be $(I + y/2)$ and it could be looked for in e^+e^- collision.

The second stage of the symmetry breaking is controlled by $\langle \phi \rangle$ and $\langle \Delta_L \rangle$. Since the $\Delta I_w = \frac{1}{2}$ mass relations are known to be obeyed experimentally it implies $v_L \ll \kappa, \kappa'$. An outstanding feature of this model is that this constraint is automatically satisfied for the parity violating minimum of the Higgs potential and we obtain $v_L \simeq \gamma(\kappa^2/v_R)$. Thus $\langle\phi\rangle$ mostly contributes to W_L, Z_L masses; but since ϕ transforms nontrivially under the left- and right-handed gauge groups it mixes the $W_L^{\pm}$ with $W_R^{\pm}$. In fact, we obtain the following eigenstates:

$$W_L \simeq W_1 = \cos\zeta W_L + e^{i\alpha}\sin\zeta W_R,$$
$$W_R \simeq W_2 = -\sin\zeta e^{i\alpha} W_L + \cos\zeta W_R, \tag{6.2.10}$$

with

$$\tan\zeta = \frac{\kappa\kappa'}{\kappa^2 + \kappa'^2 + 8v_R^2} \tag{6.2.11}$$

and

$$m_{W_L}^2 \simeq \frac{g^2}{2}(\kappa^2 + \kappa'^2) \equiv m_{W_1}^2. \tag{6.2.12}$$

Let us study the charged current weak Lagrangian in this case. The gauge interactions are given by

$$\mathscr{L}_{\text{wk}} = \frac{g}{2}(\mathbf{J}_{\mu_L}\cdot\mathbf{W}_{\mu_L} + \mathbf{J}_{\mu_R}\cdot\mathbf{W}_{\mu_R}). \tag{6.2.13}$$

The effective charged current piece at low energies can be written as

$$H_{\text{wk}}^{\text{cc}} = \frac{G_F}{\sqrt{2}}[(\cos^2\zeta + \eta\sin^2\zeta)J_{\mu_L}^{+}J_{\mu_L}^{-} + (\eta\cos^2\zeta + \sin^2\zeta)J_{\mu_L}^{+}J_{\mu_{\overline{R}}}$$
$$+ e^{i\alpha}\cos\zeta\sin\zeta(1-n)J_{\mu_L}^{+}\cdot J_{\mu_R}^{-} + \text{h.c.}],$$

where

$$\eta = (m_{W_L}/m_{W_R})^2. \tag{6.2.14}$$

The observed predominance of $V - A$ currents at low energies can be explained if $\eta \ll 1$ and $\zeta \ll 1$. This requires that $v_R \gg \kappa, \kappa'$ for the symmetry breaking parameters which, of course, guarantees $v_L \lll v_R$. Turning this around, we can use the accuracy of the low-energy weak interaction data to obtain a lower bound on the mass of the right-handed W-bosons. This is the question we address in the next section.

Before closing this section we would like to give the mass of the light neutral Z-boson. This turns out to be lighter than the mass of the Z-boson (denoted by Z_0) of the $SU(2)_L \times U(1)$ model and we obtain

$$m_{Z_L}^2 \simeq m_{Z_0}^2[1 - \eta A_W] + O(\eta^2), \tag{6.2.15}$$

where

$$A_W = \frac{\text{Cos}\, 2\theta_W}{2}(1 - \tfrac{1}{4}\tan^4\theta_W) \tag{6.2.16}$$

(for $\sin^2\theta_W \simeq 1/4$, $A_W \approx 1/4$). Thus, we see that if $\eta \simeq 1/10$, the Z-boson mass has to be measured to about 1% accuracy to disentangle the radiative corrections from the effect of right-hand currents. The interesting point, however, is that the effect of the right-handed currents is always to lower the mass of the Z-boson.

Neutral Current Interactions and the Lower Bound on m_{Z_R}

To study the neutral current interactions in the left–right models we start by writing down the gauge interactions involving the neutral gauge bosons (W_{3L}, W_{3R}, and B)

$$L_{\text{wk}}^{\text{N.C.}} = ig(J^3_{\mu_L}W^3_{\mu_L} + J^3_{\mu_R}W^3_{\mu_R} + J^{B-L}_\mu B_\mu). \tag{6.2.17}$$

After symmetry breaking the weak eigenstates get mixed and diagonalization of the mass matrix leads to the mass eigenstates Z_R, Z_L, and A given as follows:

$$\begin{aligned} A &= \sin\theta_W(W_L^3 + W_R^3) + B(\cos 2\theta_W)^{1/2}, \\ Z_L &= \cos\theta_W W_L^3 - \sin\theta_W \tan\theta_W W_R^3 - \tan\theta_W(\cos 2\theta_W)^{1/2}B, \\ Z_R &= (\cos 2\theta_W)^{1/2}\sec\theta_W W_{3R} - \tan\theta_W B \end{aligned} \tag{6.2.18}$$

Equation (6.2.17) can be rewritten in terms of the physical eigenstates

$$\begin{aligned} L_{\text{wk}}^{\text{N.C.}} &= eJ_\mu^{\text{em}}A_\mu + \frac{g}{\cos\theta_W}\{Z_{L_\mu}[J^Z_{L\mu} - C_W\eta(\sin^2\theta_W J^Z_{L\mu} + \cos^2\theta_W J^Z_{R\mu})] \\ &\quad + (\cos 2\theta_W)^{-1/2}Z_{R_\mu}(\sin^2\theta_W J^Z_{L\mu} + \cos^2\theta_W J^Z_{R\mu})\}, \end{aligned} \tag{6.2.19}$$

where we have written

$$J^Z_{L,R} = J^3_{L,R} - Q\sin^2\theta_W.$$

This leads to the following neutral current Hamiltonian

$$\begin{aligned} H_{\text{wk}}^{\text{N.C.}} &= \frac{4G_F}{\sqrt{2}}\left[J_L^Z J_L^Z + \eta\left\{\frac{\cos 2\theta_W}{\cos^4\theta_W}J_L^Z(J_L^Z\sin^2\theta_W + J_R^Z\cos^2\theta_W)\right.\right. \\ &\quad \left.\left. + \frac{1}{2}\frac{1}{\cos^4\theta_W}(J_L^Z\sin^2\theta_W + J_R^Z\cos^2\theta_W)\right]\right.. \end{aligned} \tag{6.2.20}$$

It is therefore clear that the modification to the $H_{\text{wk}}^{\text{N.C.}}$ of the standard model is of order $\eta \equiv (m_{W_L}/m_{W_R})^2$ and is small for small η.

Several groups [10] have analyzed the available neutral current data to obtain a bound on η. Barger *et al.* have done a systematic analysis of the neutral current predictions of general left–right symmetric models. Applying their results to our case where $g_{2L} = g_{2R}$ we obtain $m_{Z_R} \geq 4m_{Z_L}$ which implies $m_{W_R} \geq 220$ GeV. It is worth pointing out that this bound is independent of the nature of the neutrino (i.e., Dirac or Majorana, etc.) and is therefore of most general validity. This bound on W_R mass derived in subsequent sections of this chapter will rely on specific assumptions concerning the mass of the neutrino.

§6.3. Limits on m_{W_R} from Charged Current Weak Interactions

A model independent limit on the mass of the right-handed gauge boson can be obtained by looking for $V+A$ current effects in various low-energy processes. Three classes of probes are available:

(i) purely leptonic processes such as $\mu \to e\bar{\nu}_e\nu_\mu$;
(ii) semileptonic decays such as $n \to pe^-\bar{\nu}_e$, $K \to \pi\mu\bar{\nu}_\mu$, $\pi e\bar{\nu}_e$, etc.; and
(iii) nonleptonic processes such as $K \to 2\pi$, $K \to 3\pi$, hyperon decays, and $K_L - K_S$ masses.

We will analyze all these various processes to put bounds on the two parameters that characterize the presence of the right-handed currents, i.e., m_{W_R} and ζ. It turns out that in the processes involving neutrinos the bound on m_{W_R} depends on the nature of the right-handed component of the neutrino. The point is that, since the neutrino is an electrically neutral fermion, it can admit two kinds of mass terms: (i) Dirac mass for which

$$L_D = m_D\bar{\nu}_L\nu_R + \text{h.c.} \tag{6.3.1}$$

or (ii) Majorana masses

$$L_M = m_L\nu_L^T C^{-1}\nu_L + m_R\nu_R^T C^{-1}\nu_R + \text{h.c.} \tag{6.3.2}$$

or a combination of both. In the Dirac neutrino case both helicity components have the same mass: therefore, there is no kinematical obstacle to the manifestation of $V+A$ currents in lepton involved processes. However, in the case of the Majorana neutrinos, m_R is an arbitrary parameter and therefore there exists the possibility that $V+A$ current mediated processes involving neutrinos may be kinematically forbidden if m_R is big. In fact, theoretical preference [11] is for $m_R \approx m_{W_R}$, in which case most lepton involved processes will arise either from the pure left-handed current terms (i.e., first term in eqn. (6.2.14)) or the left–right mixing term (third term in eqn. (6.2.14)). If, however, m_R is chosen in the MeV range [12], by adjustment of parameters, some semileptonic processes may provide a useful bound on m_{W_R}. Let us proceed to discuss these bounds for each case.

Case (i). Dirac Neutrino Case

Bounds on m_{W_R} for the case of the Dirac neutrino were first obtained by Beg *et al.* [13] in 1977. An analysis of various leptonic and strangeness conserving semileptonic decays indicated that the most severe bounds on η and ζ came from the measurement of electron polarization in O^{14}-decay and the measurement of the Michael ρ-parameter in μ-decay. The two best experimental values [14, 15] are

$$\rho_{\text{expt}} = 0.7518 \pm 0.0026$$

and $P/P_{V-A} = 1.001 \pm 0.008$ in Gamow–Teller transition.

Pure $V-A$ theory implies $\rho = \frac{3}{4}$ and $P/P_{V-A} = 1$. Thus we obtain

$$\sqrt{\eta} \leq 0.35 \tag{6.3.3}$$

and

$$\tan \zeta < 0.05.$$

This corresponds to $m_{W_R} \geq 200$ GeV.

It was noted [13] that a crucial parameter in the probe of $V+A$ currents is the angular asymmetry parameter ζ in polarized muon-decay. This parameter is expected to be $+1$ in pure $V-A$ theories. In 1977, the best value was [15]

$$\zeta P_\mu = 0.972 \pm 0.013.$$

Recent experimental searchs for $V+A$ current effects by Carr *et al.* [16] in μ-decay has led to much better measurement of the value of ζP_μ

$$\zeta P_\mu > 0.9959$$

leading to bounds on $m_{W_R} \geq 380$ GeV for arbitrary ζ. At such precision levels it is not clear whether the fourth-order radiative corrections play an important role or not.

Several other searches for right-handed currents are by Corriveau *et al.* [17] who have also measured the ζ-parameter

$$\zeta P_\mu = 1.010 \pm 0.064.$$

There is also an ongoing experiment by Vanklinken *et al.* of the parameter P_F/P_{G_T} where a precision at the 10^{-3} level is expected. Finally, Yamazaki *et al.* [18] have looked for muon polarization in K_{μ_2}-decay where for $V-A$ currents $P_\mu = -1$. Their result is

$$P_\mu = -0.967 \pm 0.047,$$

which allows a 4% admixture of $V+A$ currents. The most stringent bound that remains, however, is the one from ref. [16].

Finally, analysis of the asymmetry [19] in ${}^{19}N_e$ β-decay puts a more stringent bound on ζ

$$0.005 \leq \zeta \leq 0.015.$$

although this involves assumptions that are more model-dependent than Ref. [13].

Case (ii). Light-Majorana Neutrinos ($m_{\nu_R} \leq 10$–100 MeV)

In this case, the right-handed neutrinos cannot be emitted in β-decay. It was pointed out [12] that the most stringent bounds in this case come from peak searches in π_{l_2} decay. There is now a new channel for the decay of the pion: $\pi \to e_R^- \nu_R$ with width given by

$$R = \frac{\Gamma(\pi \to e_R^- \nu_R)}{\Gamma(\pi \to e_L \nu_L)} = (|\eta|^2 + |\zeta|^2)\left(\frac{m_{\nu_R}}{m_e}\right)^2 \left(1 - \frac{m_{\nu_R}}{m_\pi^2}\right)^2, \qquad (6.3.4)$$

where $\eta = (m_{W_L}/m_{W_R})^2$ as defined before; ζ denotes the mixing between the light and heavy neutrino. Experiment searches for secondary peaks at these mass ranges for the heavy neutrino have been conducted. Lack of evidence for any secondary peak in π-decay at TRIUMF [20] and KEK [21] translates into an upper bound on $R < 4 \times 10^{-3}$ for 45 MeV $< m_{\nu_R} <$ 74 MeV, which implies

$$|\eta| \leq 10^{-3}. \qquad (6.3.5)$$

We will see in the next section that similar bounds appear from considerations of the ν_R contributions to neutrinoless double β-decay.

Case (iii). Bound on m_{W_R} from Nonleptonic Decays and $K_L - K_S$ Mass Difference

(a) *Nonleptonic Decays* [22], [23]

Here it is assumed that current algebra can be used to understand certain features of nonleptonic weak interactions for pure $V-A$ theory. The presence of a small admixture of $V+A$ currents would then lead to deviations from these features. From the extent to which these predictions are known to agree with experiment we can obtain bounds on m_{W_R}. One parameter of interest is the slope in $K \to 3\pi$ decay and another is the current algebra relation between parity conserving and parity violating nonleptonic weak Hamiltonians $H_{\text{wk}}^{(\pm)}$; for the case of no $V+A$ currents, i.e., $\eta = 0$.

$$[F_i^5, H_{\text{wk}}] = [F_i, H_{\text{wk}}]. \qquad (6.3.6)$$

For $\eta \neq 0$, $\zeta \neq 0$, we will get deviations from (6.3.6). Assuming that eqn. (6.3.6) implies agreement with experiment to within 10% we can put bounds [23]

$$\eta < 1/16$$

and

$$\zeta < 4 \times 10^{-3}. \qquad (6.3.7)$$

In obtaining this bound it is assumed that the quark mixing angles for the right-handed charged currents are the same as the angles for left-handed charged currents ($U_L = U_R$). Also, hadronic matrix elements are evaluated

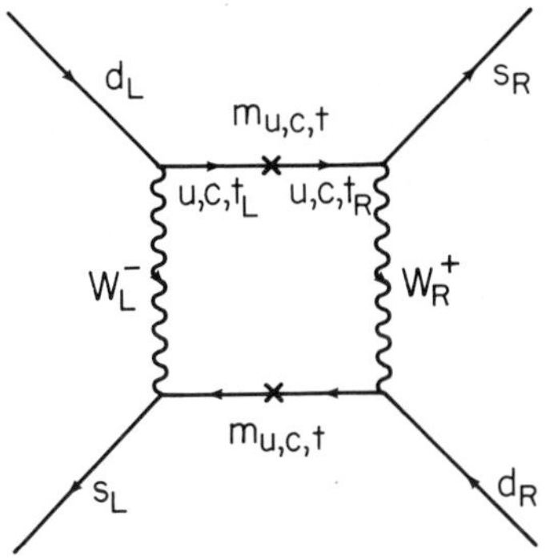

Figure 6.1

using the bag model which could have its own uncertainties. Thus, bounds in eqn. (6.3.7) are quite model independent—they, however, have the virtue that they are independent of the nature of the neutrino.

(b) $K_L - K_S$ *Mass Difference*

In 1982, Beall, Bender, and Soni [24] pointed out that short distance contributions to $K_L - K_S$ mass difference receives a huge contribution from the $W_L - W_R$ box graph (Fig. 6.1) in the case of the two-generation left–right symmetric model. Their result was that

$$\Delta m_K \simeq \frac{G_F^2 m_{W_L}}{4\pi^2}(\cos^2\theta_C \sin^2\theta_C)\left(\frac{m_c^2}{m_{W_L}^2}\right)\langle K^0|\mathrm{O}^{(1)}|\bar{K}^0\rangle$$
$$\times\left[1 + 8\eta\left(\frac{m_c^2}{m_{W_L}^2}\right)\frac{\langle K^0|\mathrm{O}^{(2)}|\bar{K}^0\rangle}{\langle K^0|\mathrm{O}^{(1)}|\bar{K}^0\rangle}\right], \tag{6.3.8}$$

where

$$\mathrm{O}^{(1)} = \bar{s}_L\gamma_\mu d_L \bar{s}_L\gamma_\mu d_L,$$
$$\mathrm{O}^{(2)} = \bar{s}_L d_R \bar{s}_R d_L. \tag{6.3.9}$$

Evaluating all matrix elements in the vacuum saturation approximation [25] we find that the second term within the bracket becomes about -430η. Since the pure left–left contribution [26] (the expression outside the bracket in eqn. (6.3.8)) explains the bulk of the $K_L - K_s^0$ mass difference at least in magnitude (presumably also in sign [27]), the second term must be small compared to 1. This implies that

$$\eta \leq \frac{1}{430} \tag{6.3.10}$$

or

$$m_{W_R} \geq 1.6 \text{ TeV}.$$

Since this bound is significantly higher than any obtained before it needs to be examined critically as has been done in several subsequent papers [28].

There exist several corrections to the expression in (6.3.8): (i) contribution due to t-quarks, and (ii) new contributions from tree graphs involving neutral Higgs exchanges. In the presence of these two contributions we can write

$$\Delta M_K = M_{LL} + M_{LR} + M_{H^0}, \tag{6.3.11}$$

where contributions characteristic of left–right models under the assumption of equal left- and right-mixing angles are given by

$$\begin{aligned} M_{LR} = {} & \frac{G_F^2 m_{W_L}^2}{4\pi^2} 8\eta \Big[\lambda_c^2 x_c(1 + \ln x_C) + \tfrac{1}{2}\lambda_c\lambda_t \frac{\sqrt{x_c x_t}}{(1 - x_t)}(4 - x_t)\ln X_t \\ & + \frac{1}{4}\lambda_t^2 \frac{x_t}{(1 - x_t)^2}\{(4 - x_t)(1 - x_t) + (4 - 2x_t)\} \ln x_t\Big] \mathrm{O}^{(2)}, \end{aligned} \tag{6.3.12}$$

$$M_{H^0} = \frac{2G_F}{m_H^2}\left(\sum_{i=u,c,t} \lambda_i m_i\right)^2 [(\bar{s}\gamma_5 d)^2 - (\bar{s}d)^2], \tag{6.3.13}$$

with

$$\lambda_u = s_1 c_1 c_3,$$

$$\lambda_c = -s_1 c_2(c_1 c_2 c_3 - s_2 s_3 e^{-i\delta}),$$

$$\lambda_t = -s_1 s_2(c_1 s_2 c_3 + c_2 s_3 e^{-i\delta}).$$

Using the vacuum saturation assumption we obtain

$$\langle(\bar{s}\gamma_5 d)^2 - (\bar{s}d)^2\rangle_{K^0\bar{K}^0} = \tfrac{4}{3}B(7.7)f_K^2 M_K. \tag{6.3.14}$$

Several ways to avoid these constraints may be suggested. For example, if m_t were much larger than m_W, the second and third terms within the brackets are positive and large which by suitable fine tuning of the parameters could cancel the negative t-quark contribution. For example, this generally happens if

$$s_2 \approx 0 \quad \text{or} \quad s_2 + s_3 e^{-i\delta} \approx 0. \tag{6.3.15}$$

In this case, the neutral Higgs exchange contribution is also small allowing for a light Higgs boson in the theory; so a light Higgs boson would appear to go hand-in-hand with a light W_R.

It has, however, subsequently been noted by Gilman and Reno [28] that experimental limits on the ratio of b-quark semileptonic decay widths, i.e.,

$$\frac{\Gamma(b \to ue^-\nu^-)}{\Gamma(b \to ce^-\nu^-)} < 0.05 \tag{6.3.16}$$

rules out (6.3.15) as a solution. Thus, it would appear that unless the dispersive [29] contributions to $K_L - K_S$ mass difference are assumed to be important, in manifest left–right symmetric models, we would have $m_{W_R} \geq 1.6$ TeV.

Table 6.1

Process	(m_{W_R}) min
Dirac ν	380 GeV
$P_\mu \zeta$	
Neutral current + triplet model	220 GeV
Majorana ν, $m_\nu \leq 100$ MeV	2 TeV
$K_L - K_S$ mass difference + manifest $L - R$ Sym (with QCD effects included)	1.6 TeV (3 TeV)

Recently, it has been noted [29a] that this bound increases by a factor of 2 to 3 after QCD corrections are taken into account.

It is, however, possible to have departures from the manifest left–right symmetric models. In fact, it is possible to show that natural manifest left–right models must be CP conserving and therefore not realistic. In fact, a natural realistic left–right model is the one where both P and CP are violated spontaneously, in which case the left- and right-mixing matrices satisfy the constraint [30]

$$U_R = U_L^*. \tag{6.3.17}$$

Harari and Leurer [28] have argued that in this case b-quark physics can be accommodated together with the fine tuning equation (6.3.15) modified appropriately to take account of (6.3.17). In the case that eqn. (6.3.15) does not hold, the Higgs contribution to $K^0 - \bar{K}^0$ matrix elements become enormous unless $M_{H^0} > 5$ TeV.

Finally, another simpler way out, suggested by several groups [31] , is to assume that the left- and right-mixing angles themselves are different from each other, in which case the constraint on m_{W_R} clearly gets weaker. The results on the bounds on m_{W_R} are summarized in Table 6.1.

Limit on ζ from $\bar{\nu}N$ Scattering

In the left–right symmetric models the deep inelastic neutrino and anti-neutrino scattering receives a contribution proportional to the $W_L - W_R$ mixing parameter ζ in addition to the usual left-handed contribution, i.e.,

$$\frac{d\sigma(\nu N)/dy}{d\sigma(\bar{\nu}N)/dy} = \frac{q(x) + \zeta^2 q(x)(1-y)^2 + \bar{q}(x)(1-y)^2 + \zeta^2 \bar{q}(x)}{\{q(x) + \zeta^2 \bar{q}(x0)\}(1-y)^2 + \{\bar{q}(x) + \zeta^2 q(x)\}}. \tag{6.3.18}$$

Using the CDHS data we can obtain a bound on $\zeta \leq 0.095$.

Decay Properties and Possible High-Energy Signature for W_R- and Z_R-Bosons

If the W_R- and Z_R-boson masses are in the 300–400 GeV range they can be produced in the high-energy machines of the coming generation in $p\bar{p}$ colliders. It may therefore be useful to list their decay modes and properties.

Assuming that $2m_{\nu_R} \ll m_{Z_R}$ we obtain [32]

$$
\begin{aligned}
\Gamma_{Z_2} &\simeq 12 \text{ GeV}, \\
B(Z_R \to 2\nu) &\simeq 1\%, \\
B(Z_R \to \mu^+\mu^-) &\simeq 2\%, \\
B(Z_R \to u\bar{u}) &\simeq 7\%, \\
B(Z_R \to d\bar{d}) &\simeq 14\%, \\
B(Z_R \to 2N_R) &\simeq 9\%.
\end{aligned}
\tag{6.3.19}
$$

For the W_R^+ the leptonic branching ratio is $B(W_R^+ \to N_\mu \mu^+) \simeq 8\%$.

It has been pointed out [32] that if neutrinos are Majorana particles then, in $p\bar{p}$ collision, we obtains $p\bar{p} \to \mu^+\mu^+ + X$ without any missing energy. This should provide a clean signature both for the W_R-boson as well as for the Majorana character of the neutrino.

In high-energy ep colliding machines (i.e., HERA, ...) possible effects of m_{W_R} as heavy as 600 GeV can be detected through its effect on the y-distribution in $ep \to eX$ processes. To detect a right-handed gauge boson in the mass range of TeV's, we need a higher energy machine such as Tevatron or the proposed SSC.

Finally, it has been suggested [33] that a good way to look for right-handed charged currents is to observe the absorption cross section for μ_L^+ at high energies, i.e., $\mu_L^+ N \to \bar{\nu}_R X$, i.e.,

$$
\frac{\sigma(\mu_L^+)}{\sigma(\mu_R^+)} = \left(\frac{m_{W_L}}{m_{W_R}}\right)^4. \tag{6.3.20}
$$

This much is easy. The main problem is to isolate a pure μ_L^+ beam from the decay of π^+ in flight. From simple kinematic considerations it follows that for pion momenta p_π^+, the decay μ^+'s with $p_\mu \simeq p_\pi$ have left-handed helicity which is of interest in the detection of right-handed currents. It must, however, be noted that for $m_{W_R} \simeq 800$ GeV, the ratio of cross sections in (6.3.20) is 10^{-4} and this means that the momentum spread of π's and μ's must be less than 10^{-4} which may not be so easy to attain.

§6.4. Properties of Neutrinos and Lepton Number Violating Processes

(a) Connection Between Small Neutrino Mass and Strength of $V + A$ Current

Our knowledge of weak interactions since the time of Fermi and Pauli has been intimately connected with our understanding of the nature of the neutrino. When $V - A$ theory was proposed by Sudarshan and Marshak, they based their argument on the assumption that neutrinos are massless. If

$m_\nu = 0$, then the neutrino spinor obeys the Weyl equation which is invariant under γ_5-transformations, i.e. $(\nu \to \gamma_5 \nu)$. They argued that, since neutrinos participate only in weak interactions, the weak Hamiltonian H_{wk} ought to be invariant under separate γ_5-transformations of the various fermions participating in it, and this leads to the successful $V-A$ theory of charged current weak processes. From this point of view, the existence of $V+A$ currents ought to be connected with a nonvanishing neutrino mass and, furthermore, as has been argued by G. Senjanovic and myself [8], the smallness of the neutrino mass (if it exists) and the suppression of $V+A$ currents should be connected. We will show in this chapter that in the framework of left–right symmetric theories this connection comes out very naturally, i.e., we show that

$$m_\nu = \gamma \frac{m_e^2}{m_{W_R}}. \tag{6.4.1}$$

In this section we analyze the implications of the left–right symmetric models for neutrino mass and lepton number nonconservation. However, before proceeding with those discussions, we wish to give a brief discussion of astrophysical constraints on neutrino masses, laboratory constraints on mixings, and make some general remarks on the structure of neutrino masses in gauge theories.

(b) Constraints on Properties of Neutrinos from Laboratory and Astrophysical Considerations

The limits on neutrino masses from laboratory experiments are

$16 \text{ eV} \leq m_{\nu_e} \leq 46 \text{ eV}$ (Lubimov *et al.* [34]),

$m_{\nu_\mu} \leq 250 \text{ keV}$ (review by Barbiellini *et al.* [35])

$m_{\nu_\tau} \leq 70 \text{ MeV}$ (Ref. 36)

As mentioned in the previous chapter, there also exist strong limits on the neutrino mixings from ν-oscillation and peak search experiments for various mass ranges of neutrinos: if we define the general charged current weak interaction Hamiltonian as

$$\mathscr{L}_{\text{wk}} = \frac{g}{2\sqrt{2}} \sum_{l=e,\mu,\tau} \bar{l}\gamma_\mu(1+\gamma_5)\nu_i U_{li} W_\mu + \text{h.c.} \tag{6.4.2}$$

and $\Delta_{ij}^2 = |m_i^2 - m_j^2|$ and θ_{ij} as the mixing parameter for the neutrinos only, then we have the following limits on the mixing parameters [37]:

ν-Oscillations

(i) Caltech–Sin-Tum

$\Delta^2 < 0.16 \text{ eV}^2$ for large θ,

$\theta^2 < 0.03$ for large L.

(ii) Fermilab

$$\theta_{e\mu}^2 < 0.001, \qquad \Delta^2 \approx 1\ \text{eV}^2.$$

Peak Searches

$$5\ \text{MeV} \le m_{\nu_h} \le 160\ \text{MeV}, \qquad |U_{eh}|^2 < 10^{-6}\text{–}10^{-4},$$

$$5\ \text{MeV} \le m_{m_{\nu_h}} \le 30\ \text{MeV}, \qquad |U_{\mu h}|^2 < 10^{-6}\text{–}10^{-4}.$$

Beam Dump

$$310\ \text{MeV} \le m_{\nu_h} \le 370\ \text{MeV}, \qquad |U_{\mu h}|^2 \le 10^{-6}.$$

The astrophysical constraints on the neutrino masses arise from known upper limits on the mass density of the universe, i.e., $\rho \le 10$–20 KeV cc^{-1}. Since stable neutrinos (i.e., $\tau_\nu \ge 10^{18}$ s) will survive to the present day and will have a number density [38] of about 100–200 cm^{-1}, this will imply $\sum_i m_{\nu_i} \le 100$ eV. For unstable neutrinos, with τ_ν such that $1 \ll \tau_\nu \ll \tau_U$, the limit gets modified to [39]

$$m_{\nu_H} < \left(\frac{\tau_U}{\tau_{\nu_H}}\right)^{1/2} (100\ \text{eV}). \tag{6.4.3}$$

For $m_{\nu_H} \ge 1$ MeV, this restricts $\tau_{\nu_H} \le 10^9$ s. It is worth pointing out that photonic decays such as $\nu_H \to \nu_L + \gamma$ usually lead to much longer lifetimes. For a recent discussion of these constraints and ways to accommodate them in left–right symmetric models, see Refs. [39a] and [39b].

(c) Majorana Mass for Neutrinos: Model-Independent: Considerations

Two kinds of mass terms are allowed by proper Lorentz transformations for a neutral spin 1/2 fermion such as the neutrino ν. If we denote ν_L and ν_R as the left- and right-handed projections of ν, we can write a Dirac mass $\mathscr{L}_D$ and a Majorana mass $\mathscr{L}_M$ as follows (we consider only one neutrino species):

$$\mathscr{L}_D = m_D \bar{\nu}_L \nu_R + \text{h.c.} \tag{6.4.4}$$

and

$$\mathscr{L}_M = m_L \nu_L^T C^{-1} \nu_L + m_R \nu_R^T C^{-1} \nu_R \tag{6.4.5}$$

(where m_D and $m_{L,R}$ are arbitary complex numbers).

First point to note is that $\mathscr{L}_D$ is invariant under a global U(1) symmetry under which $\nu \to e^{i\theta}\nu$, where $\mathscr{L}_M$ is not. This U(1) symmetry may be identified with the lepton number L with $L(\nu) = -L(\bar{\nu}) = 1$. $\mathscr{L}_M$ breaks the lepton number by two units ($\Delta L = 2$). Therefore, in the presence of $\mathscr{L}_M$, the $\Delta L = 2$ type lepton number violating processes such as neutrinoless double β-decay, $K^+ \to \pi^- e^+ e^+$, etc., will take place. Thus, observation of any such process will constitute strong evidence (though not conclusive [40], for the Majorana character of the neutrino mass. Furthermore, both types of mass terms involve as yet unobserved physics (i.e., right-handed neutrino or lepton number violation). This means that discovery of the neutrino mass is an important step in uncovering new physics beyond the standard model.

Now we would like to present a brief discussion connecting the neutrino mass matrix of a general unified theory with the C- and CP-properties of the Majorana neutrino. It is important to point out that this is mainly a question of language [41] (or basis states for Majorana particles). To clarify this point we note that there are three parts to the neutrino Lagrangian

$$\mathscr{L} = \mathscr{L}_{\text{kin}} + \mathscr{L}_{\text{mass}} + \mathscr{L}_{\text{wk}}. \tag{6.4.6}$$

In the absence of $\mathscr{L}_{\text{wk}}$ and $\mathscr{L}_{\text{mass}}$ the Hamiltonian conserves CP, C, as well as CPT; therefore, we could choose Majorana neutrino states which are eigenstates of C as well as CP. Then any possible C- or CP-violation will reside in the mass term in the propagators and interaction vertices (i.e., complex couplings or nonpositive complex masses) that appear in the calculation of definite physical processes.[1]

It is, however, customary to define a physical neutrino state as one which is an eigenstate of the mass matrix (see below) with a positive mass. For Majorana neutrinos this tends to fix the C- and CP-phase of the neutrino state (with $\mathscr{L}_{\text{wk}}$ ignored which is all right since for all known weak processes weak interaction is treated perturbatively). We will show below that in the simplest cases the CP- or C-phase of the Majorana neutrino is precisely the phase of its Majorana mass term. For more generations this connection is more complicated.

To show this let us choose the Weyl basis with the following choice of γ-matrices ($k = 1, 2, 3$):

$$\gamma_k = \begin{pmatrix} 0 & -i\sigma_k \\ i\sigma_k & 0 \end{pmatrix}, \qquad \gamma_4 = \begin{pmatrix} 0 & 1 \\ 1 & 0 \end{pmatrix}, \qquad \gamma_5 \begin{pmatrix} 1 & 0 \\ 0 & -1 \end{pmatrix}. \tag{6.4.7}$$

We can write a four-component Dirac spinor ψ in the basis as

$$\psi = \begin{pmatrix} u \\ i\sigma_2 v^* \end{pmatrix}, \tag{6.4.8}$$

where u and $i\sigma_2 v^*$ denote the left ($(1 + \gamma_5)/2$) and right ($(1 - \gamma_5)/2$) chiral projections of ψ. Ignoring $\mathscr{L}_{\text{wk}}$ we can write the field equations following from eqns. (6.4.4), (6.4.5), (6.4.6) in terms of u and v as follows [44]:

$$(\boldsymbol{\sigma}\nabla - \partial_t)\begin{pmatrix} u \\ v \end{pmatrix} + \sigma_2 \begin{pmatrix} m_L^* & m_D \\ m_D & m_R \end{pmatrix}\begin{pmatrix} u^* \\ v^* \end{pmatrix} = 0. \tag{6.4.9}$$

[1] The experts will recognize that in this language the cancellation between two neutrino states in $(\beta\beta)_{0\nu}$-decay may be thought of as between two CP-even neutrino states with opposite signs for masses [42].

(In this case, m_D can be chosen real by redefining the phase of either u or v.) This matrix is a complex symmetric matrix (to be denoted henceforth by M) connecting the two-component spinors. In a more general situation this becomes a more complicated matrix. The various cases known as Dirac, Majorana, or pseudo-Dirac neutrinos correspond to various forms for the matrix M

$$M \equiv \begin{pmatrix} m_L^* & m_D \\ m_D & m_R \end{pmatrix}. \tag{6.4.10}$$

(i) *Dirac*: $m_L = m_R = 0$; $m_D \neq 0$.
(ii) Majorana: either m_L or m_R or both nonzero; m_D arbitrary.
(iii) *Pseudo-Dirac* [43]: $m_D \neq 0$ and $m_L = m_R \ll m_D$.

As is evident case (iii) is a special case of the Majorana mass matrix. In fact, even the Dirac case is a special case, where an additional U(1) symmetry appears in the Lagrangian in the limit $m_L = m_R = 0$. In summary, a Dirac neutrino consists of two Majorana neutrinos with equal masses and opposite CP-properties. To see this note that in case (i) one mass eigenvalue is negative which can be made positive by defining $\psi = -C\bar{\psi}^T$, i.e., a state odd under CP; a pseudo-Dirac neutrino consists of two Majorana neutrinos with opposite CP-properties but slightly different masses ($m_M \pm M_D$). This case will become of interest if the neutrino mass measured in tritium decay and that obtained from $(\beta\beta)_{0\nu}$ decay happen to be different, as is sometimes thought. (This situation is similar to the $K^0 - \bar{K}^0$ mixing case and will become almost identical to it if the tiny Majorana mass is induced by weak interactions instead of being present from the beginning.)

To proceed further, we can diagonalize the complex and symmetric mass matrix M which can be done, in general, as follows:

$$UMU^{\mathrm{T}}\Lambda = M_D, \tag{6.4.11}$$

where M_D is diagonal with positive, real eigenvalues and Λ is a diagonal unitary matrix. In terms of the eigenstates

$$\chi \equiv \begin{pmatrix} \chi_1 \\ \chi_2 \end{pmatrix} \equiv \Lambda U \begin{pmatrix} u \\ v \end{pmatrix} \tag{6.4.12}$$

the field equations (6.4.9) factorize

$$(\boldsymbol{\sigma}\nabla - \partial_t)\begin{pmatrix} \chi_1 \\ \chi_2 \end{pmatrix} + \begin{pmatrix} m_1 e^{i\alpha_1}\chi_1^* \\ m_2 e^{i\alpha_2}\chi_2^* \end{pmatrix} = 0. \tag{6.4.13}$$

It therefore suffices to study one part to learn about the connection between the phase of the mass, i.e., $e^{i\alpha_1}$, and the CP-property of the Majorana spinor. Using methods already given earlier [44] we can expand χ as follows:

$$\chi = \frac{1}{\sqrt{V}} \sum_{\lambda=1,2} (A_\lambda(k)u_\lambda(k)e^{ik\cdot x} + e^{i\alpha_1}A_\lambda^*(k)v_\lambda(k)e^{-ik\cdot x}), \tag{6.4.14}$$

where

$$u_1 = v_2 = \alpha(k)$$

and

$$-v_1 = +u_2 = \frac{m\beta(k)}{E + |\mathbf{k}|}, \qquad \alpha^*\beta = 0. \tag{6.4.15}$$

Two points ought to be stressed here. (i) In the limit of $m_\nu \to 0$, $v_1 = u_2 = 0$ and, therefore, the lepton number is actually conserved with the $L = +1$ state being associated with A_1 and the $L = -1$ antiparticle state being associated with A_2. The two helicity states now correspond to particle and antiparticle states. (ii) The second point is that the phase of the Majorana mass term has permeated to the field operator expansion in eqn. (6.4.14). Let us now give the prescription for going to the four-component description. We can adopt the prescription that the Majorana mass term $me^{i\alpha}\psi_C^T C^{-1}\psi_L$be expressible in the form of a Dirac mass term, i.e., $m\bar{\psi}_R\psi_L$ requires the condition that

$$\bar{\psi}_R = e^{i\alpha}\psi_L^T C^{-1}. \tag{6.4.16}$$

Thus, in some sense, the phase of the mass term gives the C-phase of the neutrino.

At this point we wish to note the way to understand the smallness of the neutrino masses in unified gauge theories in the case of Majorana neutrinos. Since we work with chiral spinors we chose appropriate Higgs boson couplings so that the mass matrix in eqn. (6.4.10) has the following form:

$$M = \begin{pmatrix} 0 & m_D \\ m_D & m_R \end{pmatrix}. \tag{6.4.17}$$

The eigenvalues of this mass matrix are (for $m_R \gg m_L$)

$$m_N \simeq m_R \quad \text{(heavy neutrino)}$$

and

$$m_\nu \simeq -\frac{m_D^2}{m_R} \quad \text{(light neutrino)}. \tag{6.4.18}$$

Thus, the physical Majorana neutrino mass can be small without any unnatural fine tuning or parameters if there is a heavy scale m_R in the theory. We will use this mechanism to establish eqn. (6.4.1).

(d) Neutrino Masses and Left–Right Symmetry

The basic equation that leads to the connection between neutrino mass and the magnitude of $V + A$ currents (or parity violation) is the formula for the electric charge given in eqn. (6.1.1)

$$Q = I_{3_L} + I_{3_R} + \frac{B - L}{2}. \tag{6.1.1}$$

If we work at a distance scale where weak left-handed symmetry is a good symmetry, then from eqn. (6.1.1) we obtain the relation

$$\Delta I_{3_R} = -\tfrac{1}{2}\Delta(B - L) \tag{6.4.19}$$

for processes involving leptons, this implies that $\Delta L = 2\Delta I_{3R}$. Since in our case $|\Delta I_{3R}| = 1$, this means that neutrinos must be Majorana particles and their Majorana mass must be connected to the strength of parity violation.

Below we present an explicit Higgs model for this. As discussed in Section 6.2, we will use only those Higgs multiplets which are bilinears in the fundamental fermion fields of the theory, i.e.,

$$\Delta_L(1, 0, -2) + \Delta_R(0, 1, -2)$$

and

$$\phi = (\tfrac{1}{2}, \tfrac{1}{2}, 0).$$

The gauge symmetry is then broken in stages as follows (we also show the neutrino mass matrices for each stage):

$$SU(2)_L \times SU(2)_R \times U(1)_{B-L}$$

$$\downarrow$$

$$\langle \Delta_L^0 \rangle \simeq 0, \qquad \langle \Delta_R^0 \rangle = V_R \neq 0, \qquad M_\nu = \begin{pmatrix} 0 & 0 \\ 0 & V_R \end{pmatrix}$$

$$\downarrow$$

$$SU(2)_L \times U(1)$$

$$\downarrow$$

$$\langle \phi \rangle = \begin{pmatrix} k & 0 \\ 0 & k' \end{pmatrix}, \qquad M_\nu = \begin{pmatrix} 0 & hk \\ hk & fV_R \end{pmatrix}$$

$$\downarrow$$

$$U(1)_{\text{em}}.$$

Here M_ν denotes the Majorana–Dirac mass matrix for the neutrino at the different stages of symmetry breakdown, which arises from the following leptonic Higgs couplings:

$$L_Y = \bar{\psi}_L(h\phi + \tilde{h}\tilde{\phi})\psi_R + if(\psi_L^T C^{-1}\tau_2\psi_L\Delta_L + \psi_R^T C^{-1}\tau_2\tau\psi_R\Delta_R) + \text{h.c.} \tag{6.4.20}$$

By diagonalizing the above M_ν we obtain the following eigenstates and masses:[1]

[1] It actually turns out that, as the $SU(2)_L \times U(1)$ symmetry gets broken by $\kappa, \kappa' \neq 0$, the left-handed triplet field Δ_L acquires a nonzero vacuum expectation value, i.e., $\langle \Delta_L \rangle = v_L \neq 0$. It has, however, been shown by a detailed analysis of the potential that it is proportional to

$$v_L \simeq \gamma\kappa^2/v_R.$$

It also turns that γ is nonzero only when the Yukawa couplings h and f are nonzero and $\gamma \approx h^2 f^2$. Since h and f are small numbers ($\approx 10^{-2}$), γ is expected to be of order 10^{-8}. In the presence of nonzero v_L we find

$$m_\nu \approx \gamma f\kappa^2/v_R - h^2\kappa^2/fv_R, \tag{6.4.21a}$$

which reduces to eqn. (6.4.21) for $\gamma \approx 10^{-8}$ [See Ref. [45] for a natural way to obtain such a small number].

$$\nu = \nu_L \cos \xi + \nu_R \sin \xi, \qquad m_\nu \simeq \frac{h^2 k^2}{v_R},$$

$$N = -\nu_L \sin \xi + \nu_R \cos \xi, \qquad m_N \simeq v_R, \tag{6.4.21}$$

where $\tan \xi \approx (m_\nu/m_N)^{1/2}$. As claimed before

$$v_R \to \infty, \qquad m_\nu \to 0.$$

Leptonic charged currents now look as follows:

$$\begin{pmatrix} \nu \cos \xi + N \sin \xi \\ e^- \end{pmatrix}_L \quad \text{and} \quad \begin{pmatrix} -\nu \sin \xi + N \cos \xi \\ e^- \end{pmatrix}_R. \tag{6.4.22}$$

For convenience let us reparametrize the m_ν and m_N in terms of m_e and m_{W_R}

$$m_\nu \simeq \frac{r^2}{\beta} \frac{m_e^2}{m_{W_R}},$$

$$m_N \simeq \beta m_{W_R}. \tag{6.4.23}$$

r and β are free dimensionless parameters and we can obtain different values for m_N and m_ν in the electron volt range using $m_{W_R} \geq 250$ GeV–2.5 TeV for different choices of these parameters. Case (i): $r \simeq 1$; $\beta \simeq 10^{-1}$ gives $m_\nu \leq 1$–10 eV and $m_N \geq 25$–250 GeV. Case (ii): $r \simeq 1/25$; $\beta \simeq 2 \times 10^{-4}$ gives $m_{\nu_e} \leq$ 1–10 eV and $m_N \geq 50$ MeV to 500 M_W. It must, however, be noted that the second case is not strictly "natural."

If we assume r and β to be independent of generations the neutrino masses scale as square of the mass of the charged lepton of the corresponding generation leading to the following expectations for m_{ν_μ} and m_{ν_τ}:

$$m_{\nu_\mu} \leq 40\text{–}400 \text{ keV}$$

and

$$m_{\nu_\tau} \leq 20\text{–}200 \text{ MeV}. \tag{6.4.24}$$

Let us study the laboratory and astrophysical consequences of m_{ν_μ}, m_{ν_τ}, and m_N in the above mass range.

(e) Astrophysical Implications

In view of the known cosmological constraints on the neutrino masses (both stable and unstable) summarized earlier, we must study the interaction and decay rates of ν_μ, ν_τ, and N to see if $m_N \approx 50$ MeV and $m_{\nu_\tau} \leq 20$ MeV are cosmologically allowed. (We have chosen certain typical values for the masses to illustrate our point and the discussion can be easily extended for the whole range of masses in (6.4.24).)

(a) *Right-Handed Neutrino, N*; $m_N \approx 50$ MeV

For such a low value of N there will be constraints on η coming from the absence of peaks in π-decay and K-decay, i.e.,

$$\pi^- \to e_R^- N_e,$$
$$K^- \to e_R^- N_e,\ \mu_R^- N_\mu. \tag{6.4.25}$$

For example, the constraints given in Section 6.1 imply that

$$|\zeta + \eta|^2 \leq 10^{-6}. \tag{6.4.26}$$

This restricts the overall interaction rate of N. In order for this particle to satisfy the cosmological constraint its lifetime must be of order $\leq 10^5$ s.

This implies that

$$|\zeta|^2 \geq 10^{-9}. \tag{6.4.27}$$

For values of $\zeta + \eta$ below this the particle will also go out of thermal equilibrium. Thus, if cosmological constraints are taken seriously, ζ cannot be lower than $10^{-4.5}$. Thus, if peak searches or beam dump type experiments could be improved by two to three orders of magnitudes, an N_R with $10 \leq m_N \leq 100$ MeV will be ruled out (if not discovered).

(b) ν_τ: $m_{\nu_\tau} \approx 20$ MeV

Since the ν_τ has normal weak interactions it remains in thermal equilibrium in the early universe until the age of the universe is 1 s. For this particle to be consistent with cosmology we have $\tau_{\nu_\tau} \leq 10^7$ s which implies that $|U_{e\tau}|^2 \geq 3 \times 10^{-10}$. Again improvements in the precision of peak searches can help to decide the fate of a ν_τ with mass in the MeV range.

(c) ν_μ: $m_{\nu_\mu} \geq 100$–400 keV

This case merits more careful and more model-dependent consideration to meet the cosmological constraints. The reason is that, in minimal weak interaction models, this particle has only the photonic decay model, i.e., $\nu_\mu \to \nu_e \gamma$ and we obtain $\tau_{\nu_\mu} \geq 10^{20}$ s for the most optimistic choice of parameters as against the cosmological requirement of $\tau_{\nu_\mu} \leq 10^{10}$–$10^{12}$ s.

In left–right symmetric models there exist two ways out of this difficulty.

The first way was pointed out by Roncadelli and Senjanovic [46] who noted that, in the left–right symmetric models, there exists a coupling of Δ_L^0 to neutrinos as follows:

$$\mathscr{L}_{\Delta^0} = f[\sin\theta_{\mu e}\nu_{\mu_L}^T C^{-1}\nu_{e_L} + \nu_{e_L}^T C^{-1}\nu_{e_L}]\Delta_L^0 + \text{h.c.} \tag{6.4.28}$$

Thus the process $\nu_\mu \to 3\nu_e$ can occur via the exchange of the Δ-Higgs meson and gives a lifetime of ν_μ of

$$\tau_{\nu_\mu} \approx 4\left(\frac{m_\Delta^0}{f}\right)^4 \frac{1}{\sin^2\theta_{\mu e}}\tau_\mu\left(\frac{m_\mu}{m_{\nu_\mu}}\right)^5 \frac{G_F^2}{192\pi^2}. \tag{6.4.29}$$

This implies that for $\theta_{\mu e} \approx 10^{-1}$ cosmological requirements are met by the constraint

$$\left(\frac{M_{\Delta^0}}{f}\right) \leq 10^3 \text{ GeV}. \tag{6.4.30}$$

However, in the same weak multiplet as Δ^0, there exists a Higgs boson Δ^{++} that can cause $\mu \to ee\bar{e}$, whose branching ratio presently has an upper bound of 2×10^{-9}. It has been noted [47] that this implies

$$\left(\frac{m_{\Delta^{++}}}{f}\right) \geq 10^5 \text{ GeV}. \tag{6.4.31}$$

Since both Δ^0 and Δ^{++} are the same multiplet, such a mass splitting may not be easy to understand, although for $f \approx 10^{-2}$ there may be no strict phenomenological objection to it. In any case it is perhaps somewhat unnatural. Other objections to this scheme have also been raised recently [47a].

Another resolution of this problem is to use a mechanism discovered by Hosotani [48] and Schecter and Valle [48]. The point is that when the Dirac–Majorana mass matrix in eqn. (6.4.17) is generalized to include higher generations after diagonalizations, there can be large flavor changing $\nu_\mu - \nu_e$ weak neutral currents. This will give rise to lifetimes for ν_μ of order 10^7–10^8 s, which is consistent with cosmology [49]. However, this mechanism does not work if $m_R \gg m_L$.

This leaves us with the mechanism proposed in Ref. [39b] where heavier neutrinos are allowed to decay emitting massless Goldstone bosons and a light neutrino.

(f) Double β-Decay

Now we proceed to discuss lepton number violating processes, $\Delta L \neq 0$. The most important among them is neutrinoless double β-decay: $(A, Z) \to (A, Z + 2) + e^- + e^-$. We will now discuss the implications of our model for the lifetime for this process.

Neutrinoless double β-decay has been the subject of a great deal of discussion [50] in recent years because it provides a sensitive test of lepton number conservation. Due to more available phase space this decay rate is enhanced over the lepton number conserving process, $(A, Z) \to (A, Z + 2) + e^- + e^- + \bar{\nu}_e + \bar{\nu}_e$. In order to see the general orders of magnitude, if we assume the strength of $(\beta\beta)_{0\nu}$ amplitude to be $G_F^2\eta_0$ compared to that of $(\beta\beta)_{2\nu}$ as G_F^2, $T_{1/2}(2\nu) \simeq 10^{20}$ yr where as $T_{1/2}(0\nu) \simeq 10^{14}|\eta_0|^{-2}$. Thus, non-observation of the $(\beta\beta)_{0\nu}$ process implies that $\eta_0 = 10^{-4}$.

In the left–right symmetric model there are three distinct contributions (Figs. 6.2(a), (b), (c)): ν-mass, N-mass, and left–right mixing contribution. All these contributions are incoherent. The strengths of the three amplitudes are roughly given by

$$M_a \simeq G_F^2 m_\nu \left\langle \frac{e^{-m_\nu r}}{r} \right\rangle_{\text{Nuc}}. \tag{6.4.32}$$

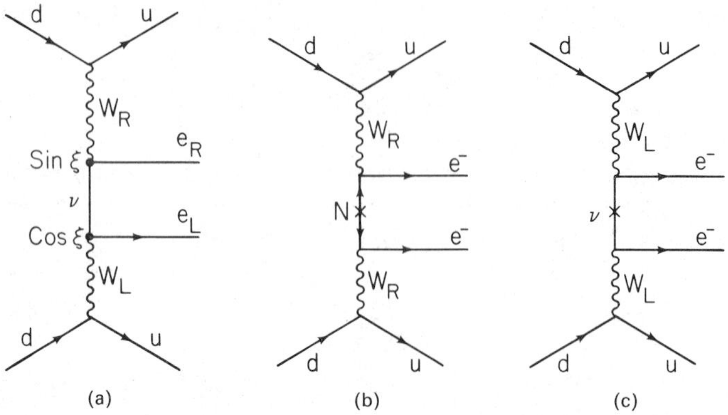

Figure 6.2

This is a $0^+ \to 0^+$ transition

$$M_b \simeq G_F^2 \eta^2 \frac{1}{m_N} \langle \delta^3(r) \rangle_{\text{Nuc}}. \tag{6.4.33}$$

This is also a $0^+ \to 0^+$ transition and is likely to be suppressed by a hard-core piece of the nuclear potential and is quite dependent on the nature of the nuclear model.

$$M_c \simeq G_F^2 \eta \sin \xi \langle |\mathbf{p}| \rangle \left\langle \frac{1}{r} \right\rangle_{\text{Nuc}}. \tag{6.4.34}$$

This is, however, a $0^+ \to 2^+$ transition and is a characteristic signature of right-handed currents. A nucleus well suited for study of this contribution is $G_e^{76} \to S_e^{76}$ which is the focus of at least two experiments [51].

To discuss the predictions for $T_{1/2}(0\nu)$ of left–right models we will consider the equivalent of η_0 in the above three cases.

Case C. $\eta_0 = \eta \sin \xi$

Phenomenologically, $\eta < \frac{1}{400}$ and for the case $m_N \approx 100$ GeV, $\sin \xi \approx (m_\nu/m_N)^{1/2} \approx 10^{-5}$ which gives $\eta_0 \leq 2\,(10^{-8})$. However, phenomenologically, m_N can be as low as 100 MeV in which case $\eta_0 \simeq 10^{-6}$. The present experiment limits on η_0 are if order 10^{-4}.

Case B.

The equivalent η_0 parameter in this case is

$$\eta_0 \simeq \eta^2 \frac{1}{m_N} \langle \delta^3(r) \rangle_{\text{Nuc}} \frac{1}{|\mathbf{p}| \langle 1/r \rangle}. \tag{6.4.35}$$

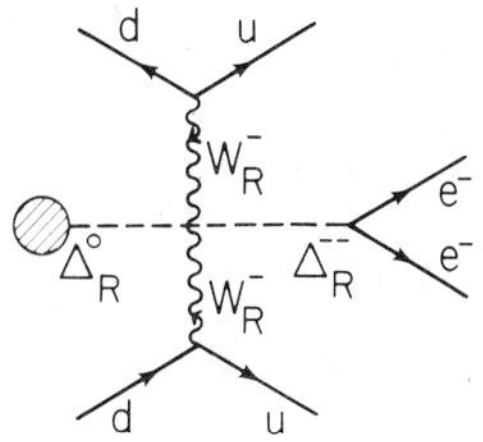

Figure 6.3

For a 100 GeV right-handed neutrino a very crude estimate for this parameter is $\eta_0 \approx 10^{-7}$ which is beyond the reach of any planned experiment.

Case A. $\eta_0 = m_\nu/|\mathbf{p}| \approx 10^{-5}$

For p (the momentum difference of the electrons) of order 2 MeV and $m_\nu \approx$ 10 eV.

In this model there exist additional contributions to $(\beta\beta)_{0\nu}$-decay which do not involve neutrinos but the exchange of doubly charged Higgs bosons [40] (Fig. 6.3), which are quite important and must be included in a detailed analysis of data. Thus, study of $(\beta\beta)_{0\nu}$-decay can provide important information about the left–right symmetry of weak interactions.

There is a variety of other processes such as $\mu \to e\gamma$, $\mu^- + (A, Z) \to e^+ + (A, Z-2)$, and muonium–antimuonium conversion which are characteristic of the left–right symmetric model. A detailed discussion of these processes is given in Ref. [52].

§6.5. Baryon Number Nonconservation and Higher Unification

(a) Selection Rules for Baryon Nonconservation

This section is devoted to the study of baryon number nonconserving processes that arise within the framework of left–right symmetric theories. The reason that these models lead to breakdown of the baryon number is due to the fact that the U(1) generator is the $B - L$ quantum number and that this U(1) symmetry is spontaneously broken. To study the selection rules for baryon number violation we note the formula for electric charge

$$Q = I_{3L} + I_{3R} + \frac{B-L}{2}. \tag{6.1.1}$$

Restricting ourselves to distance scales where $\Delta I_{3L} = 0$ we find that electric charge conservation implies

$$\Delta I_{3R} = -\tfrac{1}{2}\Delta(B - L). \tag{6.5.1}$$

Since in our case $(\Delta I_{3R}) = 1$ this implies the selection rule $|\Delta(B - L)| = 2$. This has the following implications:

(i) $\Delta L = 2, \qquad \Delta B = 0.$ (see Section 6.4)

(ii) $\Delta L = 0, \qquad \Delta B = 2.$

(iii) $\Delta L = -1, \qquad \Delta B = 1 \quad (\text{or } \Delta(B + L) = 0).$

In this section we will give the explicit Higgs model realizations of these selection rules for baryon nonconservation. First, we note that the minimal model described in the previous chapters (i.e., Higgs multiplets ϕ, $\Delta_L(3, 1, +2) + \Delta_R(1, 3, +2)$) has an extra symmetry under which $Q \to e^{i\alpha/3}Q$, which is unbroken by vacuum. Thus, even though $\langle\Delta_R\rangle \neq 0$ breaks the local $B - L$ symmetry, a final global symmetry survives in the end which can be identified with the baryon number. So, to realize the full potential of left–right models, the Higgs sector has to be increased. A more convenient framework for this is the partial unified gauge theory based on the group $SU(2)_L \times SU(2)_R \times SU(4)_C$ and the necessary Higgs multiplets to break $B - L$ symmetry [9].

(b) Partial Unification Model Based on $G_1 \equiv SU(2)_L \times SU(2)_R \times SU(4)_C$

This group was suggested by Pati and Salam [1] and the symmetry breaking pattern of interest for baryon nonconservation was first discussed by Mohapatra and Marshak [9] using the following multiplets (extending the work of Mohapatra and Senjanović [8]):

$$\phi(2, 2, 0),$$

$$\Delta_L(3, 1, 10) + \Delta_R(1, 3, 10).$$

Under $SU(2)_L \times SU(2)_R \times U(1)_{B-L} \times SU(3)_C$ the Δ's decompose as follows:

$$\Delta_L(3, 1, 10) = \{\Delta_{ll}(3, 1, -2, 1) + \Delta_{lq}(3, 1, -2/3, 3) + \Delta_{qq}(3, 1, 2/3, 6)\}_L \tag{6.5.2}$$

and similarly for Δ_R. Note that all the submultiplets of Δ and ϕ are bilinears of the fermion fields, i.e., two leptons, two quarks, or quarks and leptons. Let us denote the fermions as follows:

$$\psi \to \begin{pmatrix} u_1 & u_2 & u_3 & \nu \\ d_1 & d_2 & d_3 & e \end{pmatrix}. \tag{6.5.3}$$

The G_1-invariant Yukawas couplings can be written as

$$\mathscr{L}_Y = f(\psi_{La}^T C^{-1}\tau_2\boldsymbol{\tau}\psi_{Lb}\cdot\boldsymbol{\Delta}_{Lab}^\dagger + L \to R) + \text{h.c.}, \tag{6.5.4}$$

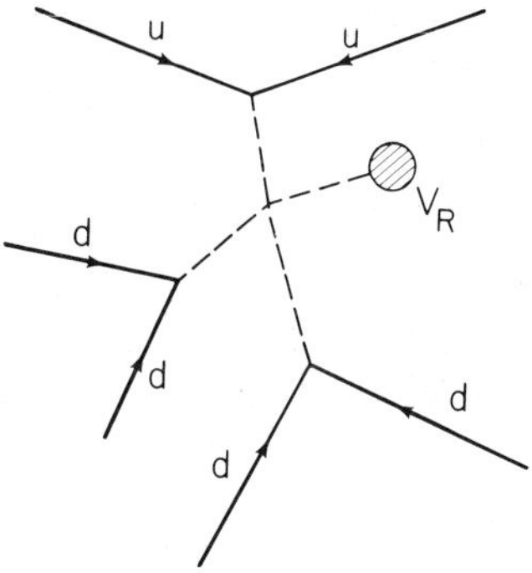

Figure 6.4

where a, $b = 1, \ldots, 4$ are the $SU(4)_C$ indices. In this notation the $SU(2)_R$ breaking occurs due to

$$\langle \Delta_{R44}^{1+i2} \rangle = v_R \neq 0. \tag{6.5.5}$$

To obtain baryon violating processes, note that, the most general Higgs potential contains a term of the form

$$V' = \lambda \varepsilon^{a_1a_2a_3a_4} \varepsilon^{b_1b_2b_3b_4} \varepsilon^{pq} \varepsilon^{p'q'} \varepsilon^{rs} \varepsilon^{r's'}$$
$$\times \Delta_{a_1b_1}^{pp'} \Delta_{a_2b_2}^{qq'} \Delta_{a_3b_3}^{rr'} \Delta_{a_4b_4}^{ss'} + \text{all permutations.} \tag{6.5.6}$$

Equations (6.5.4), (6.5.5), and (6.5.6) lead to the Feynman diagram in Fig. 6.4 which causes $\Delta B = 2$ transitions such as $N - \bar{N}$ oscillation and $p + n \to \pi$'s, etc. Let us now proceed to estimate the strength [9] of these processes: $G_{\Delta B=2}$:

$$G_{\Delta B=2} \simeq \frac{f^3 \lambda v_R}{m_{\Delta_{qq}}^6}. \tag{6.5.7}$$

This is the strength of the free-quark Hamiltonian defined at mass scale $m_{\Delta_{qq}}$. The actual low-energy $\Delta B = 2$ transition strength can be obtained from this; first, by doing renormalization group corrections to its strength [53]; and then taking account of the hadronic wave function effects [54], [55]. All these corrections have the effect of reducing the $N - \bar{N}$ transition strength from $G_{\Delta B=2}$ (defined above) to $G_{\Delta B=2} \times 10^{-4}$. Thus, it is reasonable to say that, the $N - \bar{N}$ mixing strength δm is given by

$$\delta m \approx G_{\Delta B=2} \times 10^{-4} \text{ GeV}. \tag{6.5.8}$$

To present an estimate for δm in our model we note that v_R breaks both $SU(4)_C$ as well as $SU(2)_R$. Therefore we expect $m_{\Delta_{qq}} \approx v_R$ to be consistent with the minimal fine tuning hypothesis. It can be argued from considerations of CP-violation that $m_{W_R} \leq 36$ TeV which implies 3 TeV $\leq v_R \leq 100$ TeV. If we choose $m_{\Delta_{qq}} \simeq v_R \approx 30$ TeV as a typical value and $f \approx \lambda \approx 10^{-1}$ we obtain

$$\delta m \simeq 10^{-31}\ \text{GeV}. \tag{6.5.9}$$

This corresponds to $\tau_{N-\bar{N}} \simeq h/\delta m \simeq 10^7$ s.

The phenomenon of $N - \bar{N}$ oscillation can be connected to nuclear instability from which we can get experimental information on $\delta m_{N\bar{N}}$. We present below the phenomenology of free and bound neutron oscillation to see whether these theoretical ideas can be experimentally tested. (We will define $\tau_{N-\bar{N}} = \hbar/\delta m_{N-\bar{N}}$ in what follows.)

Bounds on $\tau_{N-\bar{N}}$ from Stability of Matter

A question of crucial importance in planning experiments to detect $N - \bar{N}$ oscillation is the theoretical lower bound on $\tau_{N-\bar{N}}$ that is consistent with the known limits on matter stability. The main point is that the interaction Hamiltonian, that gives rise to $N - \bar{N}$ oscillation, can also lead to $N_1 + N_2 \to \pi$'s where $N_{1,2}$ can be protons and/or neutrons and thus lead to nuclear decay. Similarly, a neutron in the nucleus can convert into an antineutron which subsequently annihilates with a nucleon from the surrounding nuclear matter leading also to the same effect, i.e., nuclear instability. However, there already exist limits [56] on nuclear instability from the various experiments that have looked for proton decay. This puts an upper bound on the strength of neutron–antineutron mixing, as we discuss below. From a preliminary discussion of the connection between $\delta m_{N-\bar{N}}$ and the amplitude for $N + P \to \pi$'s, $A_{N+P} \to \pi$'s, i.e.,

$$\delta m_{N-\bar{N}} \approx \sqrt{M/\tau_{NP\to\pi's}} \tag{6.5.10}$$

it was suggested earlier that $\tau_{N-\bar{N}} \geq 10^5$ s. Subsequently, more refined analysis has been carried out [55], and in this section we will discuss them.

For this purpose we study $N - \bar{N}$ oscillation inside the nucleus. The $N, \bar{N}$ mass matrix in the effective nuclear potential can be written as

$$M = \begin{array}{c} \\ N \\ \bar{N} \end{array} \!\!\begin{array}{c} \begin{array}{cc} N & \quad\bar{N} \end{array} \\ \begin{pmatrix} V & \delta m \\ \delta m & \bar{V} - iW \end{pmatrix} \end{array} \tag{6.5.11}$$

where δm represents the $N - \bar{N}$ transition mass; and V and $\bar{V} - iW$ denote, respectively, the nuclear potentials seen by the neutron and the antineutron. The imaginary part of the $\bar{N}$-term represents the absorption of antineutrons by the nucleus. Diagonalization of the matrix in eqn. (6.5.11) leads to two states $|N_\pm\rangle$ with mass eigenvalues $m_\pm$ given, respectively, by

$$m_+ \simeq V + \frac{(\delta m)^2(V - \bar{V})}{(V - \bar{V})^2 + W^2} - \frac{iW(\delta m^2)}{(V - \bar{V})^2 + W^2} \tag{6.5.12}$$

and

$$m_- \simeq V - iW. \tag{6.5.13}$$

The $|N_+\rangle$ eigenstate represents the conversion of N and $\bar{N}$ and subsequent nuclear decay with $\Delta B = 2$ with a lifetime τ_+ given by

$$\tau_+^{-1} \simeq \frac{W(\delta m)^2}{(V - \bar{V})^2 + W^2}. \tag{6.5.14}$$

Since τ_+ is known from present limits on nuclear instability and V, $\bar{V}$, and W can be obtained from low-energy nuclear physics, eqn. (6.5.14) can be translated to give an upper bound on δm or lower bound on $\tau_{N-\bar{N}} \geq h/\delta m$, i.e.,

$$\tau_{N-\bar{N}} \simeq \left\{\tau_+ h \frac{W}{|V - \bar{V}|^2 + W^2}\right\}^{1/2}. \tag{6.5.15}$$

There exist lower bounds on τ_+ from existing experiments on proton decay [56]. However, we need information on $(V - \bar{V})$ and W from nuclear physics. There exist considerable uncertainties in their values, as inferred from different types of experiments.

One source of information on the $N\bar{N}$ potential is the $\bar{p}$-atom scattering experiments [57] which lead to the following values:

$$\begin{aligned} \bar{V} + iW = &\text{(a)} \quad 240 - i120 \text{ MeV} \quad (\text{Barnes } et\ al.\ [57]),\\ &\text{(b)} \quad 165 - i165 \text{ MeV} \quad (\text{Poth } et\ al.\ [57]),\\ &\text{(c)} \quad 70 - i210 \text{ MeV} \quad (\text{Robertson } et\ al.\ [57]). \end{aligned}$$

The variations could be attributed to the fact that $\bar{p}$-atom scattering is a surface effect where nuclear density is not very well known.

Another source of information on $\bar{V} - iW$ is from $\bar{p}p$ scattering data at low energies [58]. This leads to

$$\begin{aligned} W &= -850 \text{ MeV} \quad \text{for } I = 1\\ &= -659 \text{ MeV} \quad \text{for } I = 0. \end{aligned}$$

Auerbach *et al.* [59] writes a potential with comparable values for the parameters

$$\text{(d)} \quad \bar{V} + iW = 1000 + i700 \text{ MeV}.$$

If we choose $\tau_+ \geq 1 \times 10^{31}$ yr as is indicated by recent data [56] and use $V = 0$–50 MeV we obtain for various cases

$$\begin{aligned} \tau_{N-\bar{N}} &\geq 1.5 \times 10^7 \text{ s} \quad (\text{Case (a)}),\\ &\geq 2 \times 10^7 \text{ s} \quad (\text{Case (b)}),\\ &\geq 2.6 \times 10^7 \text{ s} \quad (\text{Case (c)}),\\ &\geq 0.9 \times 10^7 \text{ s} \quad (\text{Case (d)}). \end{aligned} \tag{6.5.16}$$

This corresponds to a value of $\delta m \approx 10^{-23}$ eV. Another way to obtain a lower limit on $\tau_{N-\bar{N}}$ has been discussed by Riazuddin [60]. In this method we consider the annihilation of $\bar{N}$ in nucleus via the ρ' virtual state, ρ' being the only resonance available with $J^P = 1^-$ and $I - 1$ in the 2 GeV mass region. Taking $g_{\rho' NN} \approx g_\rho$ and $g_{\rho' \bar{u}d}$ as the value of the quark–antiquark bound state wave function at the origin we obtain

$$\tau_{N-\bar{N}} \geq 3 \times 10^6 \text{ s.} \tag{6.5.17}$$

We caution that there are uncertainties in the value of $g_{\rho' NN}$.

Finally, there are two recent analyses [61], [62] where the nuclear effects have been studied with greater care and their conclusion is that the present lower limit [63] on $\tau_{N-\bar{N}}$ is between (2–8) $\times$ 10^7 s.

We point out, however, that there is an additional contribution to nuclear instability that arises from the same $\Delta B = 2$ interaction that gives rise to the $N - \bar{N}$ oscillation. This has to do with the direct decay of two nucleons in contact with pions in the presence of $\Delta B = 2$ interactions. Of course, the hard-core nature of the nuclear potential is likely to suppress this contribution. In any case, if we denote this contribution by $\Gamma_{\Delta B=2}$, the total width for nuclear decay is given by

$$\Gamma = (\Gamma_+)_{\text{ocs}} + \Gamma_{\Delta B=2}, \tag{6.5.18}$$

where $\tau_+^{-1} \equiv (\Gamma_+)_{\text{osc}}$ and $\tau_{\text{Nuc}} \simeq h/\Gamma$, and this additional contribution has the effect of increasing $\tau_{N-\bar{N}}$ somewhat. It has also been pointed out that the parameter $\delta m_{N-\bar{N}}$ may receive additional unknown nuclear renormalization that could effect the connection between free and bound $N - \bar{N}$ oscillation [64]. Let us now proceed to discuss ways of measuring $N - \bar{N}$ oscillation.

Free Neutron–Antineutron Oscillation

It was pointed out by Glashow [65] that the existence of the earth's magnetic field splits the $N - \bar{N}$ states by an amount $\Delta E = 2\mu B$, where μ is the magnetic moment of the neutron and leads to $\Delta E \approx 10^{-11}$ eV $\gg \delta m$. This would pose an immediate difficulty in searching for $N - \bar{N}$ mixing. However, it was subsequently demonstrated [66] that reducing $\Delta E \approx 10^{-14}$ eV is sufficient to simulate the conditions of a free neutron beam $((\Delta E t/h) \ll 1)$ thus making the experiment feasible. This would require degaussing the earth's magnetic field by a factor of one thousand. For a free neutron beam the probability for detecting an $\bar{N}$ starting with a beam of neutrons is given by (as can be inferred from the analyzing equation (6.5.11)) $W = 0$ and $V = \bar{V} = m_N$. For thermal neutrons we can expect to have a flight time of about $t \approx 10^{-2}$ s. If we have a neutron beam intensity of about 10^{12} s^{-1}, then we can expect about 10 antineutrons per year making experimental detection of $N - \bar{N}$ oscillation feasible [67]. Observation of $N - \bar{N}$ oscillation will provide a clear vindication of the local $B - L$ symmetry models.

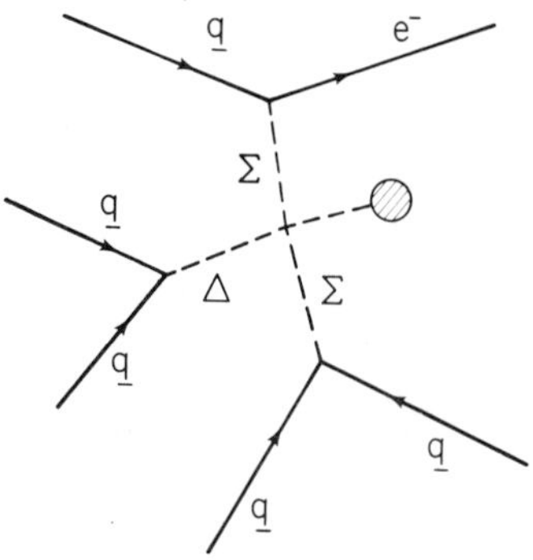

Figure 6.5

It has also been pointed out [68], [69] that by extending this model to include a (2, 2, 15) ≡ Σ Higgs multiplet we can realize the third selection rule which follows from electric charge conservation, i.e., $\Delta(B + L) = 0$. The argument is that in the presence of Σ there is a term in the Higgs potential of the form

$$V'' = \lambda' \Delta_{aa'} \Delta_{bb'} \Sigma_c^{a'} \Sigma_d^{b'} \varepsilon^{abcd}, \tag{6.5.19}$$

where for simplicity we have dropped the SU(2) indices. This gives rise to the Feynman diagram in Fig. 6.5 which obeys the $\Delta(B + L) = 0$ selection rule. The strength of this process is

$$G_{B+L=0} \approx \frac{f\lambda' h^2 V_R}{m_{\Delta qq}^2 m^4 \Sigma}. \tag{6.5.20}$$

This can lead to decays of proton of type $p \to e^- \pi^+ \pi^+$, $n \to e^- \pi^+$, etc. With a lifetime of order 10^{31}–10^{32} yr. However, in general grand unified models, this process becomes highly suppressed due to survival hypothesis.

§6.6. Sin² θ_W and the Scale of Partial Unification

In this section we would like to consider the possibility that $G_1 \equiv SU(2)_L \times SU(2)_R \times SU(4)_C$ is a partial unification group of the left–right symmetric model. G_1 has two couplings $g_2 = g_{2L} = g_{2R}$ and g_4 as against g_2, g_{B-L} and g_3 of the $SU(2)_L \times SU(2)_R \times U(1)_{B-L} \times SU(3)_C$ group. Due to this unification we can obtain, at the unification scale M_c, a relation between $\sin^2 \theta_W$ and α_s. To obtain this relation we note that

$$\frac{1}{e^2} = \frac{2}{g_2^2} + \frac{1}{g_{BL}^2}. \tag{6.6.1}$$

But at M_c, $g_{BL}(M_c) = \frac{3}{2}g_4(M_c) = \frac{3}{2}g_3(M_c)$; from this it follows that

$$\sin^2\theta_W(M_c) = \frac{1}{2} - \frac{1}{3}\frac{\alpha(M_c)}{\alpha_s(M_c)}. \tag{6.6.2}$$

In order to extrapolate this relation to $\mu = m_W$ we have to know the pattern of symmetry breaking from G_1 to $G_{123} \equiv U(1)_Y \times SU(2) \times SU(3)_C$. Let us assume the pattern

$$SU(2)_L \times SU(2)_R \times SU(4)_C$$

$$\downarrow M_c$$

$$SU(2)_L \times SU(2)_R \times U(1)_{B-L} \times SU(3)_C$$

$$\downarrow M_R$$

$$U(1)_Y \times SU(2)_L \times SU(3)_C.$$

For this case we obtain, using the techniques of Georgi, Quinn, and Weinberg, that

$$\begin{aligned}\text{Sin}^2\theta_W = \frac{1}{2} - \frac{1}{3}\frac{\alpha(m_W)}{\alpha_s(m_W)} - \frac{\alpha(m_W)}{4\pi}\Big[\{A_{2R} - A'_{2L}) + \tfrac{2}{3}(A_{BL} - A_3)\}\\ \times \ln\frac{M_c}{m_W} + \{\tfrac{5}{3}A_Y - A_{2R} - \tfrac{2}{3}A_{BL} + A_{2L} - A'_{2L}\}\ln\frac{M_R}{m_W}\Big],\end{aligned} \tag{6.6.3}$$

where the A_i's are the coefficients in the β-function given for the SU(N) group by

$$A = -\tfrac{11}{3}N + \tfrac{4}{3}N_G + \tfrac{1}{6}T(R), \tag{6.6.4}$$

where N_G is the number of generations and $T(R)$ is defined in terms of the generators θ_i of the group on the space of the Higgs multiplets

$$\text{Tr}(\theta_i\theta_j) = T(R)\delta ij. \tag{6.6.5}$$

If we keep only the fermion contribution then this equation becomes

$$\text{Sin}^2\theta_W = \frac{1}{2} - \frac{1}{3}\frac{\alpha(m_W)}{\alpha_s(m_W)} - \frac{11\alpha(m_W)}{6\pi}\left(\ln\frac{M_c}{m_W} + \ln\frac{M_R}{m_W}\right). \tag{6.6.6}$$

This implies that for $\sin^2\theta_W \simeq 0.23$, $M_c M_R \simeq m_W^2 \times 10^{29}$ and for $\sin^2\theta_W \simeq 0.25$, $M_c M_R \simeq m_W^2 \times 10^{27}$. If we choose $M_c = M_R$, this implies a partial unification scale of 10^{16}–10^{15} GeV. On the other hand, if we assume $M_c \simeq M_{Pl} \simeq 10^{19}$ GeV, this implies $M_R \simeq 10^{12}$–10^{14} GeV. In either case, partial unification implies a very high scale for M_R; thus the $N - \bar{N}$ oscillation is highly suppressed and so are the $\Delta(B + L) = 0$ processes.

It has recently been pointed out [70] that if the parity and $SU(2)_R$ breaking scales are decoupled then at $\mu \approx m_{W_R}$, $g_L \neq g_R$ and in that case both M_c and M_{W_R} can be lowered to the level of 100 TeV or so without conflicting with the observed value of $\sin^2\theta_W$.

§6.7. Conclusions

In summary, we have noted some compelling intuitive reasons for suspecting the existence of a next level of unification symmetry which is left–right symmetric. It not only restores parity to the status of a conserved quantum number of all fundamental interaction, but it also brings $B - L$ (the only other anomaly-free generator for each generation of fermions) to the level of a local symmetry like electric charge. This is very appealing since it extends the Gell-Mann–Nishijima electric charge formula from the domain of strong interactions to that of weak interactions. Extending this analogy a little further leads us to the suggestion that, just as the quark picture provides an underlying dynamical basis for the Gell-Mann–Nishijima formula for strong interactions, an underlying preonic substructure of quarks and leptons may form the basis of $B - L$ as the gauge generator of weak interactions. Indeed, most attractive preon models bear out this conjecture. Thus, left–right symmetric models may provide the big leap forward from a successful geometrical picture of weak interactions based on the $SU(2)_L \times U(1) \times SU(3)_c$ model to a dynamical model that may provide a natural basis for understanding many of the puzzles of weak interactions.

Clearly, further experimental and theoretical work is needed before this appealing scenerio receives its rightful place in the annals of particle physics. At the experimental level we believe that any or all of the following pieces of evidence would constitute manifestations of left–right symmetry:

(i) ν_e mass in the electron volt range;
(ii) $\Delta B = 2$ transitions such as nucleon + nucleon → pions and $N - \bar{N}$ oscillations; and
(iii) lepton number violating and lepton flavor changing processes such as $(\beta\beta)_{0\nu}$-decay, $\mu^- A \to e^+ A$, $\mu^- e^+ \to \mu^+ e^-$, $\mu \to 3e$.

References

[1] J. C. Pati and A. Salam, *Phys. Rev.* **D10**, 275 (1974);
R. N. Mohapatra and J. C. Pati, *Phys. Rev.* **D11**, 566, 2558 (1975);
G. Senjanovic and R. N. Mohapatra, *Phys. Rev.* **D12**, 1502 (1975).

[2] V. Lyubimov *et al.*, *Phys. Lett.* **94B**, 266 (1980).
V. Lyubimov, Talk at Leipzig Conference, 1984.

[3] R. Cowsik and J. McClelland, *Phys. Rev. Lett.* **29**, 699 (1972);
S. S. Gershtein and Ya B. Zeldovich, *JETP Lett.* **4**, 120 (1966).

[4] R. N. Mohapatra and R. E. Marshak, *Phys. Lett.* **91B**, 222 (1980);
A. Davidson, *Phys. Rev.* **D20**, 776 (1979).

[5] R. N. Mohapatra and J. C. Pati, *Phys. Rev.* **D11**, 566 (1975).

[6] D. Chang, *Nucl. Phys.* **B214**, 435 (1983);
G. Branco, J. M. Frere, and J. M. Gerard, *Nucl. Phys.* **B221**, 317 (1983).

[7] D. Chang, R. N. Mohapatra, and M. K. Parida, *Phys. Rev. Lett.* **50**, 1072 (1984); *Phys. Rev.* **D30**, 1052 (1984).

[8] R. N. Mohapatra and G. Senjanovic, *Phys. Rev. Lett.* **44**, 912 (1980); *Phys. Rev.* **D21**, 165 (1981).

[9] R. N. Mohapatra and R. E. Marshak, *Phys. Rev. Lett.* **44**, 1316 (1980).
[10] V. Barger, E. Ma, and K. Whisnant, *Phys. Rev.* **D26**, 2378 (1982);
I. Liede, J. Malampi, and M. Roos, *Nucl. Phys.* **B146**, 157 (1978);
T. Rizzo and G. Senjanovic, *Phys. Rev.* **D24**, 704 (1981);
X. Li and R. E. Marshak, *Phys. Rev.* **D25**, 1886 (1982);
For earlier work see
J. E. Kim, P. Langacker, M. Levine, and H. Williams, *Rev. Mod. Phys.* **53**, 211 (1981).
[11] M. Gell-Mann, P. Ramond, and R. Slansky, in *Supergravity* (edited by D. Freedman *et al.*), North-Holland, 1979;
T. Yanagida, KEK lectures, 1979;
R. N. Mohapatra and G. Senjanovic, *Phys. Rev. Lett.* **44**, 912 (1980).
[12] M. Gronau and S. Nussinov, Fermilab preprint, 1982;
M. Gronau and R. Yahalom, *Nucl. Phys.* **B236**, 233 (1984).
[13] M. A. B. Beg, R. Budny, R. N. Mohapatra, and A. Sirlin, *Phys. Rev. Lett.* **38**, 1252 (1977);
For a subsequent extensive analysis see
J. Maalampi, K. Mursula, and M. Roos, *Nucl. Phys.* **B207**, 233 (1982).
[14] M. Roos *et al.*, *Phys. Lett.* **111B**, 1 (1982).
[15] F. W. Koks and J. Vanklinken, *Nucl. Phys.* **A272**, 61 (1976).
[16] J. Carr *et al.*, *Phys. Rev. Lett.* **51**, 627 (1983).
[17] F. Corriveau *et al.*, *Phys. Rev.* **D24**, 2004 (1981).
[18] T. Yamazaki *et al.*, KEK preprint, 1983.
[19] B. Holstein and S. Treiman, *Phys. Rev.* **D16**, 2369 (1977).
[20] D. Bryman, Talk at Mini-Conference on Low-Energy Tests of Conservation Law, 1983.
[21] T. Yamazaki *et al.*, KEK preprint, 1983.
[22] I. I. Bigi and J. M. Frere, *Phys. Lett.* **110B**, 255 (1982).
[23] J. Donoghue and B. Holstein, *Phys. Lett.* **113B**, 383 (1982).
[24] G. Beall, M. Bender, and A. Soni, *Phys. Rev. Lett.* **48**, 848 (1982).
[25] Earliest use of vacuum saturation of short distance contribution to $K_L - K_S$ mass difference was by
R. N. Mohapatra, J. S. Rao, and R. E. Marshak, *Phys. Rev.* **171**, 1502 (1968);
B. L. Ioffe and E. Shabalin, *Sov. J. Nucl. Phys.* **6**, 328 (1967).
[26] M. K. Gaillard and B. W. Lee, *Phys. Rev.* **D10**, 897 (1974).
[27] See J. Trampetic, *Phys. Rev.* **D27**, 1565 (1983) for a discussion of this point.
[28] R. N. Mohapatra, G. Senjanovic, and M. Tran, *Phys. Rev.* **D28**, 546 (1983);
G. Ecker, W. Grimus, and H. Neufeld, CERN preprint TH-3551, 1983;
M. Hwang and R. J. Oakes, Fermilab preprint 83/38-THY, 1983;
H. Harari and M. Leurer, Fermilab preprint 83/59-Thy, 1983;
F. Gilman and M. Reno, *Phys. Rev.* **D29**, 937 (1974).
[29] L. Wolfenstein, *Nucl. Phys.* **B160**, 1979 (1981);
C. Hill, *Phys. Lett.* **97B**, 275 (1980).
[29a] I. I. Bigi and J. M. Frere, University of Michigan preprint, 1983.
[30] R. N. Mohapatra, F. E. Paige, and D. P. Sidhu, *Phys. Rev.* **D17**, 2642 (1978).
[31] A. Datta and A. Raychaudhuri, Calcutta preprint CU PP/82–7, 82–88, 1982;
F. Olness and M. E. Ebel, Wisconsin preprint, 1983;
T. Rizzo, lowa preprint, 1983.
[32] T. Rizzo and G. Senjanovic, *Phys. Rev.* **D24**, 704 (1981);
W. Y. Keung and G. Senjanovic, *Phys. Rev. Lett.* **50**, 1427 (1983).
For a more recent detailed study, see
N. G. Deshpande, J. Gunion and B. Kayser, Proceedings of the Telemark Conference (edited by V. Barger), 1984. American Institute of Physics, New York.

[33] K. Winter, CERN preprint (unpublished).
[34] V. Lubimov *et al.*, *Phys. Lett.* **94B**, 266 (1980).
[35] G. Barbiellini and R. Santoni, *Rev. del. Nuovo Cimento* (1986).
[36] F. Wagner *et al.*, Argus Collaboration (unpublished).
[37] For a review see
H. H. Williams, *Proceedings of the SLAC Summer School* (edited by M. Zipf), 1982.
[38] For a pedagogical review see
R. N. Mohapatra, *Forschritte Phys.* **31**, 185 (1983).
[39] D. Dicus, E. Kolb, V. Teplitz, and R. Wagoner, *Phys. Rev.* **D18**, 1819 (1978);
Y. Chikashige, R. N. Mohapatra, and R. Peccei, *Phys. Rev. Lett.* **45**, 1926 (1981).
[39a] S. Sarkar and A. M. Cooper, CERN preprint, 1984.
[39b] A. Kumar and R. N. Mohapatra, *Phys. Lett.* **150B**, 191 (1985).
[40] R. N. Mohapatra and J. D. Vergados, *Phys. Rev. Lett.* **47**, 1713 (1981);
C. Piccioto and M. Zahir, *Phys. Rev.* **D26**, 2320 (1982).
[41] B. Kayser, *Phys. Rev.* **D26**, 1662 (1982).
[42] L. Wolfenstein, *Phys. Lett.* **107B**, 77 (1981).
[43] J. Valle, *Phys. Rev.* **D27**, 1672 (1983);
S. Petcov, *Phys. Lett.* **110B**, 245 (1982);
M. Doi, M. Kenmoku, T. Kotani, H. Nishiura, and E. Taskasugi, Osaka preprint OS-GE-83-48, 1983.
[44] K. M. Case, *Phys. Rev.* **107**, 307 (1957).
[45] D. Chang and R. N. Mohapatra, *Phys. Rev.* **D32**, 1248 (1985).
[46] M. Roncadelli and G Senjanovic, *Phys. Lett.* **107B**, 59 (1983).
[47] P. Pal, Carnegie–Mellon preprint, 1983.
[47a] S. Pakvasa and B. McKellar, Hawaii preprint, 1982.
[48] Y. Hosotani, *Nucl. Phys.* **B191**, 411 (1981);
J. Schecter and J. W. F. Valle, *Phys. Rev.* **D25**, 774 (1982).
[49] For a recent discussion see
S. Sarkar and A. M. Cooper, CERN preprint, 1984.
[50] H. Primakoff and S. P. Rosen, *Rep. Prog. Phys.* **22**, 121 (1959); *Proc. Phys. Soc.* (London) **78**, 464 (1961);
A. Halprin, P. Minkowski, H. Primakoff, and S. P. Rosen, *Phys. Rev.* **D13**, 2567 (1976);
M. Doi, T. Kotani, H. Nishiura, K. Okuda, and E. Takasugi, *Prog. Theor. Phys.* **66**, 1765 (1981); **68**, 348 (1982) (E);
W. C. Haxton, G. J. Stehenson, Jr., and D. Strottman, *Phys. Rev. Lett.* **47**, 153 (1981); *Phys. Rev.* **D25**, 2360 (1982);
J. D. Vergados, *Phys. Rev.* **C24**, 640 (1981).
[51] F. Avignone *et al.*, Talk at Fourth Workshop on Grand Unification, held in Philadelphia, 1983;
E. Fiorini, *Proceedings of XXI International Conference on High Energy Physics*, Paris, 1982.
[52] Riazuddin, R. E. Marshak, and R. N. Mohapatra, *Phys. Rev.* **D24**, 1310 (1981).
[53] W. Caswell, J. Milutinovic, and G. Senjanovic, *Phys. Rev.* **D26**, 161 (1982);
S. Rao and R. Shrock, Stonybrook preprint, 1983.
[54] J. Pasupathy, *Phys. Lett.* **B** (to be published);
S. Rao and R. Shrock, *Phys. Lett.* **116B**, 238 (1982);
U. Sarkar and S. P. Misra, *Phys. Rev.* **D28**, 249 (1983).
[55] K. Chetyrkin *et al.*, *Phys. Lett.* **99B**, 358 (1981);
P. G. Sandars, *J. Phys.* **G6**, L161 (1980);
Riazzuddin, *Phys. Rev.* **D25**, 885 (1982);

C. Dover, M. Gal, and J. Richards, *Phys. Rev.* **D27**, 1090 (1983);
W. Alberico *et al.*, *Phys. Lett.* **114B**, 266 (1982);
A. Kerman *et al.*, MIT preprint, 1983;
For a review see
R. N. Mohapatra, *Proceedings of the Harvard Workshop on $N - \bar{N}$ Oscillation*, 1982.

[56] R. Bionta *et al.*, *Phys. Lett.* **114B**, 266 (1982);
V. L. Narasimhan *et al.*, Chapter V;
For other references see Chapter V.

[57] Barnes *et al.*, *Phys. Rev. Lett.* **29**, 1132 (1972);
Poth *et al.*, *Nucl. Phys.* **A294**, 435 (1977);
Roberson *et al.*, *Phys. Rev.* **C16**, 1945 (1977);
For a review see
C. J. Batty, Rutherford Laboratory preprint, 1981.

[58] J. Cote *et al.*, *Phys. Rev. Lett.* **48**, 13198 (1982).

[59] R. Auerbach *et al.*, *Phys. Rev. Lett.* **46**, 702 (1980).

[60] Riazzuddin, *Phys. Rev.* **D25**, 885 (1982).

[61] C. Dover, A. Gal, and J. Richards, *Phys. Rev.* **D27**, 1090 (1983).

[62] A. Kerman *et al.*, MIT preprint, 1983.

[63] L. Jones *et al.*, *Phys. Rev. Lett.* **52**, 720 (1984).

[64] P. K. Kabir, *Phys. Rev. Lett.* (1983).

[65] S. L. Glashow, Cargese lectures, 1979.

[66] R. N. Mohapatra and R. E. Marshak, *Phys. Lett.* **94B**, 183 (1980).

[67] M. Baldoceolin *et al.*, CERN preprint, 1983.

[68] A Higgs model realization of this idea has been discussed recently by
J. C. Pati, A. Salam, and U. Sarkar, University of Maryland preprint, 1983.

[69] The $\Delta(B + L) = 0$ result was first realized in a composite model by
H. Harari, R. N. Mohapatra, and N. Seiberg, *Nucl. Phys.* **B209**, 174 (1982).

[70] D. Chang, R. N. Mohapatra, J. Gipson, R. E. Marshak, and M. K. Parida, *Phys. Rev.* **D31**, 1718 (1985).

CHAPTER 7

SO(10) Grand Unification

§7.1. Introduction

The possibility of SO(10) as a grand unification group of the standard $SU(2)_L \times U(1)_Y \times SU(3)_c$ was first noted by Georgi [1] and Fritzsch and Minkowski [1]. Unlike SU(5), SO(10) is a group of rank 5 with the extra diagonal generator of SO(10) being $B - L$ as in the left–right symmetric groups. The advantages of SO(10) over SU(5) grand unification are that:

(a) only one 16-dimensional spinor representation of SO(10) has the right quantum numbers to accommodate all fermions (including the right-handed neutrino) of one generation;
(b) the gauge interactions of SO(10) conserve parity thus making parity a part of a continuous symmetry: this has the advantage that it avoids the cosmological domain wall problem associated with parity symmetry breakdown; and
(c) it is the minimal left–right symmetric grand unified model that gauges the $B - L$ symmetry and is the only other simple grand unification group that does not need mirror fermions [2]. The model does not have any global symmetries.

Before proceeding to a discussion of the model we first spell out the various maximal subgroups of SO(10) as well as the decomposition of the small-dimensional representations under the interesting subgroups

$$\mathrm{SO}(10) \begin{cases} \nearrow \mathrm{SU}(5) \times U(1), \\ \searrow \mathrm{SO}(6) \times \mathrm{SO}(4), \quad \text{or} \quad \mathrm{SU}(4) \times \mathrm{SU}(2)_L \times \mathrm{SU}(2)_R \times D. \end{cases} \tag{7.1.1}$$

Decomposition of some of the smaller SO(10) irreducible multiplets under (a) $SU(2)_L \times SU(2)_R \times SU(4)_C$ and under (b) $SU(5) \times U(1)$ are

$\{10\}$ (a) $(1, 1, 6) + (2, 2, 1)$;
(b) $\{5\} + \{\bar{5}\}$.

$\{16\}$ (a) $(2, 1, 4) + (1, 2, \bar{4})$;
(b) $\{10\} + \{\bar{5}\} + \{1\}$.

$\{45\}$ (a) $(3, 1, 1) + (2, 2, 6) + (1, 1, 15) + (1, 3, 1)$;
(b) $\{24\} + \{10\} + \{\overline{10}\} + \{1\}$.

$\{54\}$ (a) $(1, 1, 1) + (2, 2, 6) + (1, 1, 20) + (3, 3, 1)$;
(b) $\{15\} + \{\overline{15}\} + \{24\}$.

$\{120\}$ (a) $(2, 2, 15) + (3, 1, 6) + (1, 3, \bar{6}) + (2, 2, 1) + (1, 1, 20)$;
(b) $\{5\} + \{\bar{5}\} + \{10\} + \{\overline{10}\} + \{45\} + \{\overline{45}\}$.

$\{126\}$ (a) $(3, 1, 10) + (1, 3, \overline{10}) + (2, 2, 15) + (1, 1, \bar{6})$;
(b) $\{1\} + \{\bar{5}\} + \{10\} + \{\overline{15}\} + \{45\} + \{\overline{50}\}$.

$\{210\}$ (a) $(1, 1, 1) + (1, 1, 15) + (2, 2, 20) + (3, 1, 15) + (1, 3, 15) + (2, 2, 6)$;
(b) $1 + \{5\} + \{\bar{5}\} + \{10\} + \{\overline{10}\} + \{24\} + \{40\} + \{\overline{40}\} + \{75\}$.

In order to discuss the SO(10) model it is convenient to give the representations in a simple algebraic formulation, which we do in the next section, and identify the fermions in the spinor representation.

§7.2. SO(2N) in an SU(N) Basis [3]

There exist different ways [3, 4] to discuss the algebra of SO(10). A particularly convenient way is to use the spinor SU(N) basis [3] which we present below. Consider a set of N operators χ_i $(i = 1, \ldots, N)$ and their Hermitian conjugate χ_i^+ satisfying the following anticommutation relations:

$$\{\chi_i, \chi_j^+\} = \delta_{ij},$$
$$\{\chi_i, \chi_j\} = 0. \tag{7.2.1}$$

We use the symbol $\{\ ,\ \}$ to denote anticommutation and $[\ ,\ ]$ to denote the commutation operation. It is well known that the operators T_j^i defined as

$$T_j^i = \chi_i^+ \chi_j. \tag{7.2.2}$$

satisfy the algebra of the U(N) group, i.e.,

$$[T_j^i, T_l^k] = \delta_j^k T_l^i - \delta_l^i T_j^k. \tag{7.2.3}$$

Now let us define the following $2N$-operators, Γ_μ $(\mu = 1, \ldots, 2N)$:

$$\Gamma_{2j-1} = -i(\chi_j - \chi_j^+)$$

and

$$\Gamma_{2j} = (\chi_j + \chi_j^+), \qquad j = 1, \ldots, N. \tag{7.2.4}$$

It is easy to verify, using eqns. (7.2.1), that

$$\{\Gamma_\mu, \Gamma_\nu\} = 2\delta_{\mu\nu}. \tag{7.2.5}$$

Thus, the Γ_μ's form a Clifford algebra of rank $2N$ (of course, $\Gamma_\mu = \Gamma_\mu^+$). Using the Γ_μ's we can construct the generators of the SO(2N) group as follows:

$$\Sigma_{\mu\nu} = \frac{1}{2i}[\Gamma_\mu, \Gamma_\nu]. \tag{7.2.6}$$

The $\Sigma_{\mu\nu}$ can be written down in terms of χ_j and χ_j^+ as follows:

$$\Sigma_{2j-1, 2k-1} = \frac{1}{2i}[\chi_j, \chi_k^+] - \frac{1}{2i}[\chi_k, \chi_j^+] + i(\chi_j\chi_k + \chi_j^+ \chi_k^+), \quad \checkmark$$

$$\Sigma_{2j, 2k-1} = \tfrac{1}{2}[\chi_j, \chi_k^+] + \tfrac{1}{2}[\chi_k, \chi_j^+] - (\chi_j\chi_k + \chi_j^+ \chi_k^+), \quad \checkmark$$

$$\Sigma_{2j, 2k} = \frac{1}{2i}[\chi_j, \chi_k^+] - \frac{1}{2i}[\chi_k, \chi_j^+] - i(\chi_j\chi_k + \chi_j^+ \chi_k^+). \sim \tag{7.2.7}$$

It is well known that the spinor representation of SO(2N) is 2^N dimensional. To write it in terms of the SU(N) basis let us define a "vacuum" state $|0\rangle$ which is SU(N) invariant. The 2^N-dimensional spinor representation is then given in Table 7.1.

This representation can be split into the 2^{N-1}-dimensional representations under a chiral projection operator. We now proceed to construct this operator. Define

$$\Gamma_0 = i^N \Gamma_1 \Gamma_2 \ldots \Gamma_{2N}. \tag{7.2.8}$$

Also define a number operator $n_j = \chi_j^+ \chi_j$.

Using eqn. (7.2.8) Γ_0 can be written as follows:

$$\Gamma_0 = [\chi_1, \chi_1^+][\chi_2, \chi_2^+] \ldots [\chi_N, \chi_N^+]$$

$$= \prod_{j-1}^{N} (1 - 2n_j). \tag{7.2.9}$$

Using the property of the number operator $n_j^2 = n_j$ we can show that $1 - 2n_j = (-1)^{n_j}$ and so we get

$$\Gamma_0 = (-1)^n, \qquad n = \sum_j n_j. \tag{7.2.10}$$

It is then easily checked that

$$[\Sigma_{\mu\nu}, (-1)^n] = 0. \tag{7.2.11}$$

The "chirality" projection operator is therefore given by $\frac{1}{2}(1 \pm \Gamma_0)$. Each irreducible "chiral" subspace is therefore characterized by odd or even numbers of χ-particles. To make it more explicit let us consider $N = 5$ and define a column vector $|\psi\rangle$ as follows:

$$|\psi\rangle = |0\rangle\psi_0 + \chi_j^+|0\rangle\psi_j + \tfrac{1}{2}\chi_j^+ \chi_k^+|0\rangle\psi_{jk} + \tfrac{1}{12}\varepsilon^{ijklm}\chi_k^+ \chi_l^+ \chi_m^+|0\rangle\bar\psi_{ji}$$

$$+ \tfrac{1}{24}\varepsilon^{jklmn}\chi_k^+ \chi_l^+ \chi_m^+ \chi_n^+|0\rangle\bar\psi_j + \chi_1^+ \chi_2^+ \chi_3^+ \chi_4^+ \chi_5^+|0\rangle\bar\psi_0, \tag{7.2.12}$$

where $\bar{\psi}$ is not the complex conjugate of ψ but an independent vector. We will denote complex conjugate by *. The generalization of eqn. (6.7.12) to the case of arbitrary N is obvious if we write

$$\psi = \begin{pmatrix} \psi_0 \\ \psi_j \\ \psi_{jk} \\ \bar{\psi}_{jk} \\ \bar{\psi}_j \\ \bar{\psi}_0 \end{pmatrix}. \tag{7.2.13}$$

Under chirality

$$\psi = \begin{pmatrix} \psi_+ \\ \psi_- \end{pmatrix},$$

where

$$\psi_\pm = \tfrac{1}{2}(1 \pm \Gamma_0)\psi$$

and

$$\chi_+ = \begin{pmatrix} \psi_0 \\ \psi_{ij} \\ \bar{\psi}_j \end{pmatrix}, \qquad \psi_- = \begin{pmatrix} \bar{\psi}_0 \\ \bar{\psi}_{ij} \\ \psi_i \end{pmatrix}. \tag{7.2.14}$$

For the $N = 5$ case $\bar{\psi}_i$ and ψ_{ij} represent the $\bar{5}$- and 10-dimensional representation of SU(5) and ψ_0 is the singlet. All the fermions are assigned to ψ_+. It is then easy to write down the gauge interaction of the fermions. We further note that for the case of SO(10) the formula for electric charged Q is given by

$$Q = \tfrac{1}{2}\Sigma_{78} - \tfrac{1}{6}(\Sigma_{12} + \Sigma_{34} + \Sigma_{56}). \tag{7.2.15}$$

We next tackle the problem of spontaneous generation fermion masses in the SO(10) model.[1]

§7.3. Fermion Masses and the "Charge Conjugation" Operator

As is well known, in the framework of gauge theories, at the present state of art, the fermion masses arise from Yukawa couplings of fermions to Higgs bosons and subsequent breakdown of the gauge symmetry by nonzero vacuum expectation values of the Higgs mesons. In general, in grand unified

[1] It is also worth pointing out that to discuss SO($2N + 1$) groups we have to adjoint Γ_0 to the Clifford algebra generated by $\Gamma_1 \ldots \Gamma_{2N}$. There is no chirality operator in this model. So the irreducible representation of SO($2N + 1$) is 2^{2n} dimensional. The irreducible spinor representation can also be constructed in the same manner.

theories, both particles and antiparticles belong to the same irreducible representation of the gauge group. So, to generate all possible mass terms, we must write down gauge invariant Yukawa couplings of the form

$$\tilde{\psi} BC^{-1}\Gamma_\mu \psi \phi_\mu, \qquad \tilde{\psi} BC^{-1}\Gamma_\mu \Gamma_\nu \Gamma_\lambda \psi \phi_{\mu\nu\lambda}, \tag{7.3.1}$$

where $\tilde{\psi}$ stands for the transpose of ψ, B is the equivalent of the charge conjugation matrix for SO(10); and C is the Dirac charge conjugation matrix. The ϕ_μ, $\phi_{\mu\nu\lambda}$, etc. are the Higgs mesons belonging to totally irreducible antisymmetric representations of appropriate dimensions of SO($2N$), i.e., ϕ_μ is 2^N dimensional; $\phi_{\mu\nu\lambda}$ is $[2N(2N-1)(2N-2)/6]$ dimensional, etc. (for $N = 5$, ϕ_μ, $\phi_{\mu\nu\lambda}$ are, respectively, 10 and 120 dimensional). To see the need for inserting B we note that under the group transformation

$$\begin{aligned}
\delta\psi &= i\varepsilon_{\mu\nu}\Sigma_{\mu\nu}\psi, \\
\delta\psi^+ &= -i\varepsilon_{\mu\nu}(\Sigma_{\mu\nu}\psi)^+ = -i\varepsilon_{\mu\nu}\psi^+\Sigma_{\mu\nu}, \\
\delta\tilde{\psi} &= i\varepsilon_{\mu\nu}\tilde{\psi}\tilde{\Sigma}_{\mu\nu}.
\end{aligned} \tag{7.3.2}$$

Thus, $\tilde{\psi}$ does not transform like a conjugate spinor representation of SO($2N$). However, if we introduce a $2^N \times 2^N$ matrix

$$B^{-1}\tilde{\Sigma}_{\mu\nu}B = -\Sigma_{\mu\nu}. \tag{7.3.3}$$

Then

$$\delta(\tilde{\psi}B) = -i\varepsilon_{\mu\nu}\tilde{\psi}B\Sigma_{\mu\nu}. \tag{7.3.4}$$

Thus, B has the correct transformation property under SO($2N$). It is easy to see that eqn. (7.3.3) requires that

$$B^{-1}\tilde{\Gamma}_\mu B = \pm\Gamma_\mu. \tag{7.3.5}$$

We will choose the negative sign on the right-hand side. Since the Γ_μ's are represented by symmetric matrices for even μ in the spinor basis of Table 7.1, one obvious representation of B in the spinor space is (for odd N)

$$B = \prod_{\mu=\text{odd}} \Gamma_\mu. \tag{7.3.6}$$

Using eqns. (7.2.12) and (7.3.6) we conclude that

$$B\begin{pmatrix} \psi_0 \\ \psi_{ij} \\ \bar{\psi}_i \\ \psi_i \\ \bar{\psi}_{ij} \\ \bar{\psi}_0 \end{pmatrix} = \begin{pmatrix} \bar{\psi}_0 \\ -\bar{\psi}_{ij} \\ \psi_i \\ -\bar{\psi}_i \\ \psi_{ij} \\ \psi_0 \end{pmatrix}. \tag{7.3.7}$$

Since

$$\tilde{\psi}BC^{-1}\Gamma_\mu\psi = \langle\psi^*|BC^{-1}\Gamma_\mu|\psi\rangle.$$

The Yukawa coupling (for $N = 5$) of f

$$\phi_\mu \langle \psi^* | BC^{-1}\Gamma_\mu | \psi \rangle = \sum_{j=\pm} \langle \psi_j^* | BC^{-1}\Gamma_\mu | \psi_{\bar{\mu}} \rangle \phi_\mu, \tag{7.3.8}$$

where all quantities are listed in this and the previous sections. Note that in writing eqn. (6.7.22) we used the fact that $[\Gamma_0, B\Gamma_\mu] = 0$. To get fermion masses all we have to do is set $\langle \phi_\mu \rangle \neq 0$ for appropriate μ and evaluate $\langle \psi_+^* | BC^{-1}\Gamma_\mu | \psi_+ \rangle$ using the anticommutation relations of χ_j's and the fact that $\chi_j|0\rangle = 0$. In the next section we give explicit examples for the case of the SO(10) grand unified group.

Fermion Masses in SO(10); an Application

As an explicit application of our techniques we will calculate the fermion masses for SO(10) theory with Higgs mesons belonging to both 10-dimensional (ϕ_μ) and 120-dimensional ($\phi_{\mu\nu\lambda}$) representations. Before doing that we would like to identify the various particle states belonging to the 16-dimensional spinor representation of SO(10). We identify

$$\psi_0 = \nu_L^c, \qquad \bar{\psi}_i = \begin{pmatrix} d_1^c \\ d_2^c \\ d_3^c \\ e^- \\ \nu \end{pmatrix}, \qquad \psi_{ij} = \begin{pmatrix} 0 & u_3^c & -u_2^c & u_1 & d_1 \\ & 0 & u_1^c & u_2 & d_2 \\ & & 0 & u_3 & d_3 \\ & & & 0 & e^+ \\ & & & & \end{pmatrix}. \tag{7.3.9}$$

We remind the reader that $\bar{\psi}_i$ and ψ_{ij} are the usual SU(5) representations of Georgi and Glashow (see Section 5.4).

(a) *10-Dimensional Higgs*

Since we want color symmetry unbroken the components of ϕ_μ, which can acquire v.e.v.'s, are ϕ_9 and ϕ_{10}. Let us set

$$\langle \phi_9 \rangle = v_1, \qquad \langle \phi_{10} \rangle = v_2. \tag{7.3.10}$$

We have to evaluate

$$\mathscr{L}_{\text{mass}} = -i\kappa v_1 \langle \psi_+^* | B(\chi_5 - \chi_5^\dagger) | \psi_+ \rangle + \kappa_2 \langle \psi_+^* | B(\chi_5 + \chi_5^\dagger) | \psi_+ \rangle. \tag{7.3.11}$$

Using eqns. (7.2.12) and (7.3.6), and after some algebra, we obtain

$$\mathscr{L}_{\text{mass}} = \kappa(v_2 - v_1)(\bar{d}_L d_R + \bar{e}_L e_R) + \kappa(v_2 + v_1)(\bar{u}_L u_R + \bar{\nu}_L \nu_R). \tag{7.3.12}$$

We thus see that

$$m_d = m_e, \qquad m_u = m_\nu. \tag{7.3.13}$$

It is also easily seen that, if there are more than one-spinor multiplets of fermions corresponding to different families of particles, the mass matrix is symmetric. Also, we remind the reader that the relevant neutrino mass here is the Dirac mass.

(b) *120-Dimensional Case*

The invariant Yukawa coupling of O(10) spinor fermions ψ_+ to the 120-dimensional Higgs field $\phi_{\mu\nu\lambda}$ can be written down as follows:

$$\mathscr{L}_Y^{ab} = \kappa_{ab}\tilde{\psi}_a BC^{-1}\Gamma_\mu\Gamma_\nu\Gamma_\lambda\psi_b\phi_{\mu\nu\lambda}, \tag{7.3.14}$$

where a and b stand for the different generations of fermions. Using the fact that

$$\tilde{B} = -B \quad \text{and} \quad \tilde{C} = -C,$$

we find that

$$\mathscr{L}_Y^{ab} = -\mathscr{L}_Y^{ab}. \tag{7.3.15}$$

So, if we restrict ourselves to only one generation, it does not contribute to the fermion masses. It will, however, contribute to mixings between various generations. To analyze the kind of mixing pattern that this representation generates, we note that under SU(5), $\phi_{\mu\nu\lambda}$ breaks up as follows:

$$\{120\} = \{45\} + \{45^*\} + \{10\} + \{10^*\} + \{5\} + \{5^*\}. \tag{7.3.16}$$

Thus, we can choose either of the following patterns of vacuum expectation values:

(i) A linear combination of $\{45\}$ and $\{5\}$ acquires v.e.v.
This means that we have

$$\langle\phi_{789}\rangle \neq 0, \qquad \langle\phi_{7810}\rangle \neq 0. \tag{7.3.17}$$

Inserting this into the Yukawa couplings and proceeding with the calculation as in the case of $\{10\}$-dimensional Higgs, we find that the mixing between the various generations of quarks and leptons are related as follows:

$$m_{d_a} = 3m_{E_aE_b}, \tag{7.3.18}$$

$$m_{u_au_b} = 3m_{\nu_a\nu_b}, \tag{7.3.19}$$

where $m_{\nu_a\nu_b}$ stands for mixing terms in the mass matrix between generations a and b; d_a means the $-1/3$ charged quark of an ath generation and similarly for u, E^-, ν.

(ii) Only the SU(5) $\{45\}$-dimensional Higgs acquires v.e.v.
This means the following fields acquire v.e.v.'s:

$$\langle\phi_{789}\rangle = -3\langle\phi_{2k-1,2k,9}\rangle, \qquad k = 1, 2, 3,$$

$$\langle\phi_{7890}\rangle = -3\langle\phi_{2k-01,2k,10}\rangle, \qquad k = 1, 2, 3. \tag{7.3.20}$$

In this case the miximg pattern is very different from case (i); we get

$$\begin{aligned} m_{E_aE_b} &= 3m_{d_ad_b}, \\ m_{\nu_a\nu_b} &= 0, \\ m_{u_au_b} &\neq 0. \end{aligned} \tag{7.3.21}$$

(c) *126-Dimensional Case*

The invariant Yukawa coupling in this case involves five Γ matrices

$$\mathscr{L}_Y^{ab} = \kappa_{ab}\tilde{\psi}_a BC^{-1}\Gamma_\mu\Gamma_\nu\Gamma_\lambda\Gamma_\sigma\Gamma_\alpha\psi_b\phi_{\mu\nu\lambda\sigma\alpha}. \tag{7.3.22}$$

We first note that $\mathscr{L}^{ab} = \mathscr{L}^{ba}$. Thus, this makes a symmetric contribution to various masses. We may choose the following vacuum expectation values for the Higgs field to be consistent with the local color SU(3) symmetry remaining exact

$$\langle\phi_{1278\mu}\rangle = \langle\phi_{3478\mu}\rangle = \langle\phi_{5678}\mu\rangle \neq 0, \tag{7.3.23}$$

where $\mu = 9$ or 10. Substituting this into the Yukawa couplings we get for one generation ($a = b = 1$) the following kind of mass relations:

$$m_e = 3m_d, \qquad m_u = 3m_\nu. \tag{7.3.24}$$

Neutrino Mass in the SO(10) Model

As we saw from eqn. (7.3.13) the smallness of the neutrino mass is not easy to understand. The solution to this was first suggested by Gell-Mann, Ramond, and Slansky [6], using the {126}-dimensional representation of SO(10). Their idea was to give nonzero vacuum expectation value to the SU(5) singlet part of {126} (note the decomposition of {126} under SU(5)):

$$\{126\} = \{1\} + \{5\} + \{10\} + \{10^*\} + \{50\} + \{50^*\}. \tag{7.3.25}$$

Under $SU(2)_L \times SU(2)_R \times SU(4)$ its decomposition can be written as

$$\{126\} = (3, 1, 10) + (1, 3, 10) + (2, 2, 15) + (1, 1, 6). \tag{7.3.26}$$

The Higgs representation of eqn. (7.3.26), that acquires vacuum expectation values in this case, is the right-handed triplet (1, 3, 10) as in the left–right symmetric models. Thus, as in this case, the smallness of the neutrino mass can be understood as a consequence of suppression of $V + A$ currents.

In our notation the SU(5) singlet component of $\phi_{\mu\nu\lambda\sigma\alpha}$ has the form $\chi_1^+\chi_2^+\chi_3^+\chi_4^+\chi_5^+$ or $\chi_1\chi_2\chi_3\chi_4\chi_5$. Substituting this in eqn. (7.3.22) we note that this gives a Majorana mass only to the right-handed neutrino $N_R^TCN_R$, whereas Dirac masses arise from the introduction of {10}-dimensional Higgs representations as in eqn. (7.3.13). Thus, in the SO(10) model, we predict (as in eqn. (6.4.18))

$$m_\nu = \frac{m_u^2}{m_{N_R}}. \tag{7.3.27}$$

Table 7.1 Construction of the states belonging to the spinor representation of SO(N) dimensionality.

SO($2N$) spinor state	SU(N) dimension
$\|0\rangle$	1
$\chi_j^+ \|0\rangle$	N
$\chi_j^+ \chi_k^+ \|0\rangle$	$\frac{N(N-1)}{2}$
$\chi_j^+ \chi_k^+ \chi_l^+ \|0\rangle$	$\frac{N(N-1)(N-2)}{6}$
$\vdots$	
$\chi_1^+ \chi_2^+ \ldots \chi_N^+ \|0\rangle$	1
Total	$2N$

An important point worth noting here is that, in deriving eqn. (7.3.27). the mass matrix for neutrinos ν, N is assumed to be of the form in eqn. (6.4.17). But in mass realistic models, minimization of the potential leads to a mass matrix of the form:

$$M = \begin{pmatrix} f\dfrac{k^2}{V_R} & m_D \\ m_D & fV_R \end{pmatrix} \tag{7.3.28}$$

This upsets the see-saw mechanism as was noted in Chapter 6. This can be cured by breaking the D-parity symmetry present in SO(10) model at the GUT scale separately from $SU(2)_R$ symmetry by introducing a $\{210\}$- or $\{45\}$-dimensional Higgs multiplet [5a]. In this case fk^2/V_R in eqn. (7.3.28) is replaced by $fk^2 V_R/M_{\text{GUT}}^2$ which tiny if $V_R \ll M_{\text{GUT}}$.

§7.4. Symmetry Breaking Patterns and Intermediate Mass Scales

The SO(10) group has rank 5 where the low-energy standard electro-weak group has rank 4. Because of this there exist several intermediate symmetries through which SO(10) can descend to the $SU(3)_c \times SU(2)_L \times U(1)_Y$ group (G_{321}). To list these various possibilities let us recall that SO(10) has two maximal subgroups which contain the G_{321} group: (i) SU(5) $\times$ U(1); and (ii) $SU(2)_L \times SU(2)_R \times SU(4)_c \times D$; where D is a discrete symmetry which transforms [6], [7]

$$q_L \xrightarrow{D} q_L^c. \tag{7.4.1}$$

In terms of the SO(10) generators thus is given by

$$D = \Sigma_{23}\Sigma_{67}. \tag{7.4.2}$$

The breaking of D-parity has important cosmological, as well as physics [8], implications.

The subsequent stages of symmetry breaking depend on whether we are in Case (i) or Case (ii).

Case (i)

$$SO(10) \to \mathrm{SU}(5) \times \mathrm{U}(1) \to \mathrm{SU}(5) \xrightarrow{M_x} \mathrm{SU}(3)_c \times \mathrm{SU}(2)_L \times \mathrm{U}(1)_Y.$$

For Case (ii) there exists a large variety of possibilities since $\mathrm{SU}(2)_R \times \mathrm{SU}(4)_c \times D$ can break down to $\mathrm{U}(1)_Y \times \mathrm{SU}(3)_c$ in many ways. We see that, unlike the SU(5) grand unified model, the SO(10) model has many ways in which it can appear. So, How do we experimentally test SO(10) grand unification? This question has been addressed in Refs. [7]–[10] where the low-energy constraints on $\sin^2\theta_W$ and α_s are imposed to isolate the values of the intermediate mass scales corresponding to various symmetry breaking chains. There exist two distinct possibilities in Case (ii) depending on whether D-parity and $\mathrm{SU}(2)_R$ local symmetry are broken together or at separate scales.

Let us first deal with Case (i). It has been shown in Case (i) that the proton lifetime is predicted to be less than that of the SU(5) model. Thus, the $\mathrm{SU}(5) \times \mathrm{U}(1)$ intermediate symmetry is ruled out by experiment.

Case (ii) leads to two distinct symmetry breaking patterns if we assume that D-parity and $\mathrm{SU}(2)_R$ break down simultaneously:

$$\text{(iiA)} \quad \mathrm{SO}(10) \xrightarrow[M_U]{} G_{224D} \xrightarrow[M_c]{} G_{2213D} \xrightarrow[M_{R^+}]{} G_{2113} \xrightarrow[M_{R^0}]{} G_{213};$$

$$\text{(iiB)} \quad \mathrm{SO}(10) \xrightarrow[M_U]{} G_{224D} \xrightarrow[M_{R^+}]{} G_{214} \xrightarrow[M_c]{} G_{2113} \xrightarrow[M_{R^0}]{} G_{213}.$$

In both these cases we can write down the formulas for $\sin^2\theta_W(m_W)$ and $\alpha_s(m_W)$ following the same procedure as in Chapter 5 and we find the following for Case (iiA).

Case (iiA)

$$\begin{aligned}
\sin^2\theta_W(m_W) = \frac{3}{8} - \frac{\alpha(m_W)}{48\pi}\Bigg[&(110 + 3T_Y - 5T_L)\ln\frac{M_{R^0}}{m_W} \\
&+ (110 + 3T_{R^0} + 2T_{BL} - 5T_L)\ln\left(\frac{M_{R^+}}{M_{R^0}}\right) \\
&+ (44 + 3T_R + 2T_{B-L} - 5T_L)\ln\left(\frac{M_c}{M_{R^+}}\right) \\
&+ (-44 + 3T_R + 2T_4 - 5T_L)\ln\left(\frac{M_U}{M_c}\right)\Bigg],
\end{aligned} \tag{7.4.3}$$

$$\frac{\alpha(m_W)}{\alpha_s(m_W)} = \frac{3}{8} - \frac{\alpha(m_W)}{16\pi}\Bigg[(66 + T_L + T_Y - \tfrac{8}{3}T_s)\ln\left(\frac{M_{R^0}}{m_W}\right)$$
$$+ (66 + T_L + T_{R^0} + \tfrac{2}{3}T_{BL} - \tfrac{8}{3}T_s)\ln\left(\frac{M_{R^+}}{M_{R^0}}\right)$$
$$+ (44 + T_L + T_R + \tfrac{2}{3}T_{BL} - \tfrac{8}{3}T_s)\ln\left(\frac{M_C}{M_{R^+}}\right)$$
$$+ (44 + T_L + T_R - 2T_4)\ln\left(\frac{M_U}{M_C}\right)\Bigg]. \tag{7.4.4}$$

The T_i's in eqns. (7.4.3) and (7.4.4) denote the contributions of the Higgs boson multiplets to the β-functions for the ith symmetry group. Their contributions have to be included in accordance with the minimal fine tuning and extended survival hypotheses [11]. Already a very important conclusion can be drawn looking at eqns. (7.4.3) and (7.4.4). Note that if we ignore the Higgs contributions to these equations in the range $M_{R^0} < \mu < M_{R^+}$ then M_{R^0} drops out of both equations. (In fact, all the T_i's are small, of order unity, in this mass range.) Thus, an important feature of the SO(10) grand unified models is that M_{R^0} can be as light as possible without affecting the values of $\sin^2\theta_W(m_W)$ and $\alpha_s(m_W)$. This point has been emphasized before in Refs. [10] and [11]. In fact, as was noted in Ref. [10], if all Higgs contributions are ignored then we find

$$M_U = M_5\left(\frac{M_5}{M_{R^+}}\right)^{1/2}, \tag{7.4.5}$$

which is even independent of the scale M_c (where M_5 is the unification scale in the SU(5) model). From this we can conclude that if M_{R^+} is lower than the SU(5) scale, M_U for SO(10) becomes higher, making the proton lifetime longer, i.e.,

$$\tau_{10} = \tau_5\left(\frac{M_5}{M_{R^+}}\right)^2. \tag{7.4.6}$$

Thus, a longer proton lifetime in the context of SO(10) grand unification implies that there must exist an intermediate left–right symmetric scale. Furthermore, we can relate the prediction for $\sin^2\theta_W$ in the SO(10) model to that of the SU(5) model to get an idea of how small M_{R^+} can be. Ignoring the Higgs contributions we find

$$\Delta(\sin^2\theta_W) = \sin^2\theta_{10} - \sin^2\theta_5 = \frac{11\alpha(m_W)}{24\pi}\left[\ln\left(\frac{M_U^2 M_5^5}{M_C^4 M_{R^+}^3}\right)\right]. \tag{7.4.7}$$

We conclude that if $M_C = M_U$ and $M_{R^+} \simeq 1$ TeV then $\Delta(\sin^2\theta_W) \simeq 0.06$ which, on using the SU(5) result for $\sin^2\theta_W$, implies $\sin^2\theta_{10} \simeq 0.27$ [12] which is in conflict with experimental data.

Let us now look at the symmetry breaking pattern (iiB) where $M_{R^+} > M_C$. The equations for $\sin^2\theta_W$ and α_s can be written as follows:

$$\sin^2\theta_W(m_W) = \frac{3}{8} - \frac{\alpha(m_W)}{48\pi}\Bigg[(110 + 3T_Y - 5T_L)\ln\left(\frac{M_{R^0}}{M_W}\right)$$
$$+ (110 + 3T_{R^0} + 2T_{BL} - 5T_L)\ln\left(\frac{M_C}{M_{R^0}}\right)$$
$$+ (22 + 3T_{R^0} + 2T_4 - 5T_L)\ln\left(\frac{M_{R^+}}{M_C}\right)$$
$$+ (-44 + 3T_R + 2T_4 - 5T_L)\ln\left(\frac{M_U}{M_{R^+}}\right)\Bigg], \tag{7.4.8}$$

$$\frac{\alpha(m_W)}{\alpha_s(m_W)} = \frac{3}{8} - \frac{\alpha(m_W)}{16\pi}\Bigg[(66 + T_L + T_Y - \frac{8}{3}T_s)\ln\left(\frac{M_C}{M_{R^0}}\right)$$
$$+ (66 + T_L + T_{R^0} + \tfrac{2}{3}T_{BL} - \tfrac{8}{3}T_s)\ln\left(\frac{M_C}{M_{R^0}}\right)$$
$$+ (66 + T_L + T_{R^0} - 2T_4)\ln\left(\frac{M_{R^+}}{M_C}\right)$$
$$+ (44 + T_L + T_R - 2T_4)\ln\left(\frac{M_U}{M_{R^+}}\right)\Bigg]. \tag{7.4.9}$$

Again, as in chain (iiA), we note that, ignoring the Higgs contributions, the equations become independent of M_{R^0} and, therefore, a low M_{R^0} is allowed by the second symmetry breaking chain without conflicting with low-energy data. However, as far as the proton lifetime is concerned, the result is sensitive to the Higgs contributions and no umambiguous prediction can be made; nevertheless, no low values of M_C or M_{R^+} can be obtained.

We conclude that, within the framework of conventional SO(10) grand unification, the only interesting physics—beyond the predictions of the standard subgroup G_{123}–is a low mass (≈ 300 GeV to 1 TeV) right-handed neutral Z_R-boson [9]. The charged right-handed W-boson W_{R^+} has to be extremely heavy ($\simeq 10^{11}$ GeV) due to the constraints of grand unification.

§7.5. Decoupling Parity and $SU(2)_R$ Breaking Scales

So far, in our discussion of symmetry breaking in the left–right symmetric models, we have broken parity and $SU(2)_R$ symmetry at the same scale. But, in a recent paper, Chang, Mohapatra, and Parida have suggested [7] an alternative approach where, by including a real parity odd $SU(2)_R$ singlet σ-field in the theory, we can decouple the parity and $SU(2)_R$ breaking scales.

The method can be illustrated in the case of the left–right symmetric model by choosing the following set of Higgs multiplets:

$$\Delta_L(3, 1, +2) + \Delta_R(1, 3, +2), \qquad (7.5.1)$$
$$\phi(2, 2, 0),$$
$$\sigma(1, 1, 0).$$

The relevant part of the Higgs potential can be written as

$$V' = -\mu^2\sigma^2 + \lambda\sigma^4 + m\sigma(\Delta_L^+\Delta_L - \Delta_R^+\Delta_R) + \mu_\Delta^2(\Delta_L^+\Delta_L + \Delta_R^+\Delta_R) + \text{quartic terms}. \qquad (7.5.2)$$

From eqn. (6.7.48) we observe that parity symmetry is broken by $\langle\sigma\rangle = \mu/\sqrt{2\lambda}$ which then makes the Δ_L, Δ_R mass terms asymmetric. By choosing $m > 0$, we see that, Δ_R mass can be negative if $\mu_4^2 < m\mu/\sqrt{2\lambda_2}$ and this, then, triggers $SU(2)_R$ breaking at a different (lower) scale $M_R = (\mu_4^2 - m\mu/\sqrt{2\lambda})/\gamma$ where γ is a function of scalar couplings. Thus, parity and $SU(2)_R$ scales are decoupled.

This has the following implications: (i) the Higgs masses are parity asymmetric even above the $SU(2)_R$ breaking scale; and (ii) this asymmetry leads to $g_L \neq g_R$ at the scale M_R which, therefore, has important implications for low-energy phenomenology.

It is interesting that this idea can be embedded in an SO(10) grand unified model. To do this, we have to search for an irreducible representation of SO(10) which has an $SU(2)_R$ gauge singlet parity odd field. Such a representation turns out to be the {210}-dimensional representation of SO(10). The {210}-dimensional representation is denoted by the totally antisymmetric fourth rank tensor $\phi_{\mu\nu\lambda\sigma}$. The relevant D-parity odd component is

$$\sigma = \phi_{78910}. \qquad (7.5.3)$$

Thus, if we break $SO(10) \xrightarrow[M_u]{\{210\}} SU(2)_L \times SU(2)_R \times SU(4)_C$, we break parity symmetry which changes the pattern of gauge symmetry breaking. These have been studied in detail in Refs. [7] and [8]. A major outcome of detailed analysis of various SO(10) breaking chains is that we can now lower both the $SU(4)_C$ as well as W_R, Z_R scales, which then imply observable right-handed current effects at low energies. For detailed discussion of these ideas we refer the reader to Refs. [7] and [8]; but we simply note the symmetry breaking chain favorable for low-energy phenomenology

$$SO(10) \xrightarrow[\{54\}]{M_u} SU(2)_L \times SU(2)_R \times SU(4)_C \times P$$
$$M_P \downarrow \{210\}$$
$$SU(2)_L \times SU(2)_R \times SU(4)_C$$
$$M_C \downarrow \{210\}$$
$$SU(2)_L \times U(1)_R \times U(1)_{B-L} \times SU(3)_C$$
$$M_{R^0} \downarrow \{126\}$$
$$SU(2)_L \times U(1)_Y \times SU(3)_c.$$

This leads to $M_{R^+} \simeq M_C \simeq 10^5$ GeV, $M_{R^0} \simeq 1$ TeV for $M_U \simeq 10^{16}$ GeV and $\sin^2\theta_W \simeq 0.227$ for a D-parity breaking scale of 10^{14} GeV. In this analysis [8] the two-loop β-function effects have been taken into account. This leads to testable experimental predictions such as $K_L^0 \to \mu\bar{\mu}$ decay with $B(K_L^0 \to \mu\bar{\mu}) \simeq 10^{-9}$ and neutron–antineutron oscillation with mixing time $\tau_{N-\bar{N}} \simeq 10^7$–$10^8$ s. The proton lifetime is now predicted to be $\simeq 10^{35\pm 2}$ yr which may barely be within reach of the on-going experiments. If proton decay can be measured a further check of this symmetry breaking pattern comes from measurement of the branching ratios, i.e.,

$$\Gamma(p \to e^+\pi^0) = \Gamma(p \to \bar{\nu}\pi^+), \tag{7.5.4}$$

which is different from the SU(5) prediction of

$$\tfrac{2}{5}\Gamma(p \to e^+\pi^0) = \Gamma(p \to \bar{\nu}\pi^+). \tag{7.5.5}$$

References

[1] H. Georgi, in *Particles and Fields* (edited by C. E. Carlson), A.I.P., 1975; H. Fritzsch and P. Minkowski, *Ann. Phys.* **93**, 193 (1975).
[2] Y. Tosa and S. Okubo, *Phys. Rev.* **D23**, 2486 (1981).
[3] R. N. Mohapatra and B. Sakita, *Phys. Rev.* **D21**, 1062 (1980).
[4] F. Wilczek and A. Zee, *Phys. Rev.* **D25**, 553 (1982).
[5] M. Gell-Mann, P. Ramond, and R. Slansky, in *Supergravity* (edited by D. Freedman *et al.* North-Holland, Amsterdam, 1980;
See also T. Yanagida, K.E.K. preprint, 1979;
R. N. Mohapatra and G. Senjanovic, *Phys. Rev. Lett.* **44**, 912 (1980).
[5a] D. Chang and R. N. Mohapatra, *Phys. Rev.* **D32**, 1248 (1985).
[6] T. W. B. Kibble, G. Lazaridis, and Q. Shafi, *Phys. Rev.* **D26**, 435 (1982).
[7] D. Chang, R. N. Mohapatra, and M. K. Parida, *Phys. Rev. Lett.* **52**, 1072 (1984); *Phys. Rev.* **D30**, 1052 (1984).
[8] D. Chang, R. N. Mohapatra, J. Gipson, R. E. Marshak, and M. K. Parida, *Phys. Rev.* **D31**, 1718 (1985).
[9] R. Robinett and J. L. Rosner, *Phys. Rev.* **D25**, 3036 (1982).
[10] Y. Tosa, G. C. Branco, and R. E. Marshak. *Phys. Rev.* **D28**, 1731 (1983).
[11] R. N. Mohapatra and G. Senjanovic, *Phys. Rev.* **D27**, 1601 (1983);
F. del Aguila and L. Ibanez, *Nucl. Phys.* **B177**, 60 (1981).
[12] T. Rizzo and G. Senjanovic, *Phys. Rev.* **D25**, 235 (1982).

CHAPTER 8

Technicolor and Compositeness

§8.1. Why Compositeness?

In the previous chapters we have emphasized that while the success of the $SU(2)_L \times U(1)_Y \times SU(3)_C$ model has indicated that the unified gauge theories are perhaps the right theoretical framework for the study of quark–lepton interactions, it still leaves a lot of questions unanswered. Some of the outstanding questions are:

(a) the nature of the Higgs bosons and the origin of electro-weak symmetry breaking;
(b) the apparent superfluous replication of quarks and lepton (and even Higgs bosons if electro-weak symmetry is higher); and
(c) the origin of fermion masses which are much smaller than the scale of electro-weak symmetry breaking: for instance, $m_{e,u,d} \sim 10^{-5}\,\Lambda_W$.

Historically, this is similar to the situation that existed in the domain of strong interactions in the early 1960s before Gell-Mann and Zweig introduced the quarks as the substructure of baryons and mesons. Most of the observed hadronic spectrum received a simple and elegant explanation in terms of bound states of the quarks and antiquarks. Crucial to the success of the quark idea was the introduction of the concept of color and non-abelian gauge theory of color interaction which provided a solid theoretical framework for quark interactions. We may extrapolate this idea to the domain of weak interactions and ask whether the Higgs bosons, quarks, and leptons (perhaps even the W, Z, etc.) are composites of a new set of fermions and bosons (which we will call preons). There must, then, exist a new kind of strong interaction that binds the preons to form the composites, i.e., quarks, leptons, and Higgs bosons. Since we do not know anything about this hypothetical interaction (which has been called by a variety of names in the literature such as metacolor, technicolor, hypercolor, etc.; we will use the word hypercolor) we will assume that its properties are very similar to that of QCD.

(a) It (QHCD) is described by an unbroken non-abelian gauge symmetry under which preons are nonsinglets.
(b) Like QCD, it confines nonsinglet composites and only asymptotic states correspond to a singlet of QHCD. Quarks, leptons, and Higgs bosons will be singlets under QHCD.

Following the existing literature we will present our discussion in two stages. First, we will discuss partial compositeness where only Higgs bosons are treated as composites of new kinds of fermions (to be called techni-quarks) and quarks and leptons are left elementary. This has generally gone under the name of technicolor. Second, we will introduce total compositeness when quarks, leptons, Higgs (techni-quarks), etc., will all be treated as composites of preons.

§8.2. Technicolor and Electro-Weak Symmetry Breaking

The goal of dynamical symmetry breaking is to remove all scalar bosons from electro-weak physics and replace them by a new set of fermions. In this process we may learn about the deeper structure of weak interactions. The Higgs fields, which led to such a successful picture of electro-weak symmetry breaking, are therefore to be thought of as bilinear composites of these new sets of fermions. For purposes of easy identification, and to distinguish them from ordinary quarks and leptons, we will call them techni-fermions. For the formation of composites we need a strong binding force. We cannot use the known strong interactions for this purpose since the new effective composites must be present at an energy scale of 100 GeV or more, whereas the usual quantum chromodynamic interactions become strong only at $\Lambda_{QCD} \sim (1-2)$ 10^{-1} GeV. We, therefore, need to introduce a force (to be called technicolor) which becomes strong in a scale of 1 TeV or so. This fact can be represented by the following expressions:

$$\frac{g_S^2}{4\pi}(\Lambda_{QCD} \approx 100 \text{ MeV} \geq 1, \tag{8.2.1}$$

where

$$\frac{g_{TC}^2}{4\pi}(\Lambda_{TC} \leq 1 \text{ TeV}) \geq 1. \tag{8.2.2}$$

This is the general picture of dynamical symmetry breaking introduced by Susskind [1] and Weinberg [1]. To realize these ideas in the building of electro-weak models we start by giving the gauge group, and the assignment of fermions under it, as follows (we consider one generation of fermions). The gauge group is

$$G = G_{WK} \times G_{St} \times G_{TC}. \tag{8.2.3}$$

Let us choose $G_{St} \times G_{WK} = \mathrm{SU}(3)_C \times \mathrm{SU}(2)_L \times \mathrm{U}(1)_Y$. We keep $G_{TC} = \mathrm{SU}(N_{TC})$, $N_{TC} > 3$. Let us denote by $\psi_L \equiv \binom{\nu}{e^-}_L$; e_R^-; $Q_L \equiv \binom{u}{d}_L$, u_R, d_R the observed quarks and leptons of first generation. Let Q_T denote the techni-quarks. We can choose their transformation properties under $G \equiv \mathrm{SU}(N_{TC}) \times \mathrm{SU}(3)_C \times \mathrm{SU}(2)_L \times \mathrm{U}(1)_Y$ as follows (this model has anomalies and will therefore need to be modified; but this will not effect the essential points of the ensuing discussion)

$$\begin{aligned}
\psi_L &\equiv (1, 1, 2, -1),\\
e_R^- &\equiv (1, 1, 1, -2),\\
Q_L &\equiv (1, 3, 2, 1/3),\\
u_R &\equiv (1, 3, 1, 4/3),\\
d_R &\equiv (1, 3, 1, -2/3),\\
Q_{TL} &\equiv (N, 1, 2, 1/3),\\
U_{TR} &\equiv (N, 1, 1, 4/3),\\
D_{TR} &\equiv (N, 1, 1, -2/3).
\end{aligned} \tag{8.2.4}$$

We then assume that the technicolor interactions behave in a manner very similar to quantum chromodynamics.

First of all, since $\mathrm{SU}(N)_{TC}$ is an unbroken non-abelian gauge symmetry it is asymptotically free. Like QCD, it is expected to have the property of confinement which means that all technicolor nonsinglet states are confined and do not appear in collisions or as decay products. Finally, like QCD, it will break chiral symmetry of techni-quarks by formation of condensates, i.e.,

$$\langle \bar{Q}_T Q_T \rangle = \Lambda_{TC}^3 \neq 0. \tag{8.2.5}$$

It is now clear that since Q_{TL} transforms as a nonsinglet under $\mathrm{SU}(2)_L \times \mathrm{U}(1)_Y$ it will break the electro-weak gauge symmetry down to $\mathrm{U}(1)_{\mathrm{em}}$ at the scale Λ_{TC}. This will give rise to the masses for the W- and Z-bosons. Before we discuss how this explicitly comes about in a model without elementary scalars, it is worth pointing out that since we would expect $M_W \simeq g\Lambda_{TC}$ observations indicate $\Lambda_{TC} \approx 300$ GeV. This implies that if there is grand unification of both QCD and QTCD (technicolor dynamics), i.e., $g_S = g_{TC}$ at a scale $\mu \gg \Lambda_{TC}, \Lambda_{\mathrm{QCD}}$, then the group $\mathrm{SU}(N)_{TC}$ must be bigger than SU(3) and the number of techni-fermions must be smaller than the number of ordinary fermions.

Let us try to discuss the Higgs–Kibble mechanism for the case with no elementary Higgs bosons. The W-bosons will couple to Q_{TC} as follows:

$$\mathscr{L}_{W,T} = \mathbf{W}_\mu\left(-\frac{ig}{2}\bar{Q}_{TL}\gamma_\mu\boldsymbol{\tau}Q_{TL}\right)$$

$$-\frac{ig'}{6}B_\mu(\bar{Q}_{TL}\gamma_\mu Q_{TL} + 4\ \bar{U}_{TR}\gamma_\mu U_{TR} - 2\bar{D}_{TR}\gamma_\mu D_{TR}). \tag{8.2.6}$$

As a result of symmetry breakdown there will be composite techni-Goldstone bosons which will have odd parity and we can denote them by $\boldsymbol{\pi}_T$. By using the ideas of PCAC they will dominate the axial techni-current, i.e.,

$$\tfrac{1}{2}\bar{Q}_T\boldsymbol{\tau}\gamma_\mu\gamma_5 Q_T = F_\pi\partial_\mu\boldsymbol{\pi}_T + \cdots. \tag{8.2.7}$$

Where $F_\pi \approx \Lambda_{TC}$ up to some numerical factor of order 1. We can therefore write

$$\mathscr{L}_{W,T} = \mathbf{W}_\mu\left(-\frac{igF_\pi}{2}\partial_\mu\boldsymbol{\pi}_T\right) - \frac{ig'}{6}B_\mu(3F_\pi\partial_\mu\pi_T^3) + \cdots \tag{8.2.8}$$

$$\equiv -\frac{igF_\pi}{2}W_\mu^+\partial_\mu\pi_T^- + \text{h.c.} - \frac{iF_\pi}{2}\partial_\mu\pi_T^3(gW_\mu^3 + g'B_\mu) + \cdots, \tag{8.2.9}$$

where ellipses indicate the rest of the interactions. The $W - \pi_T$ bilinears in eqn. (8.2.9) will contribute to the self-energy of the fields W_μ^+ and $Z_\mu = (1/\sqrt{g^2 + g'^2})(gW_\mu^2 + g'B_\mu)$ through the diagram of Fig. 8.1. Let the massless propagator for the W^+ be chosen as

$$\Delta_{v,0}^{W^+} = \frac{\delta_{\mu\nu} - q_\mu q_\nu/g^2}{q^2 + i\varepsilon}. \tag{8.2.10}$$

Since each vertex with $W - \pi_T$ contributes $(gF_\pi/2)q_\mu$ we can evaluate the modification to the propagator and find

$$\Delta_{\mu\nu}^{W^+} = \frac{\delta_{\mu\nu} - q_\mu q_\nu/q^2}{q^2 + g^2F_\pi^2/4}. \tag{8.2.11}$$

Thus, the W-boson propagator picks up a pole in q^2 at $q^2 = -g^2F_\pi^2/4$. This gives a mass to the W-boson

$$M_W = \frac{gF_\pi}{2}. \tag{8.2.12}$$

Similarly, we find

$$M_Z = \sqrt{g^2 + g'2}\frac{F_\pi}{2}. \tag{8.2.13}$$

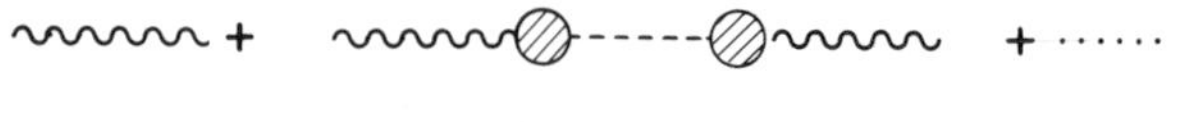

Figure 8.1

If we identify $F_\pi = v$ in eqn. (3.1.12), Chapter 3, these are precisely the gauge boson masses in the elementary Higgs picture. Thus, the electro-weak symmetry breaking has been achieved without the need for elementary scalars. The correct mass relation between W and Z with $\rho_W = 1$ is also achieved. Thus, replacing the Higgs bosons by an underlying strongly interacting fermion leads to the same picture of electro-weak symmetry breaking. The Goldstone bosons, which are now composite, will be absorbed by the gauge bosons. The underlying strong interaction will, however, give its own composite spectrum which are technicolor singlets but consist of the techni-quarks Q. We will call them techni-hadrons. Typically, their masses will be of the order of the scale of technicolor interactions, Λ_T, i.e., 1 TeV or so, unless they are protected by a symmetry to have zero or lighter mass. This will, in general, depend on the global symmetry of the technicolor theory. However, one model-independent statement can be made, i.e., since we expect the chiral symmetry of the techni-world to be spontaneously broken there will be *no* light baryonic techni-composites (i.e., $M_B \approx 1$ TeV). There will, however, be light mesonic techni-composites of which we give some examples in a subsequent section. Using similarity with QCD we may expect the techni-baryon (B_T) mass to be of order

$$M_{B_T} \sim \frac{N}{3} \times \frac{\Lambda_T}{\Lambda_{\text{QCD}}} M_{\text{proton}}. \tag{8.2.14}$$

§8.3. Techni-Composite Pseudo-Goldstone Bosons

In the previous section we chose only one doublet (U, D) of techni-quarks to break the electro-weak symmetry. As a result, in the limit of vanishing electro-weak coupling, we were left with the chiral symmetry $\text{SU}(2)_L \times \text{SU}(2)_R$ which broke down by techni-condensates (eqn. (8.2.5)) to SU(2) leading to three (π_L) Goldstone bosons. All of them were absorbed by the electro-weak gauge bosons $W^\pm$, Z. Thus, no light scalar bosons remains in this theory. If, however, for some reason, the underlying technicolor theory has N ($N > 2$) techni-quarks, then the chiral symmetry group of the techni-world will be $\text{SU}(N)_L \times \text{SU}(N)_R$. In analogy with QCD, this chiral symmetry will be expected to undergo spontaneous symmetry breakdown to $\text{SU}(N)_{L+R}$ leading to $N^2 - 1$ pseudo-scalar massless bosons. Of these, three will be absorbed by the $W_L^\pm$- and Z-bosons leaving $N^2 - 4$ massless bosons. The question we would address now is the following: Are these $N^2 - 4$ pseudo-scalar bosons massless? If not, what are their masses? All these $N^2 - 4$ bosons are actually pseudo-Goldstone bosons since the electro-weak gauge interactions do not respect the entire chiral symmetry $\text{SU}(N)_L \times \text{SU}(N)_R$ prior to spontaneous symmetry breaking. Their masses arise at the one-loop level due to gauge interactions and are expected to be of order

$$M_P^2 \approx \frac{\alpha}{\pi}\Lambda_{TC}^2. \tag{8.3.1}$$

Naively, this implies $M_P \approx 30$ GeV for $\Lambda_{TC} \approx 1$ TeV. These particles have the interesting property that they are both P- and CP-odd like the pion. If the Higgs bosons were elementary then, in extended electro-weak models, they do not have definite P- and CP-properties. In the standard $SU(2)_L \times U(1)$ model with one Higgs doublet the physical Higgs boson has even parity and CP. Thus, experimental study of the P- and CP-properties of the scalar bosons, when they are discovered, can distinguish the technicolor picture from the elementary Higgs picture.

An Example. Let us consider an anomaly-free extension [2] of the toy technicolor model in eqn. (8.2.4) by considering the techni-fermions (within the brackets we give their $SU(N)_H \times SU(3)_C \times SU(2)_L \times U(1)_Y$ quantum numbers)

$$\begin{aligned}
Q_{CL} &\equiv \begin{pmatrix} U_a \\ D_a \end{pmatrix}_L \equiv (N, 3, 2, 1/3), \qquad a = 1, 2, 3,\\
\psi_L &\equiv \begin{pmatrix} N \\ E \end{pmatrix}_L \equiv (N, 3, 2, -1),\\
U_{aR} &\equiv (N, 3, 1, 4/3), \qquad N_R \equiv (N, 1, 1, 0),\\
D_{aR} &\equiv (N, 3, 1, -2/3), \qquad E_R \equiv (N, 1, 1, -2),
\end{aligned} \tag{8.3.2}$$

In the limit of zero electro-weak gauge couplings the chiral symmetry of the techi-world is $SU(8)_L \times SU(8)_R$. This symmetry is spontaneously broken, by the following condensates, down to $SU(8)_{L+R}$:

$$\langle \bar{U}_a U_a \rangle = \langle \bar{D}_a D_a \rangle = \langle \bar{E}E \rangle = \langle \bar{N}N \rangle \neq 0. \tag{8.3.3}$$

This leads to 63 massless pseudo-scalar particles of which 60 are the pseudo-Goldstone bosons according to our previous counting arguments. We can identify them by giving their color and isospin quantum numbers

$$\begin{aligned}
\theta_p^i &\sim \bar{Q}\gamma_5 \lambda \tau^i Q, \qquad \begin{cases} p = 1, \ldots, 8, & \text{color indices,} \\ i = 1, 2, 3, & \text{isospin,} \end{cases}\\
\theta_p &\sim \bar{Q}\gamma_5 \lambda_p Q,\\
T_a^i &\sim \bar{Q}_a \gamma_5 \tau_i L, \qquad a = 1, \ldots, 3, \quad \tau = 1, \ldots, 3,\\
T_a &\sim \bar{Q}_a \gamma_5 L,\\
P_i &\sim (\bar{Q}\gamma_5 \tau^i Q - 3\bar{L}\gamma_5 \tau_i L),\\
P_0 &\sim (\bar{Q}\gamma_5 Q - 3\bar{L}\gamma_5 L).
\end{aligned} \tag{8.3.4}$$

These add up to the 60 pseudo-Goldstone bosons predicted by the theory. We would expect their masses to be smaller than 1 TeV and presumably somewhere in the range of 100 GeV or so. For instance, we expect the masses of $P^{\pm}$ to be

$$\begin{aligned} M_{P^{\pm}}^2 &\approx \frac{3\alpha}{4\pi} M_Z^2 \ln\left(\frac{\Lambda_{TC}^2}{M_Z^2}\right) \\ &\approx (5\text{–}10)\ \text{GeV}^2. \end{aligned} \tag{8.3.5}$$

This is among the lightest pseudo-scalar particles predicted by technicolor theories. The pseudo's with $SU(3)_C$ such as θ's will have higher mass due to the strong interaction corrections. Roughly, their masses will be

$$M_\theta \approx \frac{\sqrt{\alpha_{\text{strong}}\,(1\ \text{TeV})}}{\alpha_{\text{em}}\,(1\ \text{TeV})} M_P.$$

The existence of these lower mass states makes it possible to test these ideas in the current generation of accelerators. Their detailed decay properties and other implications [3] have been studied in the literature.

§8.4. Fermion Masses

In the standard electro-weak model discussed in Chapter 3 the elementary Higgs boson serves two purposes: it breaks the electro-weak symmetry and generates fermion masses. To achieve this purpose in technicolor theories we have to extend the scope of the technicolor models. The reason is that the analog of the Yukawa coupling in this case is a four-Fermi operator

$$\mathscr{L}_{\text{eff}} = \frac{1}{\Lambda_E^2} \bar{q} q \bar{Q} Q. \tag{8.4.1}$$

This can lead to fermion masses M_f, on substituting eqn. (8.2.5), i.e.,

$$M_f \approx \frac{\Lambda_T^3}{\Lambda_E^2}. \tag{8.4.2}$$

An inspection of the models discussed in the previous sections makes it clear that there is no scope for generating such interactions without further extensions. These extensions [3] are called extended technicolor models.

The idea in these models is to introduce new broken gauge interactions between the known fermions and techni-fermions, with the scale of breaking of the new interactions being of order $\Lambda_E > \Lambda_T$. To give an example, consider an $SU(8) \times SU(2)_L \times U(1)_Y$ gauge theory with techni- and known particles

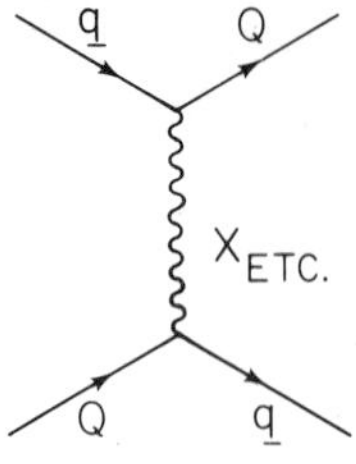

Figure 8.2

of one generation transforming under it as follows:

$$\left.\begin{pmatrix} U_1 & D_1 \\ U_2 & D_2 \\ U_3 & D_3 \\ N & E \end{pmatrix}_L \right\} \text{technicolor} \left\{ \begin{pmatrix} U_1 \\ U_2 \\ U_3 \\ N \end{pmatrix}_R \begin{pmatrix} D_1 \\ D_2 \\ D_3 \\ E \end{pmatrix}_R \right.$$

$$\left.\begin{pmatrix} u_1 & d_1 \\ u_2 & d_2 \\ u_3 & d_3 \\ \nu & e \end{pmatrix}_L \right\} \text{color} \left\{ \begin{pmatrix} u_1 \\ u_2 \\ u_3 \\ \nu \end{pmatrix}_R \begin{pmatrix} d_1 \\ d_2 \\ d_3 \\ e^- \end{pmatrix}_R \right.$$

$SU(2)_L$.

Let us assume that SU(8) gauge theory is broken down to $SU(4)_{TC} \times SU(3)_C$ by some unknown mechanism at a scale Λ_{ETC}. Then the graphs of Fig. 8.2 will induce four-Fermi interactions

$$\mathscr{L}_{\text{eff}} = \frac{1}{\Lambda_{\text{ETC}}^2} \bar{q}\gamma_\mu Q \bar{Q}\gamma_\mu q. \tag{8.4.3}$$

On Fierz rearrangement this gives rise to eqn. (8.4.1) which after techni-condensates form can lead to fermion masses. Clearly, this picture can be extended by including more light quarks by suitable extension of the extended technicolor group G_{ETC}. For instance, by choosing $G_{\text{ETC}} \equiv SU(12)$, we can include two generations of fermions. By appropriate adjustment of the masses of the extended techni-gauge bosons that connect Q with the first and the second generations, we have enough free parameters to generate fermion mixings and masses. It has, however, been argued that these models predict large amounts of flavor violation in weak interactions, e.g., $K_1 - K_2$ mass differences receive too much contribution from the extended technicolor interactions [4].

Regardless of the detailed phenomenological implications the technicolor models have introduced a new concept into physics, i.e., the existence of new strong interactions with a higher scale which can simulate many aspects of low-energy electro-weak interactions without introducing elementary scalars. This picture of "partial compositeness" may be extended to a domain where not only the Higgs bosons but also the quarks, leptons, W-bosons, and Higgs bosons all may be thought of as composites bound by a new strong interaction force analogous to technicolor. This philosophy may be dubbed one of "total compositeness" and will be the subject of the subsequent sections.

§8.5. Composite Quarks and Leptons

Even though there are theoretical reasons to suspect that the quarks and leptons may be composite, as yet, there is no experimental evidence for their compositeness. On the other hand, based on new effects predictable on the basis of compositeness, we can put an upper bound on the compositeness size (or a lower bound on the mass scale of compositeness, which we will call, Λ_H). This is also of great interest to experimentalists as they build accelerators with higher and higher energies to look for new structures in particle interactions.

In order to identify effects due to compositeness we must isolate the effects of known physics on the experimental parameters being studied to a certain level of accuracy. One such parameter where effects of known physics have been computed to a high degree of accuracy, is the $(g - 2)$ of the electron and the muon. Experimentally, $(g - 2)$ is known to a great precision

$$(g - 2)_\mu = (2331848.0 \pm 16.3) \times 10^{-9}. \tag{8.5.1}$$

Theoretical value, which takes into account QED effects to eighth order in the electric charge and other weak effects, for this parameter is given by [6]

$$(g - 2)_{\text{th}} = (2331840.4 \pm 4.0) \times 10^{-9}. \tag{8.5.2}$$

The hadronic contribution to eqn. (8.5.2) is $(140.4 \pm 3.8) \times 10^{-9}$ whereas the electro-weak contribution is $(3.90 \pm 0.02) \times 10^{-9}$.

If there is any effect due to compositeness it must therefore be less than

$$(\delta a)_\mu = [(g - 2)_{\text{expt}} - (g - 2)_{\text{th}}] < 1.5 \times 10^{-8}. \tag{8.5.3}$$

Attempts have been made [7] to estimate the contribution of compositeness to (δa). A crude estimate can be made on the basis of chiral symmetry and dimensional arguments: the latter says that, since the anomalous magnetic moment operator has dimension of inverse mass, we can write

$$\frac{1}{m_\mu}(\delta a)_\mu = f(m_\mu, \Lambda_H) = 0. \tag{8.5.4}$$

Furthermore, since the $(g - 2)$ vanishes in the limit of chiral symmetry we conclude that

$$\lim_{m_\mu \to 0} f(m_\mu, \Lambda_H) = 0. \tag{8.5.5}$$

This implies that the leading contribution to $(\delta a)_\mu$ from compositeness can be written as

$$(\delta a)_\mu \simeq \left(\frac{m_\mu}{\Lambda_H}\right)^2 \leq 10^{-8}. \tag{8.5.6}$$

This implies that $\Lambda_H \geq 1$ TeV. The corresponding value $(\delta a)_e$ for the electron is known to be mush less, i.e., $(\delta a)_e \leq 3.2 \times 10^{-10}$. However, since the analogous formula for $(\delta a)_e$ involves m_e^2, the bound on Λ_H is not as good.

Other sources of information on Λ_H are Bhabha scattering $e^+e^- \to e^+e^-$ off the forward direction. To study these contributions, we may note that in a composite theory, low-energy structure of the interaction Hamiltonian for quarks and leptons is given by the chiral symmetry of the theory. For one family of quarks and leptons this low-energy chiral symmetry is expected to be $SU(2)_L \times SU(2)_R \times SU(4)_C$ [8] or $SU(2)_R \times SU(2)_L \times SU(4)_L \times SU(4)_R$ [9]. Therefore, all four-Fermi operators consistent with these symmetries should appear at low energies. A more conservative approach may be simply to assume that the only electro-weak symmetry below the compositeness scale is $SU(2) \times U(1) \times SU(3)_C$. It may be worth pointing out that at this stage there is no compelling argument for it. In other words, we may assume either $SU(2)_L \times U(1) \times SU(3)_C$ or $SU(2)_L \times SU(2)_R \times U(1)_{B-L} \times SU(3)_C$ or $SU(2)_L \times SU(2)_R \times SU(4)_C$ to constrain the nature of the four-Fermi interactions below the compositeness scale. Below we give a list of possible four-Fermi operators consistent with the $SU(2)_L \times SU(2)_R \times SU(4)_C$ symmetries for the case of two generations: the extensions to other cases are straightforward.

(i) $SU(2)_L \times SU(2)_R \times SU(4)_C$

Let the fermion multiplet F_a be denoted by (a denotes generation).

$$F_a = \begin{pmatrix} u_1 & u_2 & u_3 & \nu \\ d_1 & d_2 & d_3 & e \end{pmatrix}_a, \tag{8.5.7}$$

$$F_{a_L} \equiv (2, 1, 4) \quad \text{and} \quad F_{a_R} \equiv (1, 2, 4),$$

under the $SU(2)_L \times SU(2)_R \times SU(4)_C$ group. The possible invariant four-Fermi operators are then given by

$$O_{1L}^{ab,cd} = (\bar{F}_{a_L}\gamma_\mu\tau_i \times \lambda_p F_{b_L})(\bar{F}_{c_L}\gamma_\mu\tau_i \times \lambda_p F_{d_L}). \tag{8.5.8}$$

Similarly,

$$O_{1L}^{ab,cd} = O_{1(L\leftrightarrow R)}^{ab,cd}, \tag{8.5.9}$$

$$O_{2\varepsilon,\varepsilon'}^{ab,cd} = (\bar{F}_{a_\varepsilon}\gamma_\mu\lambda_p F_{b_\varepsilon})(\bar{F}_{c_{\varepsilon'}}\gamma_\mu\lambda_p F_{d_{\varepsilon'}}) \tag{8.5.10}$$

where $\varepsilon, \varepsilon' = L, R$; $p = 0, \ldots, 15$ and denotes the SU(4) index; τ_i denotes the flavor SU(2) matrices, and λ_p are SU(4) matrices. The effective four-Fermi interaction at low energies can therefore be written as

$$\mathscr{L}_{\text{eff}} = \frac{1}{\Lambda_H^2} \sum_i C_i O_i, \tag{8.5.11}$$

where i goes over all the operators in eqns. (8.5.9) and (8.5.10). Of these, we can pick out the combination, involving electron fields only, that will contribute to Bhabha scattering off the forward angle. If all C_i are chosen to be of order one, then, for a particular value of momentum transfer Q, we can probe the compositeness scale $\Lambda_H \sim Q/\sqrt{\alpha_{\text{em}}}$. This kind of analysis [10] leads to bounds on

$$\Lambda_H \geq 1.5 \text{ TeV}. \tag{8.5.12}$$

The Hamiltonian in eqn. (8.5.11), in general, also leads to flavor changing effects. For instance, if we choose $a = c = 1$ and $b = d = 2$ in eqn. (8.5.10) we can get an operator of the form (for $p = 0$)

$$\mathscr{L}_{\text{eff}}^{\Delta S=2} \equiv \sum_{\varepsilon,\varepsilon'} \frac{C_{2\varepsilon\varepsilon'}}{\Lambda_H^2} d_\varepsilon \gamma_\mu s_\varepsilon \bar{d}_{\varepsilon'} \gamma_\mu s_{\varepsilon'}. \tag{8.5.13}$$

This would contribute to the $K_1 - K_2$ mass difference with a dominant part coming from $\varepsilon = L$, $\varepsilon' = R$. A simple analysis following the literature on the subject [11] can be found

$$\Lambda_H^2 \geq \frac{32\pi^2}{G_F^2 \sin^2 \theta_C M_C^2}. \tag{8.5.14}$$

This implies

$$\Lambda_H \geq 5.5 \times 10^3 \text{ TeV}. \tag{8.5.15}$$

This appears to be the strongest bound. Furthermore, from eqn. (8.5.10), taking $p = 9, \ldots, 14$ and $a = b = 1$ and $c = d = 2$ we find an interaction of type

$$\mathscr{L}_{\text{eff}} = \frac{\tilde{C}}{\Lambda_H^2} \bar{d}\gamma_\mu e \bar{\mu} \gamma_\mu s. \tag{8.5.16}$$

This would contribute to the process $K_L^0 \to \mu\bar{e}$ decay. The present experimental upper limit on the branching ratio for this decay model is [12]

$$B(K_L^0 \to \mu\bar{e}) \leq 10^{-8}. \tag{8.5.17}$$

This implies that $\Lambda_H \geq 100$ TeV. Again we see that these limits on the scale of compositeness are much more stringent than those derived previously.

It must, however, be pointed out that if there exists a family symmetry in the preon model, these limits will be considerably altered and will depend on unknown parameters corresponding to breaking of these symmetries. Thus, the two latter limits on Λ_H are to be taken with a degree of caution.

In summary, the present low-energy and high-energy scattering experiments indicate the scale of compositeness to be somewhere between 1–100 TeV.

§8.6. Light Quarks and Leptons and 't Hooft Anomaly Matching

As we saw in the last section, the scale of compositeness is at least of order 1–100 TeV. On the other hand, the quarks and leptons have masses ranging from 10 MeV to 40 GeV. It is therefore to be expected that the fermion masses owe their origin to a different mechanism than compositeness and, strictly in the absence of those "unknown mechanisms," they should vanish. The disappearance of fermion masses is generally guaranteed by the existence of chiral symmetries such as $\psi \to e^{i\gamma_5\alpha}\psi$, etc. operating on the preons as well as their composites. In other words, to understand the masslessness of quarks and leptons we must ensure that the chiral symmetries of the preonic world must remain unbroken as the composites form. This situation is in contrast to that existing in the case of QCD, where the chiral symmetry of the quarks is dynamically broken by $\bar{q}q$ condensates, which then leads to nucleon mass of order $k\Lambda_{\text{QCD}}$ where k is a number of order 1–10.

It was pointed out by 't Hooft in a classic paper [13] that the requirement that chiral symmetry remain unbroken at the composite level imposes strong constraints on the possible models for compositeness. The origin of the constraints can be understood as follows.

Let there be a global chiral symmetry $G_L \times G_R$ of the preon dynamics. Let $J_{\mu, i_{L,R}}$ be the currents corresponding to this global symmetry group. It is well known [14], [15] that the three point functions involving these currents have a singularity in the q^2-plane at $q^2 = 0$. To be more specific let us define

$$\Gamma^{ijk}_{\mu\nu\lambda}(q, k) = \int \exp(ik \cdot x + iq \cdot y)\, d^4x\, d^4y\, \langle 0|\, T(V^i_\mu(x) V^j_\nu(y) A^k_\lambda(0))|0\rangle. \tag{8.6.1}$$

The internal indices i, j, k can be factored out

$$\Gamma^{ijk}_{\mu\nu\lambda} = d^{ijk}\Gamma_{\mu\nu\lambda}, \tag{8.6.2}$$

where

$$d^{ijk} = T_r[\lambda^i\{\lambda^j, \lambda^k\}_+], \tag{8.6.3}$$

λ's being the generators of the appropriate group. Lorentz invariance implies that

$$\begin{aligned}\Gamma_{\mu\nu\lambda}(q, k) = {} & A_1 q^\tau \varepsilon_{\tau\mu\nu\lambda} + A_2 k^\tau \varepsilon_{\tau\mu\nu\lambda} + A_3 q_\nu q^\sigma k^\beta \varepsilon_{\sigma\beta\mu\lambda} + A_4 k_\nu q^\sigma k^\beta \varepsilon_{\sigma\beta\mu\lambda} \\ & + A_5 q_\mu q^\sigma k^\beta \varepsilon_{\sigma\beta\nu\lambda} + A_6 k_\mu q^\sigma k^\beta \varepsilon_{\sigma\beta\nu\lambda}.\end{aligned} \tag{8.6.4}$$

Following Ref. [15] let us restrict ourselves to the case when $k^2 = q^2$. The invariant amplitude then depends on $(q + k)^2 = k_1^2$ and $k_2^2 = q^2 = k^2$. Bose symmetry implies

$$A_1 = -A_2, \qquad A_3 = -A_6,$$

and

$$A_4 = -A_5. \tag{8.6.5}$$

Now let us impose the vector and axial vector Ward identities, i.e.,

$$\partial_\mu V_\mu = 0$$

and

$$\partial_\mu A_\mu = m\bar{\psi}\gamma_5\psi. \tag{8.6.6}$$

This implies

$$A_1 = \frac{k_1^2 - 2k_2^2}{2} A_4 + k_2^2 A_3,$$

$$A_1 - A_2 = 2mB, \tag{8.6.7}$$

where B is the Fourier transform similar to that in eqn. (8.6.1) with A_μ replaced by $\bar{\psi}\gamma_5\psi$. If the amplitude is assumed to be regular at k_1^2, $k_2^2 \to 0$, then eqn. (8.6.7) implies that for $m \to 0$, $A_1 = A_2 = 0$. Then keeping k_1^2 fixed and letting $k_2^2 \to 0$ we get $A_3(k_1^2, 0) = 0$, and taking $k_1^2 \to 0$ we get $A_3(0, k_2^2) = A_4(0, k_2^2)$. Thus at $k_1^2 = k_2^2 = 0$ the whole amplitude vanishes. This is not acceptable. This implies that either the vector or the axial vector current will not satisfy the Ward identities.

To study the nature of A_i further we assume unsubtracted dispersion relations in k_1^2, i.e.,

$$A_i^{(u)}(k_1^2, k_2^2) = \frac{1}{\pi}\int \frac{\mathrm{Im}\, A_i(s, k_2^2)}{s - k_1^2}\, ds. \tag{8.6.8}$$

It is then clear from eqn. (8.6.7) that, in order to satisfy the Ward identities, A_1 must have a subtraction with subtraction constant

$$A_1 = A_1^{(u)} - \frac{1}{2\pi}\int \mathrm{Im}\, A_4(s, k_2^2)\, ds. \tag{8.6.9}$$

Now going to the point $k_2^2 = 0$, eqns. (8.6.7) and (8.6.8) imply that

$$k_1^2\, \mathrm{Im}\, A_4(k_1^2, 0) = 0$$

and

$$A_1(k_1^2, 0) = -\frac{1}{2\pi}\int \mathrm{Im}\, A_4(s, k_2^2 = 0)\, ds. \tag{8.6.10}$$

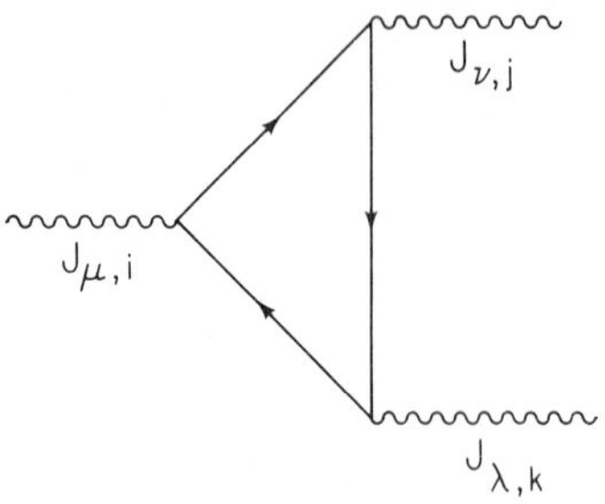

Figure 8.3

Equation (8.6.10) requires that

$$\operatorname{Im} A_4(k_1^2, 0) = c\delta(k_1^2). \tag{8.6.11}$$

This is the 't Hooft anomaly equation. The constant c is obtained by looking at the triangle graphs with internal fermion lines (Fig. 8.3) and picking out the Clebesh–Gordan coefficients. In order to apply it to constrain the confining models of weak interaction, we can choose a normalization such that we get c as follows:

$$c = \sum_f T_r[\lambda_a\{\lambda_b, \lambda_c\}]_f, \tag{8.6.12}$$

where f denotes the number of fermions with some internal global chiral quantum numbers going around in Fig. 8.3.

In our derivation of the 't Hooft anomaly equation (8.6.11) we did not assume anything about the underlying field theory. In other words, the same singularity structure should obtain, both in the component preonic field theory as well as in the composite theory. In the composite theory the singularity can come about in two ways:

(i) by spontaneous breakdown of chiral symmetry so that the resulting Goldstone bosons lead to the k_1^2 singularity of A_3;
(ii) by massless chiral fermions in which case c (defined in eqn. (8.6.12)) must be the same as in the preonic theory, i.e.,

$$c_{\text{preons}} = c_{\text{composite}}. \tag{8.6.13}$$

This is a strong constraint on the composite models. We give examples of this in the following sections.

§8.7. Examples of 't Hooft Anomaly Matching

In this section we give some examples of composite theories and show the impact of anomaly constraints.

(a) Two-flavor QCD

Let us consider the usual quantum chromodynamic theory with two flavors of quarks u, d. In the limit of $m_u, m_d = 0$ the Lagrangian can be written as

$$\mathscr{L}_{\text{QCD}} = -\sum_{\psi=u,d} \bar{\psi}\gamma_\mu\left(\partial_\mu - \frac{ig_s}{2}\boldsymbol{\lambda}_c\cdot\mathbf{B}_\mu\right)\psi - \tfrac{1}{4}\mathbf{f}_{\mu\nu}\cdot\mathbf{f}_{\mu\nu}. \tag{8.7.1}$$

This theory has the global symmetry $\mathrm{SU}(2)_L \times \mathrm{SU}(2)_R \times \mathrm{U}(1)_B \times \mathrm{U}(1)_B^5$. Of these, the $\mathrm{U}(1)_B^5$ which gives the axial baryonic symmetry is broken down to Z_2 by instanton effect. Thus, for our purpose, the continuous global symmetry is $\mathrm{SU}(2)_L \times \mathrm{SU}(2)_R \times \mathrm{U}(1)_B$. Let us now calculate the anomaly coefficient c (eqn. (8.6.12)) for this quark theory. The only three-point functions that are anomalous are $\mathrm{U}(1)_B[\mathrm{SU}(2)_L]^2$, $\mathrm{U}(1)_B[\mathrm{SU}(2)_R]^2$

$$C_{q_R} = C_{q_L} = 3 \times B_q = 1, \tag{8.7.2}$$

where B_q is the baryon number of the quarks and is 1/3.

The composites of this theory are the $\mathrm{SU}(3)_c$ singlet three-quark states, i.e., proton, $P = (uud)$ and neutron, $n = (udd)$. They form representations of the chiral $\mathrm{SU}(2)_L \times \mathrm{SU}(2)_R \times \mathrm{U}(1)_B$. Now there are two ways to satisfy the anomaly constraints:

(i) if the proton and the neutron are massless, then since their baryon number is 1, they imply $C_{N_L} = C_{N_R} = 1$; or
(ii) if chiral symmetry is broken by quark–antiquark condensates. In this case the corresponding zero mass Goldstone bosons, the pions, provide the singularity of the three-point function.

Thus, the 't Hooft anomaly does not prefer one possibility over the other in the case of two-flavor QCD. However, it rules out the possibility considered seriously in the early days whereby a massive parity odd nucleon was supposed to provide the chiral partner of the nucleon in the exact symmetry limit.

It is easy to construct theories where 't Hooft anomaly matching will require that the component global chiral symmetry be spontaneously broken.

(b) Three-flavor QCD

As an example of this kind consider three-flavor QCD with massless flavors u, d, s. The quark theory has the chiral symmetry $G_q = \mathrm{SU}(3)_L \times \mathrm{SU}(3)_R \times \mathrm{U}(1)_B$. The anomalous three-point functions in this case are:

(i) $\mathrm{U}(1)_B[\mathrm{SU}(3)_L]^2$;
(ii) $\mathrm{U}(1)_B[\mathrm{SU}(3)_R]^2$;
(iii) $[\mathrm{SU}(3)_L]^3$;
(iv) $[\mathrm{SU}(3)_R]^3$.

The corresponding values of C at the quark level are

$$c_{\mathrm{i}} = 1 = c_{\mathrm{ii}},$$
$$c_{\mathrm{iii}} = c_{\mathrm{iv}} = d_{abc}. \tag{8.7.3}$$

Let us now look at the composite theory. If the chiral symmetry is not spontaneously broken the only source of singularities for the three-point functions (i)–(iv) is the massless spin 1/2 baryon octet (p, n, Σ^+, Σ^-, Σ^0, Λ, Ξ^0, Ξ^-). Obviously, their contributions to $c_{(\mathrm{i})-(\mathrm{iv})}$ are very different. For instance, $c_{\mathrm{i}} = c_{\mathrm{ii}} = 3$. Thus, the anomalies do not match unless the chiral symmetry is spontaneously broken down to $\mathrm{SU}(2)_L \times \mathrm{SU}(2)_R \times \mathrm{U}(1)_B$, in which case in a manner analogous to Case (a) the anomalies match. It can therefore be argued that the 't Hooft anomaly implies that dynamics will require spontaneous breakdown of chiral symmetry in this case.

(c) Color Flavor Factorized Preon Model

We present a realistic preon model of quarks and leptons and show how for a particular kind of hypercolor binding force for the preons, the 't Hooft anomaly constraints match, implying massless quarks and leptons.

The model is based on the observation of Pati and Salam [16] that the symmetry of the quark–lepton world is $\mathrm{SU}(2)_L \times \mathrm{SU}(2)_R \times \mathrm{SU}(4)_C$ operating on them as followings:

$$\begin{array}{c} \rightarrow \mathrm{SU}(4)_C \\ F = \begin{pmatrix} u_1 & u_2 & u_3 & \nu \\ d_1 & d_2 & d_3 & e \end{pmatrix} \begin{array}{l}\downarrow \\ \mathrm{SU}(2)_L \times \mathrm{SU}(2)_R. \end{array} \end{array}$$

It then is suggestive to consider the quarks and leptons as bound states [17] of two-flavor fermions (F_u, F_d) and four-color bosons ($\phi_1\phi_2\phi_3\phi_l$). Let the binding force be given by an $\mathrm{SU}(N)_H$-hypercolor force analogous to QCD. Assume that, under $\mathrm{SU}(N)_H \times \mathrm{SU}(3)_C \times \mathrm{U}(1)_{\mathrm{em}}$,

$$\begin{pmatrix} F_u \\ F_d \end{pmatrix} \equiv \begin{pmatrix} N & 1 & 1/2 \\ N & 1 & -1/2 \end{pmatrix} \tag{8.7.4}$$

and

$$\phi_i \equiv (\bar{N}, 3, 1/6),$$
$$\phi_l \equiv (\bar{N}, 1, -1/2). \tag{8.7.5}$$

The anomaly constraints on this model were first analyzed by Barbieri, Mohapatra, and Masiero [8]. For this let us note that the preonic theory has the global symmetry

$$G_p^0 = \mathrm{SU}(2)_L \times \mathrm{SU}(2)_R \times \mathrm{U}(1)_F \times \mathrm{U}(4)_\phi. \tag{8.7.6}$$

In order to satisfy the anomalies we must break the $\mathrm{U}(1)_\phi$ either explicitly or spontaneously so that the surviving symmetry is

$$G_p = \mathrm{SU}(2)_L \times \mathrm{SU}(2)_R \times \mathrm{U}(1)_F \times \mathrm{SU}(4)_\phi. \tag{8.7.7}$$

There are only two sets of preonic anomalies in this case corresponding to (i) $U(1)_F[SU(2)_L]^2$ and (ii) $U(1)_F[SU(2)_R]^2$, with the following values for C (choose F's to have fermion number 1):

$$C_{\text{i}} = C_{\text{ii}} = N_H. \tag{8.7.8}$$

At the composite level we have the composites listed earlier as F. Their contribution to anomaly are (if only one set of composites appear)

$$C_{\text{i}} = 4 = C_{\text{ii}}. \tag{8.7.9}$$

This implies that for one family of fermions $N_H = 4$.

It is then easy to note that we could get Ng families by choosing $N_H = 4$ Ng. Thus, in this case, masslessness of quarks and leptons is understood as a consequence of 't Hooft anomaly constraints.

§8.8. Some Dynamical Constraints on Composite Models

In this section we summarize some other dynamical constraints which have been discussed in connection with composite models with vector-like non-abelian forces binding the constituents.

(a) Constraints Due to Heavy Constituent Decoupling

In his original paper on anomaly matching 't Hooft also proposed another kind of constraint on composite models. He argued that if one of the preonic constituents becomes heavier than the scale of compositeness then it should decouple from the entire theory [18]. This would therefore change the global chiral symmetry from, say, $SU(N)_L \times SU(N)_R$ down to $SU(N-1)_L \times SU(N-1)_R$. Then again, in the resulting theory, the anomalies should match. This argument can be applied successively to move preons being made heavy and made to decouple. The anomalies for chiral groups at each stage should match. Thus, anomaly matching should basically be independent of N. Clearly, this is a much stronger constraint on the models and makes it extremely difficult to construct composite models. However, it has since been argued [19] that the nonperturbative effects may not be analytic in the preon mass so that for the large and small preon masses the theories may be quite difficult, thus avoiding this additional constraint.

(b) Mass Inequalities in Vector-like Non-abelian Fermionic Constituent Type Theories

Recently, Weingarten [20], Witten [20], and Nussinov [20] have derived inequalities between composite fermions and scalar boson masses in vector-like confining theories with purely fermionic constituents. Their results can simply be stated as follows: if there is a vector-like confining theory with non-abelian $SU(N)_H$ gauge interaction providing binding of $SU(N)_H$ non-singlet fermionic constituents to form singlet composites, then in this theory the composite fermions and scalar bosons must satisfy the inequality

$$M_B \geq aM_S, \tag{8.8.1}$$

where B and S stand for baryons and scalar composites and a is a number depending on N_H; for instance, for QCD, $a = 3/2$. Let us discuss its implications for composite models. Naively speaking, it implies that if quarks and leptons are to be identified with the composite baryons, eqn. (8.8.1) implies that there must be spin 0 particles lighter than them in the composite spectrum. Of course, if such particles are neutral under the strong and electro-weak $U(1)_Y \times SU(2)_L \times SU(3)_C$ symmetry, then they are harmless provided some weak constraints are satisfied by their couplings [21]. However, in general, they carry color as well as electric charge. Therefore, they are in obvious conflict with experiment. This would therefore rule out most [22] fermionic type vector-like composite quark and lepton models. No such theorem exists for theories with scalar constituents. Therefore, combining this with the philosophical elegance of the color-flavor factorized model would make the fermion–boson type models of the previous section quite interesting candidates for composite models of quarks and leptons. The presence of scalar bosons may, of course, indicate an underlying supersymmetry in nature at the preonic level.

(c) Theorem of Weinberg and Witten [23]

A result of great significance for the composite models is a theorem due to Weinberg and Witten, which says that the only possible massless states in field theory carrying internal quantum numbers must have spin $\leq \frac{1}{2}$ unless there are guage symmetries in the theory. This is important for composite model building since it says, for instance, that the composite spectrum cannot have massless states with spin $J \geq 3/2$, which would, in principle, be allowed by chiral symmetries needed to understand the smallness of quark and lepton masses.

§8.9. Other Aspects of Composite Models

In this section we raise some general issues encountered in building composite models:

(a) the existence of real Goldstone bosons;
(b) composite versus elementary gauge bosons;
(c) possibilities of baryon and lepton nonconservation;
(d) CP-violation and baryon asymmetry of the universe.

(a) *Massless Goldstone Bosons.* We have discussed in the previous sections that global symmetries play a vital role in understanding the small masses of quarks and leptons. The nature of the global symmetry is dictated by the preonic structure of the theory. It may therefore happen (and it often does in actual model building) that there may be global symmetries which have to be broken spontaneously in order to satisfy 't Hooft anomaly constraints. In such a case there will appear real Goldstone bosons. It is important to point out that these Goldstone bosons are similar to the Majoron [21] discussed in Chapter 2. As pointed out there, since they lead to spin-dependent forces in the nonrelativistic limit, the phenomenological constraints on their couplings are relatively mild. To see this we note that the general form of the Goldstone boson coupling to light fermions is of the form

$$\mathscr{L}_{ffG} = \frac{M_f}{V} \bar{f} \gamma_5 f G, \tag{8.9.1}$$

where V is the scale of the symmetry breaking. The strength of the spin-dependent potential generated by it is $\approx (M_f/V)^2$ and is bounded phenomenologically to be less than one part in a million. For $M_f \approx 1$ GeV, this imples $V \geq$ TeV, since $V \approx \Lambda_H$, the compositeness scale, this is quite an acceptable constraint. The astrophysical constraint is, however, more stringent and may require [24] $\Lambda_H \approx 10^8$ GeV. In this case more careful analysis of the model is required to seek additional suppressions. This phenomenon is similar to the one discussed in Chapter 2 in connection with Majoron and is not repeated here.

(b) A second important aspect of composite models is the interpretation of the W-, Z-bosons, photon, color-gauge bosons, etc. There are two general points of view with regard to this: one [25] is to treat the massive gauge bosons W, Z as the composites of preons in much the same manner as the ρ, ω mesons are the quark–antiquark composites in QCD. The masses of the W and Z are then expected to be related to the compositeness scale, Λ_H. The photon and color-gauge bosons are associated with unbroken gauge theories $U(1)_{em} \times SU(3)_C$. In this picture there is no $SU(2)_L \times U(1)$ spontaneously broken gauge theory describing weak interactions. The structure of the four-Fermi weak interaction Lagrangian is given by the residual global $SU(2)_L$ symmetry of weak interaction together with neutrino charge radius interaction [26] as follows.

If $\mathbf{J}_{\mu_L}$ is denoted as the fermionic current associated with weak $SU(2)_L$ then the $SU(2)_L$ invariant current–current interaction is given by

$$\mathscr{L}_{wk}^{(0)} = 4\frac{G_F}{\sqrt{2}} \mathbf{J}_{\mu_L} \mathbf{J}_{\mu_L} \tag{8.9.2}$$

and the neutrino charge–radius piece can be parametrized as follows:

$$\mathscr{L}_{wk}^{(\nu)} = -2\frac{G_F}{\sqrt{2}} \sin^2\theta_W \bar{\nu}\gamma_\mu(1+\gamma_5)\nu J_\mu^{em}. \tag{8.9.3}$$

The combination of these two terms are, of course, well known [26] to give the correct structure of low-energy weak interaction.

The low-energy effective Lagrangian can also be written in terms of $\mathbf{W}_{\mu_L}$-bosons which belong to the weak isospin 1 representation of $SU(2)_L$ along with an additional $\gamma - W^3$ mixing term. Hung and Sakurai [26] have demonstrated that this also gives the correct description of electro-weak physics at low energy. In this picture the Weinberg angle is replaced by the combination $e\lambda/g$ where e is the electric charge, g is the coupling of the $\mathbf{W}$-boson to the weak isospin current $\mathbf{J}_{\mu_L}$, and λ is the $W^3 - \gamma$ mixing parameter.

Several questions have been raised concerning this very interesting alternative approach to weak interactions.

(i) λ is the analog of the $\rho - \gamma$ coupling in QCD which is known to be a very small number ($\sim 10^{-2}$), whereas neutral current data says that $e\lambda/g \approx .23$ implying $\lambda \approx 1$. The question then arises that if QHCD is so similar to QCD, how can we understand this discrepancy? It has been argued that this may be related to the question of chiral symmetry breaking of QCD and nonbreaking in the case of QHCD [27].

(ii) In this model M_W and M_Z are unpredictable. How then can we understand the discovery of W- and Z-bosons with mass values exactly predicted by the standard gauge model? A phenomenological answer to this question is to assume that, exactly like the ρ-dominance of isospin current in strong interactions, the W-boson dominates [28] the weak current. We can then prove that

$$\lambda \simeq \frac{e}{g}, \tag{8.9.4}$$

which leads to a prediction of the W- and Z-masses exactly as in the standard electro-weak model. This also explains the universality of weak interactions.

This model is philosophically different from the standard gauge theory approach and is under a great deal of discussion [29] due to the new events from the CERN $p\bar{p}$ collider experiments such as anomalous $Z \to e^+e^-\gamma$, $\mu^+\mu^-\gamma$, etc.

The other approach to the W-, Z-boson, photon, color boson, etc. is to assume that the preons carry the spontaneously broken electro-weak gauge theory which is then transmitted to their composites such as quarks and leptons in much the same way that the electro-weak interactions are transmitted from quarks to nucleons.

(c) An important conceptual triumph of grand unified theories (or a geometrical picture of unification) is that they provide a natural framework for baryon and lepton number violation. Combined with CP-violating interactions they meet all the basic requirements for understanding a great cosmological riddle—the origin of matter in the universe. If composite models are to provide a unified description of physics, not only at small distances but also on large scales such as the universe, they must provide a basis for understanding the origin of matter. This requires the introduction of baryon-violating interactions as well as *CP*-violation into composite models. While there exist no satisfactory realistic models with the above features, some general intuitive inferences can be listed.

(i) There are two possible sources of *CP*-violation in composite models: (a) the fermion condensates that break chiral symmetries; and (b) the *CP*-violating interactions arising out of hypercolor anomalies. For the former possibility an effective Lagrangian can be written down [30] that provides a realistic superweak model of *CP*-violation and provides a mechanism for baryon generation [31].

(ii) Baryon-violating interactions could also arise through multifermion condensates along with *CP*-violation. The other possibility is the grand unified composite models where hypercolor and color interactions unify at a superheavy scale. This may provide a mechanism for baryon generation of the universe.

References

[1] L. Susskind, *Phys. Rev.* **D20**, 2619 (1979);
S. Weinberg, *Phys. Rev.* **D19**, 1277 (1979).

[2] E. Fahri and L. Susskind, *Phys. Rev.* **D20**, 3404 (1979);
S. Dimopoulos, *Nucl. Phys.* **B169**, 69 (1980).

[3] S. Simopoulos and L. Susskind, *Nucl. Phys.* **B155**, 237 (1979);
E. Eichten and K. Lane, *Phys. Lett.* **90B**, 125 (1980).

[4] S. Dimopoulos and J. Ellis, *Nucl. Phys.* **B182**, 505 (1981).

[5] S. Raby, S. Dimopoulos, and G. Kane, *Nucl. Phys.* **B182**, 77 (1981);
J. Ellis, M. K. Gaillard, D. Nanopoulos, and P. Sikivie, *Nucl. Phys.* **B182**, 529 (1981);
M. A. B. Beg, H. D. Politzer, and P. Ramond, *Phys. Rev. Lett. 43*, 1701 (1979).

[6] J. Calmet, S. Narison, M. Perrottet, and E. DeRafael, *Rev. Mod. Phys.* **49**, 21 (1977);
T. Kinoshita and W. B. Lindquist, *Phys. Rev. Lett.* **41**, 1573 (1981);
T. Kinoshita, B. Nizic, and Y. Okamoto, *Phys. Rev. Lett.* **53**, 717 (1984).

[7] R. Barbieri, L. Maiani, and R. Petronzio, *Phys. Lett.* **96B**, 63 (1980);
S. J. Brodsky and S. D. Drell, *Phys. Rev.* **D22**, 2236 (1980).

[8] See, for instance, models of O. W. Greenberg and J. Sucher, *Phys. Lett.* **99B**, 339 (1981);
R. Barbieri, R. N. Mohapatra, and A. Masiero, *Phys. Lett.* **105B**, 369 (1981).

[9] O. W. Greenberg, R. N. Mohapatra, and M. Yasue, *Phys. Rev. Lett.* **51**, 1737 (1983).

[10] E. J. Eichten, K. D. Lane, and M. E. Peshkin, *Phys. Rev. Lett.* **50**, 811 (1983).

[11] M. K. Gaillard and B. W. Lee, *Phys. Rev.* **D10**, 897 (1974);
G. Beall, M. Bender, and A. Soni, *Phys. Rev. Lett.* **48**, 848 (1982).
These bounds have been discussed in composite model framework by I. Bars, *Nucl. Phys.* **B198**, 269 (1982);
and for a recent discussion of the conditions under which these bounds may be evaded, see

O. W. Greenberg, R. N. Mohapatra, and S. Nussinov, *Phys. Lett.* **148B**, 465 (1984).

[12] B. Weinstein, in *TSIMESS Workshop Proceedings, 1983* (edited by T. Goldman *et al.*), American Institute of Physics, New York, 1983.

[13] G. 't Hooft, in *Recent Developments in Gauge Theories* (edited by ???), Plenum, New York, 1980, p. 135.

[14] S. L. Alder, *Phys. Rev.* **177**, 2426 (1969);
R. Jackiw and J. S. Bell, *Nuovo Cimeno*, **60A**, 47 (1969);
S. L. Alder and W. Bardeen, *Phys. Rev.* **182**, 1517 (1969).

[15] Y. Frishman, A. Schwimmer, T. Banks, and S. Yankielowicz, *Nucl. Phys.* **B177**, 157 (1981).

[16] J. C. Pati and A. Salam, *Phys. Rev.* **D10**, 275 (1974).

[17] J. C. Pati, O. W. Greenberg, and J. Sucher (Ref. [8]).

[18] T. Applequist and J. Carrazone, *Phys. Rev.* **D11**, 2856 (1975).

[19] J. Preskill and S. Weinberg, Texas preprint, 1981.

[20] D. Weingarten, *Phys. Rev. Lett.* **51**, 1830 (1983);
E. Witten, *Phys. Rev. Lett.* **51**, 2351 (1983);
S. Nussinov, *Phys. Rev. Lett.* **51**, 2081 (1983).

[21] They are analogous to the massless Goldstone–Majoron boson suggested by Y. Chikashige, R. N. Mohapatra, and R. D. Peccei, *Phys. Lett.* **98B**, 265 (1981).

[22] For an apparent exception to this argument see an $E(6)$ hypercolor model by Y. Tosa, J. Gibson, and R. E. Marshak, Private communication, 1984.

[23] S. Weinberg and E. Witten, *Phys. Lett.* **96B**, 59 (1980);
See also E. C.G. Sudarshan, *Phys. Rev.* **D** (1981).

[24] D. A. Dicus, E. Kolb, V. Teplitz, and R. Wagoner, *Phys. Rev.* **D17**, 1529 (1978);
M. Fukugita, S. Watamura, and M. Yoshimura, *Phys. Rev. Lett.* **48**, 1522 (1982).

[25] L. Abbott and E. Farhi, *Phys. Lett.* **101B**, 69 (1981);
H. Fritzsch and G. Mandelbaum, *Phys. Lett.* **102B**, 319 (1981);
R. Barbieri, R. N. Mohapatra, and A. Masiero, *Phys. Lett.* **105B**, 369 (1981);
For a review see R. N. Mohapatra, *Proceedings of the Telemark Neutrino Mass Mini-Conference*, 1982, American Institute of Physics, New York, 1982.

[26] J. D. Bjorken, *Phys. Rev.* **D19**, 335 (1979);
P. Q. Hung and J. J. Sakurai, *Nucl. Phys.* **B143**, 81 (1978).

[27] R. Barbieri and R. N. Mohapatra, *Phys. Lett.* **120B**, 195 (1982).

[28] D. Schildknecht, in *Proceedings of the Europhysics Study Conference on Electroweak Effects at High Energies* (edited by H. Newman), Plenum, New York, 1983.

[29] U. Baur, H. Fritzsch, and H. Faissner, *Phys. Lett.* **135B**, 313 (1984).

[30] A. Masiero, R. N. Mohapatra, and R. D. Peccei, *Nucl. Phys.* **B192**, 66 (1981).

[31] A. Masiero and R. N. Mohapatra, *Phys. Lett.* **103B**, 343 (1981).

CHAPTER 9

Global Supersymmetry

§9.1. Supersymmetry

Ultimate unification of all particles and all interactions is the eternal dream of theoretical physicists. The unified gauge theories have taken us a step closer to realizing the second goal. However, since known "elementary" particles consist of both fermions (q, l) and bosons (photons γ, W, Z color octet of gluons) their ultimate unification would require them either to be composites of some basic set of fermions which can be unified within a Lie group framework or that there must exist a new symmetry that transforms bosons to fermions. In this chapter, we begin discussion of this latter kind of symmetry [1], known as supersymmetry. Supersymmetry was invented in 1973 by Wess and Zumino [2] and earlier in a nonlinear realization by Volkov and Akulov [3].

To see why symmetry between bosons and fermions may be of interest to the study of elementary particle physics, we point out that renormalizable quantum field theories with scalar particles (such as the Higgs sector of unified gauge theories) have a very disturbing feature in that the scalar masses have quadratic divergences in one- and higher-loop orders. Unlike the logarithmic divergences associated with fermion masses, which can be eliminated by taking advantage of chiral symmetries, there is no apparent symmetry that can control the divergences associated with scalar field masses. On the other hand, if we have a theory that couples fermions and bosons the scalar masses have two sources for their quadratic divergences, one from a scalar loop which comes with a positive sign and another from a fermion loop with a negative sign. It is then suggestive that, if there was a symmetry that related the couplings and masses of fermions and bosons, all divergences

{c a, b} = cab + bca
c{a, b} - [c,b] a
cab + cba - cba + bca. ✓

from scalar field masses could be eliminated. Supersymmetry provides such an opportunity.

One of the first requirements of supersymmetry is an equal number of bosonic and fermionic degrees of freedom in one multiplet. We demonstrate this with a simple example. Consider two pairs of creation annihilation operators: $(a, a^\dagger)$ and $(b, b^\dagger)$ with a being bosonic and b being fermionic. They satisfy the following commutation and anticommutation relations, respectively,

$$[a, a^\dagger] = \{b, b^\dagger\} = 1. \tag{9.1.1}$$

The Hamiltonian for this system can be written in general as

$$H = \omega_a a^\dagger a + \omega_b b^\dagger b. \tag{9.1.2}$$

$\{b^\dagger a + a^\dagger b, b^\dagger\}$
$\{b^\dagger a, b^\dagger\} = \{b^\dagger$

If we define a fermionic operator

$\{b^\dagger a, b^\dagger\} = b^\dagger a b^\dagger + b^\dagger b^\dagger a$ $b^\dagger a b^\dagger +$

$$Q = b^\dagger a + a^\dagger b, \tag{9.1.3}$$

then (assume $[a^\dagger, b^\dagger] = 0$ but $\{a, b^\dagger\} = 0$ & perms)

$[b^\dagger a + a^\dagger b, a^\dagger]$
$= [b^\dagger a, a^\dagger] + [a^\dagger b, a^\dagger]$

$b^\dagger|0\rangle$

$$[Q, a^\dagger] = +b^\dagger, \tag{}$$

$$\{Q, b^\dagger\} = a^\dagger. \checkmark \tag{9.1.4}$$

Thus, if $a^\dagger|0\rangle$ represent bosonic and fermionic states, respectively, Q will take bosons to fermions and vice versa. Moreover

$b^\dagger [a, a^\dagger] + [b^\dagger, a^\dagger]a$

$[b^\dagger a + a^\dagger b, \omega_a a^\dagger a + \omega_b b^\dagger b]$ $+ a^\dagger [b, a^\dagger] + [a^\dagger, a^\dagger]b$

$$[Q, H] = (\omega_a - \omega_b)Q. \tag{9.1.5}$$

So, for $\omega_a = \omega_b$ (i.e., equal energy for the bosonic and fermionic states), H is supersymmetric. Furthermore, in this case

$\omega_a [b^\dagger a + a^\dagger b, a^\dagger a] + [a^\dagger b, a^\dagger a]\omega_a$

$\omega_b [b^\dagger a, b^\dagger b] + [b^\dagger a, b^\dagger b]\omega_b$

$\omega_a \{ b^\dagger [a, a^\dagger a] + [b^\dagger a, a^\dagger a] + a^\dagger [b, a^\dagger a] + [a^\dagger, a^\dagger a]b \}$

$$\{Q, Q^\dagger\} = \frac{2}{\omega} H. \tag{9.1.6}$$

$-\omega_a [-b^\dagger a - a^\dagger b]$

Thus, the algebra of Q, $Q^\dagger$, and H closes under anticommutation. If there is more than one a and b, then there must be an equal number of them, otherwise eqns. (9.1.5) and (9.1.6) cannot be satisfied together.

Another point which distinguishes this symmetry from other known symmetries is that the anticommutator of Q, $Q^\dagger$ involves the Hamiltonian; for any other bosonic symmetry the charge commutation never involves the Hamiltonian. This, as we will see later, has important implications for physics. Now we proceed to the consideration of field theories with supersymmetry.

How does he figure

$[a, b^\dagger] = 0$

$\{a, b^\dagger\} = 0$

§9.2. A Supersymmetric Field Theory

Consider the free Lagrangian for the massless complex bosonic field ϕ

$$\phi = \frac{1}{\sqrt{2}}(A + iB)$$

cab $\neq$ cba ; cba + bca

$\{Q, b^+\} = \{b^+a + a^+b, b^+\}$

$= \{b^+a, b^+\} + \{a^+b, b^+\}$

§9.2. A Supersymmetric Field Theory

189

and a Majorana field ψ ($\psi = c\bar{\psi}^T$)

$b^+\{a, b^+\} - [a^+, b^+] b$

$$\mathscr{L} = -(\partial_\mu \phi^*)(\partial_\mu \phi) - \tfrac{1}{2}\bar{\psi}\gamma_\mu \partial_\mu \psi. \qquad (9.2.1)$$

The action for this Lagrangian (not the Lagrangian itself) is invariant under the transformations

$+\{a+b, b^+\}$

$$\delta A = \bar{\varepsilon}\psi,$$

$a + \{b, b^+\} - [a^+, b^+] b$

$$\delta B = -i\bar{\varepsilon}\gamma_5 \psi,$$

$$\delta\psi = \gamma_\mu(\partial_\mu A - i\gamma_5 \partial_\mu B)\varepsilon, \qquad (9.2.2)$$

where ε is an anticommuting, Majorana spinor which is an element of the Grassmann algebra. The current corresponding to this symmetry is given by

$$\bar{\varepsilon}S_\mu = -\partial_\mu A\bar{\varepsilon}\psi + i\partial_\mu B\bar{\varepsilon}\gamma_5\psi - \bar{\varepsilon}\gamma_\nu(\partial_\nu A - i\gamma_5\partial_\nu B)\gamma_\mu\psi. \qquad (9.2.3)$$

Deriving the charge that generates supersymmetry we can look for the algebra of supersymmetry charges

$$S = -i\int d^3x\, S_4 = +\int d^3x\, \partial_t A\psi - i\int d^3\, \partial_t B\gamma_5\psi. \qquad (9.2.4)$$

We note that S is a Majorana spinor operator with $S = C\bar{S}^T$. To obtain the supersymmetry algebra we consider the commutator of successive supersymmetry operations, i.e.,

$$[\delta_{\varepsilon_1}, \delta_{\varepsilon_2}]A = ?.$$

Let us evaluate this commutator explicitly using eqn. (9.2.2)

$$\delta_{\varepsilon_1}\delta_{\varepsilon_2}A - \delta_{\varepsilon_2}\delta_{\varepsilon_1}A = \delta_{\varepsilon_1}\bar{\varepsilon}_2\psi - \delta_{\varepsilon_2}\bar{\varepsilon}_1\psi$$

$$= \bar{\varepsilon}_2\gamma_\mu(\partial_\mu A - i\gamma_5\partial_\mu B)\varepsilon_1 - \bar{\varepsilon}_1\gamma_\mu(\partial_\mu A - i\gamma_5\partial_\mu B)\varepsilon_2$$

$$= 2\bar{\varepsilon}_2\gamma_\mu\varepsilon_1\partial_\mu A, \qquad (9.2.5)$$

since

$$\bar{\varepsilon}_2\gamma_\mu\gamma_5\varepsilon_1 = \bar{\varepsilon}_1\gamma_\mu\gamma_5\varepsilon_2$$

are these Fierz identities?

and

$$\bar{\varepsilon}_2\gamma_\mu\varepsilon_1 = -\bar{\varepsilon}_1\gamma_\mu\varepsilon_2.$$

We can therefore write

$$[\bar{\varepsilon}_1 S, \bar{S}\varepsilon_2]A = +2\bar{\varepsilon}_1\gamma_\mu\varepsilon_2\partial_\mu A. \qquad (9.2.6a)$$

Similarly, we can show that

$$[\bar{\varepsilon}_1 S, \bar{S}\varepsilon_2]B = 2\bar{\varepsilon}_1\gamma_\mu\varepsilon_2\partial_\mu B. \qquad (9.2.6b)$$

Finally, let us calculate $[\delta_{\varepsilon_1}, \delta_{\varepsilon_2}]\psi$. We obtain

$$(\delta_{\varepsilon_1}\delta_{\varepsilon_2} - \delta_{\varepsilon_2}\delta_{\varepsilon_1})\psi = \delta_{\varepsilon_1}\gamma_\mu(\partial_\mu A - i\gamma_5\partial_\mu B)\varepsilon_2 - \delta_{\varepsilon_2}\gamma_\mu(\partial_\mu A - i\gamma_5\partial_\mu B)\varepsilon_1$$

$$= \gamma_\mu\varepsilon_2\bar{\varepsilon}_1\partial_\mu\psi - \gamma_\mu\gamma_5\varepsilon_2\bar{\varepsilon}_1\gamma_5\partial_\mu\psi$$

$$- \gamma_\mu\varepsilon_1\bar{\varepsilon}_2\partial_\mu\psi + \gamma_\mu\gamma_5\varepsilon_1\bar{\varepsilon}_2\gamma_5\partial_\mu\psi. \qquad (9.2.7)$$

To evaluate the last term we use the identity for products of four-component spinors

$$\bar{\psi}_1 M\psi_3\bar{\psi}_4 N\psi_2 = -\tfrac{1}{4}\sum_j \sigma(j)\bar{\psi}_1 O_j\psi_2\bar{\psi}_4 N O_j M\psi_3,$$

where

$$\sigma(j) = \begin{cases} +1 \\ -1 \\ -2 \end{cases} \text{for} \quad O_j = \begin{cases} 1, \gamma_5, \gamma_\mu, \\ \gamma_5\gamma_\mu, \\ \tfrac{1}{4}[\gamma_\mu, \gamma_\nu]. \end{cases} \tag{9.2.8}$$

To make the evaluation of eqn. (9.2.7) simpler, we first note that $[\delta_{\varepsilon_1}, \delta_{\varepsilon_2}]$ is antisymmetric in the interchange of ε_1 and ε_2 and only antisymmetric products involving ε_1 and ε_2 are $\bar{\varepsilon}_1\gamma_\mu\varepsilon_2$ and $\bar{\varepsilon}_1\sigma_{\mu\nu}\varepsilon_2$. We then get

$$\begin{aligned}[\delta_{\varepsilon_1}\delta_{\varepsilon_2} - \delta_{\varepsilon_2}\delta_{\varepsilon_1}]\psi \\ = -\tfrac{1}{4}\sum_j \sigma(j)\bar{\varepsilon}_1 O_j\varepsilon_2[\gamma_\mu O_j\partial_\mu\psi - \gamma_\mu\gamma_5 O_j\gamma_5\partial_\mu\psi] - (\varepsilon_1 \leftrightarrow \varepsilon_2),\end{aligned} \tag{9.2.9}$$

where $O_j = \gamma_\alpha$ or $\sigma_{\alpha\beta}$, but $\gamma_5\sigma_{\alpha\beta}\gamma_5 = \sigma_{\alpha\beta}$ so that the two terms within the bracket cancel. Thus we get

$$\begin{aligned}[\delta_{\varepsilon_1}, \delta_{\varepsilon_2}]\psi &= -\bar{\varepsilon}_1\gamma_\sigma\varepsilon_2\gamma_\mu\gamma_\sigma\partial_\mu\psi \\ &= -2\bar{\varepsilon}_1\gamma_\mu\varepsilon_2\partial_\mu\psi + \bar{\varepsilon}_1\gamma_\sigma\varepsilon_2\gamma_\sigma\gamma\cdot\partial\psi.\end{aligned} \tag{9.2.10}$$

If we use field equations the last term is zero and we get

$$[\delta_{\varepsilon_1}, \delta_{\varepsilon_2}]\psi = -2\bar{\varepsilon}_1\gamma_\mu\varepsilon_2\partial_\mu\psi. \tag{9.2.11}$$

Noting that $\delta_{\varepsilon_1} \equiv i\bar{\varepsilon}_1 S$ and $\delta_{a_\mu} \equiv ia_\mu P_\mu \equiv a_\mu\partial_\mu$ we obtain

$$\{S, \bar{S}\} = -2i\gamma_\mu P_\mu. \tag{9.2.12}$$

Thus, the algebra of the supersymmetry generators closes with the algebra of translation. We can, therefore, write down the super-Poincaré algebra as one that includes (9.2.12) and the following equations:

$$\begin{aligned}[M_{\mu\nu}, M_{\alpha\beta}] &= \delta_{\nu\alpha}M_{\mu\beta} - \delta_{\mu\alpha}M_{\nu\beta} - \delta_{\nu\beta}M_{\mu\alpha} + \delta_{\mu\beta}M_{\nu\alpha}, \\ [M_{\mu\nu}, P_\alpha] &= \delta_{\nu\alpha}P_\mu - \delta_{\mu\alpha}P_\nu, \\ [P_\alpha, P_\beta] &= 0, \\ [P_\alpha, \bar{\varepsilon}S] &= 0, \\ [M_{\mu\nu}, \bar{\varepsilon}S] &= \bar{\varepsilon}\sigma_{\mu\nu}S.\end{aligned} \tag{9.2.13}$$

An important point to note is that it was essential for the closure of the algebra to use the field equation for ψ. This raises the question as to whether supersymmetry leaves the Lagrangian invariant for arbitrary values of the fields. This can actually be achieved by adding two more fields, F and G. They are called auxiliary fields and are required to transform under supersymmetry transformations as:

$$\delta F = \bar{\varepsilon}\gamma_\mu \partial_\mu \psi$$

and

$$\delta G = i\bar{\varepsilon}\gamma_5\gamma_\mu\partial_\mu\psi, \tag{9.2.14}$$

$$\delta\psi = \gamma_\mu(\partial_\mu A - i\gamma_5\partial_\mu B)\varepsilon + (F + i\gamma_5 G)\varepsilon. \tag{9.2.15}$$

It is then easily verified that the extra term in eqn. (9.2.7) cancels and also

$$[\delta_{\varepsilon_1}, \delta_{\varepsilon_2}](F, G) = -2\bar{\varepsilon}_1\gamma_\mu\varepsilon_2(\partial_\mu F, \partial_\mu G). \tag{9.2.16}$$

Thus, supersymmetry can indeed be a symmetry for arbitrary field configurations. The deeper reason for the necessity of the auxiliary fields is that supersymmetry requires the fermionic and bosonic degrees of freedom to be equal. Since ψ has four fermionic degrees of freedom A and B are not enough. However, by imposing field equations two fermionic degrees of freedom are removed. If we do not impose field equations two more degrees of freedom F and G must be added.

Another point to note is that the parameter ε has mass dimension $-\frac{1}{2}$ so that F and G have mass dimension two. Therefore, their kinetic energy terms will have mass dimension 6 and cannot, lead to consistent equal-time commutation relations; as a result, they cannot be dynamical fields but must be expressible in terms of other fields of the theory. In fact, in the presence of F and G, the supersymmetric Lagrangian of eqn. (9.2.1) gets modified as follows:

$$\mathscr{L} = -\tfrac{1}{2}[(\partial_\mu A)^2 + (\partial_\mu B)^2 + \bar{\psi}\gamma_\mu\partial_\mu\psi - F^2 - G^2], \tag{9.2.17}$$

which leads to new field equations $F = G = 0$ in addition to the usual ones for A, B, and ψ.

§9.3. Two-Component Notation

We saw in the previous section that the parameter ε, which is needed to describe supersymmetry transformations, is a four-component Majorana spinor, i.e., it has four independent Grassmann variables which transform as an irreducible representation of the SO(3, 1) group. But SO(3, 1) is locally isomorphic to the SL(2, C) group consisting of arbitrary complex 2×2 matrices, Z with determinant one, i.e.,

$$Z = \begin{pmatrix} Z_{11} & Z_{12} \\ Z_{21} & Z_{22} \end{pmatrix}, \tag{9.3.1}$$

with

$$Z_{11}Z_{22} - Z_{12}Z_{21} = 1$$

or

$$\varepsilon^{ab}Z_a^c Z_b^d = \varepsilon^{cd}, \tag{9.3.2}$$

where $\varepsilon^{ab} = -\varepsilon_{ab}$ (with $\varepsilon_{ab}\varepsilon^{bc} = \delta_a^c$) is the antisymmetric Levi-Civita symbol. Equation (9.3.2) implies that this is invariant under the SL(2, C) group.

The fundamental representation of the SL(2, C) group act on the two-component complex column vector ϕ_a, i.e.,

$$\phi_a \to \phi'_a = Z_a^b \phi_b. \tag{9.3.3}$$

If we call ϕ_a a covariant vector, the contravariant vector ϕ^a can be defined using ε^{ab} as follows:

$$\phi^a = \varepsilon^{ab}\phi_b. \tag{9.3.4}$$

The complex conjugate of ϕ_a, i.e., ϕ_a^* transforms as follows:

$$\phi_a^* \to \phi_a^{*\prime} = Z_a^{b*}\phi_b^*. \tag{9.3.5}$$

The matrices Z and Z^* are not equivalent. Thus, ϕ_a^* belongs to a different two-dimensional representation of SL(2, C) and we will denote its indices by $\bar{\phi}_{\dot{a}}$ to distinguish it from ϕ_a.

Now, we are ready to write the Majorana spinor in terms of two-component vectors: to see this let us write an arbitrary complex four-componet Dirac spinor ψ as $\psi = \begin{pmatrix}\phi_1\\ \phi_2\end{pmatrix}$ and impose the Majorana condition

$$\psi = C\bar{\psi}^T. \tag{9.3.6}$$

In the choice of γ-matrices

$$\gamma_k = \begin{pmatrix} 0 & -i\sigma_k \\ i\sigma_k & 0 \end{pmatrix}, \qquad \gamma_4 = \begin{pmatrix} 0 & -1 \\ -1 & 0 \end{pmatrix}$$

and

$$\gamma_5 = \begin{pmatrix} 1 & 0 \\ 0 & -1 \end{pmatrix}, \tag{9.3.7}$$

we can write

$$\psi = \begin{pmatrix} \phi \\ i\sigma_2\phi^* \end{pmatrix}. \tag{9.3.8}$$

Noting that $i\sigma_2 \equiv \varepsilon^{ab}$ we can write a Majorana spinor ψ as

$$\psi = \begin{pmatrix} \phi_a \\ \bar{\phi}^{\dot{a}} \end{pmatrix}, \tag{9.3.9}$$

where the overbar implies its transformation as a complex two-dimensional representation.

To see the connection with Lorentz transformations consider the four-vector σ_μ, i.e.,

$$\sigma_\mu = (\sigma_k, -i) \tag{9.3.10}$$

and the Lorentz scalar $\sigma_\mu P_\mu$

$$\sigma_\mu P_\mu = \begin{pmatrix} P_0 + P_3 & P_1 - iP_2 \\ P_1 + iP_2 & P_0 - P_3 \end{pmatrix}. \tag{9.3.11}$$

Note that Det $\sigma_\mu P_\mu = (P_0^2 - P_1^2 - P_2^2 - P_3^2) = -P^2$ and is invariant under the transformation

$$\sigma_\mu P_\mu \to Z \sigma_\mu P_\mu Z^\dagger.$$

If we write

$$\sigma_\mu P'_\mu = Z \sigma_\mu P_\mu Z^\dagger, \tag{9.3.12}$$

then $P'^2 = P^2$, and therefore Z, is a Lorentz transformation. We also note that the matrices σ_μ, P_μ transform as $(\frac{1}{2}, \frac{1}{2})$ under the SL(2, C) transformations, i.e., the SL(2, C) indices of σ_μ and P_μ are $\sigma_{\mu, a\dot{a}}$ and $P_{\mu, a\dot{a}}$.

The ϕ and $\bar{\chi}$ are Weyl spinors and have the following SL(2, C) invariants:

$$\phi\chi \equiv \phi_a \chi^a = \varepsilon^{ab} \phi_b \chi_a$$

and

$$\bar{\phi}\bar{\chi} = \bar{\phi}_{\dot{a}} \bar{\chi}^{\dot{a}} = \varepsilon^{\dot{a}\dot{b}} \bar{\phi}_{\dot{a}} \bar{\chi}_{\dot{b}}. \tag{9.3.13}$$

The familiar Dirac bilinear covariants can be expressed in terms of the $SL(2, C)$ invariant products. For instance, for an arbitrary Dirac spinor ψ

$$\begin{aligned} \bar{\psi}\psi &= (\bar{\phi}_{\dot{a}} \chi^a) \begin{pmatrix} 0 & 1 \\ 1 & 0 \end{pmatrix} \begin{pmatrix} \phi_a \\ \bar{\chi}^{\dot{a}} \end{pmatrix} \\ &= (\chi\phi + \bar{\phi}\bar{\chi}). \end{aligned} \tag{9.3.14}$$

Similarly,

$$\begin{aligned} \bar{\psi}\gamma_k\psi &= (\bar{\phi}_{\dot{a}} \chi^a) \begin{pmatrix} -i\sigma_k & 0 \\ 0 & i\sigma_k \end{pmatrix} \begin{pmatrix} \phi_a \\ \bar{\chi}^{\dot{a}} \end{pmatrix} \\ &= -i\bar{\phi}_{\dot{a}} \sigma_k^{a\dot{a}} \phi_a + i\chi^a \sigma_{k, a\dot{a}} \bar{\chi}^{\dot{a}}, \\ \bar{\psi}\gamma_4\psi &= \bar{\phi}_{\dot{a}} (1)^{\dot{a}a} \phi_a + \chi^a (1)_{a\dot{a}} \bar{\chi}^{\dot{a}}. \end{aligned} \tag{9.3.15}$$

We can group them to write

$$\bar{\psi}\gamma_\mu\psi = +i\chi\sigma_\mu\bar{\chi} - i\bar{\phi}\bar{\sigma}_\mu\phi,$$

where $\bar{\sigma}_\mu = (\sigma_k, +i)$.

§9.4. Superfields

Soon after the discovery of supersymmetry by Wess and Zumino [2], Salam and Strathdee [4] proposed the concept of the superfield as the generator of supersymmetric multiplets. This is a very profound concept and has far-reaching implications. To discuss superfields we introduce the conept of superspace, which is an extension of ordinary space–time by the inclusion of additional fermionic coordinates. We would like to maintain symmetry between ordinary space and fermionic space, so we introduce four extra fermionic coordinates to match the four space–time dimensions. We can describe the fermionic coordinates as elements of a Majorana spinor or as a pair of two-component Weyl spinors. Points in superspace are then identified by the coordinates

$$z^M = (x^\mu, \theta^a, \bar{\theta}_{\dot{a}}). \tag{9.4.1}$$

where θ's are anticommuting spinors with the following properties:

$$\{\theta^a, \theta^b\} = \{\bar{\theta}_{\dot{a}}, \bar{\theta}_{\dot{b}}\} = \{\theta^a, \bar{\theta}_{\dot{a}}\} = 0,$$
$$[x^\mu, \theta^a] = [x^\mu, \bar{\theta}_{\dot{a}}] = 0. \tag{9.4.2}$$

The indices a and $\dot{a}$ go over 1 and 2 and $\theta^{a*} = \bar{\theta}^{\dot{a}}$. The scalar product of θ's is given as follows:

$$\theta\theta = \theta^a\theta_a = \varepsilon^{ab}\theta_b\theta_a = \theta^1\theta_1 + \theta^2\theta_2 = -\theta_1\theta^1 - \theta_2\theta^2. \tag{9.4.3}$$

The summation convention for $\bar{\theta}$ will be $\overline{\theta\theta} = \bar{\theta}_{\dot{a}}\bar{\theta}^{\dot{a}}$; the product of more than two θ's vanishes since $(\theta_1)^2 = (\theta_2)^2 = 0$. It also follows that

$$\theta^a\theta^b = -\tfrac{1}{2}\varepsilon^{ab}\theta\theta,$$
$$\bar{\theta}_{\dot{a}}\bar{\theta}_{\dot{b}} = -\tfrac{1}{2}\varepsilon_{\dot{a}\dot{b}}\overline{\theta\theta}. \tag{9.4.4}$$

Salam and Strathdee [4] proposed that a function $\Phi(x, \theta, \bar{\theta})$ of the superspace coordinates, called superfield, which has a finite number of terms in its expansion in terms of θ and $\bar{\theta}$ due to their anticommuting property, be considered as the generator of the various components of the supermultiplets. Φ could belong to arbitrary representation of the Lorentz group but, for the moment, we will consider only scalar superfields. We can now expand Φ in a power series in θ and $\bar{\theta}$ to get the various components of the supermultiplet

$$\Phi(x, \theta, \bar{\theta}) = \phi(x) + \theta\psi(x) + \bar{\theta}\bar{\chi}(x) + \theta^2 M(x) + \bar{\theta}^2 N(x) + \theta\sigma_\mu\bar{\theta}V_\mu(x)$$
$$+ \theta^2\overline{\theta\lambda}(x) + \bar{\theta}^2\theta\alpha(x) + \theta^2\bar{\theta}^2 D(x), \tag{9.4.5}$$

ϕ, ψ, $\bar{\chi}$, M, N, V_μ, $\bar{\lambda}$, α, and D are the component fields. There are sixteen real bosonic and sixteen fermionic degrees of freedom.

To exhibit the supersymmetry transformations of various fields we define an element of the supergroup as (we will use the Pauli metric)

$$\sigma(x_\mu, \theta, \bar{\theta}) = \exp[i(-x_\mu P_\mu + \theta^a Q_a + \bar{\theta}_{\dot{a}}\bar{Q}^{\dot{a}})]. \tag{9.4.6}$$

Here we have taken the supersymmetry generators in two-component notation. Supersymmetry transformation will be defined as a translation in superspace, of the form $\sigma(0, \varepsilon, \bar{\varepsilon})$. To study the effect of supersymmetry transformations we consider

$$\sigma(0, \varepsilon, \bar{\varepsilon})\sigma(x_\mu, \theta, \bar{\theta}) = \exp[i(\varepsilon Q + \bar{\varepsilon}\bar{Q})] \exp[i(-x_\mu P_\mu + \theta Q + \bar{\theta}\bar{Q})]. \qquad (9.4.7)$$

We then use the Baker–Campbell–Hausdorf formula

$$e^A e^B = \exp(A + B + \tfrac{1}{2}[A, B] + \cdots). \qquad (9.4.8)$$

To obtain any result from eqn. (9.4.7) we need the commutation in terms of Q and $\bar{Q}$. To obtain this from (9.2.12) we write

$$S = \begin{pmatrix} Q_a \\ \bar{Q}^{\dot{a}} \end{pmatrix}$$

and we obtain

$$\left\{\begin{pmatrix} Q_a \\ \bar{Q}^{\dot{a}} \end{pmatrix}, (\bar{Q}_{\dot{b}} Q^b)\right\} = 2(-i)\begin{pmatrix} +i\sigma_{\mu,a\dot{b}} & 0 \\ 0 & -i\bar{\sigma}_\mu^{\dot{a}b} \end{pmatrix} P_\mu \qquad (9.4.9)$$

leading to

$$\{Q_a, \bar{Q}_{\dot{b}}\} = 2\sigma_{\mu,a\dot{b}} P_\mu$$

and

$$\{Q_a, Q_b\} = \{\bar{Q}_{\dot{a}}, \bar{Q}_{\dot{b}}\} = 0. \qquad (9.4.10)$$

To familiarize the reader with manipulations involving σ_μ and $\bar{\sigma}_\mu$ we see that eqn. (9.3.9) also implies

$$\{\bar{Q}^{\dot{a}}, Q^b\} = -2\bar{\sigma}_\mu^{a\dot{b}} P_\mu. \qquad (9.4.11)$$

Let us derive eqn. (9.4.11) from eqn. (9.3.10), which can be written as

$$\{\bar{Q}_{\dot{b}}, Q_a\} = 2\sigma_{\mu,\dot{a}b} P_\mu. \qquad (9.4.10')$$

To raise indices we multiply both sides by $\varepsilon^{\dot{c}\dot{b}}\varepsilon^{ba}$ and the right-hand side of eqn. (9.4.10′) then gives

$$\begin{aligned} \varepsilon^{ba}\sigma_{\mu,a\dot{b}}\varepsilon^{\dot{c}\dot{b}} &= (\sigma_2\sigma_\mu\sigma_2)^{b\dot{b}} \\ &= -(\sigma_k^T, i)^{\dot{b}b} = -\bar{\sigma}_\mu^{\dot{b}b}, \end{aligned} \qquad (9.4.12)$$

which leads to eqn. (9.4.11). From eqns. (9.4.7) and (9.4.8) it follows that

$$\sigma(0, \varepsilon, \bar{\varepsilon})\sigma(x_\mu, \theta, \bar{\theta}) = \exp\{i[-(x_\mu + \xi_\mu)P_\mu + (\theta + \varepsilon)Q + (\bar{\theta} + \bar{\varepsilon})\bar{Q}]\},$$

where

$$\xi_\mu = -i\varepsilon\sigma_\mu\bar{\theta} + i\bar{\varepsilon}\bar{\sigma}_\mu\theta. \qquad (9.4.13)$$

Thus, under supersymmetry transformation

$$\Phi(x, \theta, \bar{\theta}) \to \Phi(x_\mu - i(\varepsilon\sigma_\mu\bar{\theta} - \bar{\varepsilon}\bar{\sigma}_\mu\theta), \theta + \varepsilon, \bar{\theta} + \bar{\varepsilon}). \qquad (9.4.14)$$

To find the supersymmetry transformation of the component fields we can

Taylor expand the right-hand side and compare coefficients of θ, $\bar{\theta}$, etc.

$$\Phi(x_\mu + \xi_\mu, \theta + \varepsilon, \bar{\theta} + \bar{\varepsilon}) = \Phi(x, \theta, \bar{\theta}) + \xi_\mu \partial_\mu \Phi + \varepsilon \frac{\partial \Phi}{\partial \theta} + \bar{\varepsilon} \frac{\partial \Phi}{\partial \bar{\theta}}. \tag{9.4.15}$$

We get

$$\begin{aligned}
\delta\phi(x) &= \varepsilon\psi + \overline{\varepsilon\chi}, \\
\delta\psi &= -i\bar{\varepsilon}\bar{\sigma}_\mu \partial_\mu \phi + \varepsilon M, \\
\delta\bar{\chi} &= +i\varepsilon\sigma_\mu \partial_\mu \phi + \bar{\varepsilon} N, \\
\delta M &= \frac{i}{2}\bar{\varepsilon}\bar{\sigma}_\mu \partial_\mu \psi + \bar{\varepsilon}\bar{\lambda}, \\
\delta N &= -\frac{i}{2}\varepsilon\sigma_\mu \partial_\mu \bar{\chi} + \varepsilon\alpha, \\
\delta V_\mu &= i\varepsilon\delta_\mu \bar{\lambda} + \varepsilon\sigma_\mu \partial_\mu \bar{\chi} + \bar{\varepsilon}\bar{\sigma}_\mu \partial_\mu \psi, \\
\delta\bar{\lambda} &= \bar{\varepsilon} D + (\partial_\mu V_\nu - \partial_\nu V_\mu)\bar{\varepsilon}\bar{\sigma}_\mu \sigma_\nu, \\
\delta\alpha &= \varepsilon D + (\partial_\mu V_\nu - \partial_\nu V_\mu)\sigma_\mu \bar{\sigma}_\nu \varepsilon, \\
\delta D &= \frac{i}{4}\varepsilon\sigma_\mu \partial_\mu \bar{\lambda} - \frac{i}{4}\bar{\varepsilon}\bar{\sigma}_\mu \partial_\mu \alpha.
\end{aligned} \tag{9.4.16}$$

Now we are ready to write down the generators of supersymmetry transformations Q_a and $\bar{Q}_{\dot{a}}$ as differential operators in the superspace

$$\begin{aligned}
Q_a &= \frac{\partial}{\partial \theta^a} - i\sigma_{\mu, a\dot{a}} \bar{\theta}^{\dot{a}} \partial_\mu, \\
\bar{Q}_{\dot{a}} &= \frac{\partial}{\partial \bar{\theta}^{\dot{a}}} - i\theta^a \sigma_{\mu, a\dot{a}} \partial_\mu.
\end{aligned} \tag{9.4.17}$$

It can be checked that $\bar{Q}_{\dot{a}} = (Q_a)^*$. Using these equations we can then verify eqn. (9.4.10).

Before proceeding further, we also like to note that the product of two superfields $\Phi_1(x, \theta, \bar{\theta})$ and $\Phi_2(x, \theta, \bar{\theta})$ is also a superfield with components

$$\begin{aligned}
&(\phi_1\phi_2, \phi_1\psi_2 + \phi_2\psi_1, \phi_1\bar{\chi}_2 + \phi_2\bar{\chi}_1, \phi_1 M_2 + \phi_2 M_1 - \tfrac{1}{2}\psi\psi, \\
&\phi_1 N_2 + \phi_2 N_1 - \tfrac{1}{2}\bar{\chi}\bar{\chi}, \phi_1 V_{2\mu} + \phi_2 V_{1\mu} - i\psi\sigma_\mu \bar{\chi}_\mu, \\
&\phi_1 \bar{\lambda}_2 + \phi_2 \bar{\lambda}_1 + \bar{\chi}_1 M_2 + \bar{\chi}_2 M_1).
\end{aligned} \tag{9.4.18}$$

§9.5. Vector and Chiral Superfields

The scalar multiplet discussed in the previous section is a reducible multiplet and we can take subsets of the component fields in eqn. (9.3.5) to make irreducible multiplets. The first irreducible multiple we can construct is, by demanding reality of Φ, i.e.,

$$\Phi = \Phi^*. \tag{9.5.1}$$

Since $\theta^* = \bar{\theta}$, we find the various component fields in eqn. (9.4.5) related to each other, i.e.,

$$\phi = \phi^*, \quad \psi = \bar{\chi}, \quad M = N^*, \quad V_\mu = V_\mu^*, \quad \bar{\lambda} = \alpha, \quad D = D^*. \tag{9.5.2}$$

This multiplet is also closed under multiplication. We will call this the vector multiplet and denote it by the symbol $V(x, \theta, \bar{\theta})$.

We will now discuss another irreducible supermultiplet called the chiral superfield. The important thing here is to note that there exists an operator

$$D_a = \frac{\partial}{\partial \theta^a} + i\sigma_{\mu,a\dot{a}}\bar{\theta}^a \partial_\mu,$$

$$\bar{D}_{\dot{a}} = +\frac{\partial}{\partial \bar{\theta}^{\dot{a}}} + i\theta^a \sigma_{\mu,a\dot{a}}\partial_\mu, \tag{9.5.3}$$

which commute with Q_a and $\bar{Q}_{\dot{a}}$ and with all the Lorentz generators. Therefore we can impose the requirement that the superfields Φ_+ and Φ_- (respectively the chiral and antichiral superfields) satisfy the following supersymmetric invariant constraints,

$$D_a \Phi_- = 0 \quad \text{and} \quad \bar{D}_{\dot{a}} \Phi_+ = 0. \tag{9.5.3a}$$

Examining eqn. (9.5.3) we can work out its implications. The solution to eqn. (9.5.3a) is that

$$\Phi_-(\bar{\theta}, x_\mu - i\theta\sigma_\mu\bar{\theta}) \equiv \Phi_-(\bar{\theta}, y),$$

where

$$y = x_\mu - i\theta\sigma_\mu\bar{\theta}.$$

Since the Φ_- depends only on $\bar{\theta}$ it has only the following components:

$$\begin{aligned} \Phi_- &= A(y) + \sqrt{2}\bar{\theta}\bar{\psi}(y) + \bar{\theta}^2 F(y) \\ &\equiv \exp(-i\theta\sigma_\mu\bar{\theta}\partial_\mu)(A_-(x) + \sqrt{2}\bar{\theta}\bar{\psi}_-(x) + \bar{\theta}^2 F_-(x)). \end{aligned} \tag{9.5.4}$$

Under the supersymmetry transformation ($\theta \to \theta + \varepsilon, \bar{\theta} \to \bar{\theta} + \bar{\varepsilon}$) the components of the chiral field transform as follows:

$$\begin{aligned} \delta A &= \sqrt{2}\varepsilon\psi, \\ \delta\psi &= \sqrt{2}\varepsilon F + \sqrt{2}i(\sigma_\mu\partial_\mu)_{a\dot{a}}\bar{\varepsilon}^{\dot{a}} A, \\ \delta F &= -i\sqrt{2}\partial_\mu\psi^a(\sigma_\mu)_{a\dot{a}}\bar{\varepsilon}^{\dot{a}}. \end{aligned} \tag{9.5.4a}$$

The chiral superfield Φ_+ is similarly given by Φ_+

$$\Phi_+(\theta, x_\mu + i\theta\sigma_\mu\bar{\theta}) = \exp(+i\theta\sigma_\mu\bar{\theta}\partial_\mu)(A_+(z) + \sqrt{2}\theta\psi_+ + \theta^2 F_+). \tag{9.5.5}$$

It is clear that $\Phi_+^\dagger = \Phi_-$. It is easy to check that the product of two chiral (two antichiral) fields is also a chiral (antichiral) field. However, products of a chiral and an antichiral field give a general vector field. Note further that we could isolate multiplets by imposing the D or $\bar{D}$ operators more often, i.e.,

$$DD\Phi = 0 \quad \text{or} \quad \bar{D}\bar{D}\Phi = 0. \tag{9.5.6}$$

This multiplet is called the complex linear multiplet L and has the following component expansion for $\bar{D}\bar{D}L = 0$, i.e.,

$$L = C(y) + \theta\beta(y) + \bar{\theta}\bar{\sigma}(y) + \theta^2 g(y) + \theta\sigma_\mu\bar{\theta}V_\mu(y) + \theta^2\bar{\theta}\bar{\rho}(y), \tag{9.5.7}$$

where C, V_μ, and g are complex bosonic fields and β, σ, and ρ are Weyl spinors and $y = x_\mu + i\theta\sigma_\mu\bar{\theta}$. By demanding $L = L^*$ we can obtain a real linear multiplet for which we have $C = C^*$, $\beta = \sigma$, $g = 0$, $V_\mu = V_\mu^*$ with $\partial_\mu V_\mu = 0$, $\bar{\rho} = \beta(\sigma_\mu)\partial_\mu$. In fact, we show below that Φ does contain precisely such a multiplet which satisfies both constraints in eqn. (9.5.6).

We now wish to note some properties of the D and $\bar{D}$ operators

$$D_a D_b D_c = \bar{D}_{\dot{a}} \bar{D}_{\dot{b}} \bar{D}_{\dot{c}} = 0$$

and

$$D^a \bar{D}^2 D_a = \bar{D}_{\dot{b}} D^2 \bar{D}^{\dot{b}}. \tag{9.5.8}$$

Equation (9.5.8) is easily proved

$$\begin{aligned}
D^a\bar{D}^2D_a &= D^a\bar{D}_{\dot{b}}\bar{D}^{\dot{b}}D_a \\
&= D^a\bar{D}_{\dot{b}}[2i(\sigma\partial)_a^{\dot{b}} - D_a\bar{D}^{\dot{b}}] \\
&= [2i(\sigma\partial)_{\dot{b}}^{a} - \bar{D}_{\dot{b}}D^a][2i(\sigma\partial)_a^{\dot{b}} - D_a\bar{D}^{\dot{b}}] \\
&= -8\Box + \bar{D}_{\dot{b}}D^2\bar{D}^{\dot{b}} - 2i(\sigma\partial)_a^{\dot{b}}\{D^a, \bar{D}_{\dot{b}}\} \\
&= -8\Box + \bar{D}_{\dot{b}}D^2\bar{D}^{\dot{b}} - 2i(\sigma\partial)_a^{\dot{b}}2i(\sigma\cdot\partial_{\dot{b}}^{a}) \\
&= \bar{D}_{\dot{b}}D^2\bar{D}^{\dot{b}}.
\end{aligned} \tag{9.5.9}$$

Some other identities involving D and $\bar{D}$ are

$$D^2\bar{D}^2 + \bar{D}^2D^2 - 2D^a\bar{D}^2D_a = 16\Box, \tag{9.5.10a}$$

$$D^2\bar{D}^2D^2 = 16\Box D^2, \tag{9.5.10b}$$

$$\bar{D}^2D^2\bar{D}^2 = 16\Box\bar{D}^2, \tag{9.5.10c}$$

Using eqns. (9.5.8) and (9.5.10a–c) we can define the projection operators

$$\pi_{0+} = \frac{\bar{D}^2D^2}{16\Box}, \qquad \pi_{0-} = \frac{D^2\bar{D}^2}{16\Box}, \quad \text{and} \quad \pi_{1/2} = -\frac{2D^a\bar{D}^2D_a}{16\Box}, \tag{9.5.11}$$

$$\pi_{0+} + \pi_{0-} + \pi_{1/2} = 1. \tag{9.5.12}$$

In fact, operating on a scalar superfield π_{0+} and π_{0-} project out the chiral and antichiral parts and $\pi_{1/2}$ projects out a piece called the linear multiplet. It is now clear that the linear multiplet $L \equiv \pi_{1/2}\Phi$ satisfies the constraints in eqn. (9.5.6). Because it satisfies both the constraints in eqn. (9.5.6) it can be written as two real linear multiplets with the following independent components each $L \equiv (C, \beta, B_\mu)$ with $\partial_\mu B_\mu = 0$ and C a real field and β a Majorana spinor as pointed out.

We are now ready to give the supersymmetric generalization of the gauge transformation. Before doing that, let us give the various components of the field. For that purpose, we first realize that, V must transform in some way as $\Phi_+ \cdot \Phi_-$.

Let us study the effect of the following transformation on the real vector multiplet

$$V \to V + \Lambda_+ + \Lambda_+^\dagger. \tag{9.5.13}$$

If we write

$$\begin{aligned} V = c + i\theta\chi - i\bar{\theta}\bar{\chi} + \frac{i}{2}\theta^2(M + iN) - \frac{i}{2}\bar{\theta}^2(M - iN) \\ - \theta\sigma_\mu\bar{\theta}V_\mu + i\theta\theta\bar{\theta}\left[\bar{\lambda} + \frac{i}{8}\bar{\sigma}_\mu\partial_\mu\chi\right] - i\bar{\theta}\bar{\theta}\theta\left[\lambda + \frac{i}{8}\sigma_\mu\partial_\mu\bar{\chi}\right] \\ + \frac{1}{2}\theta^2\bar{\theta}^2\left[D + \frac{1}{16}\Box C\right] \end{aligned} \tag{9.5.14}$$

and

$$\begin{aligned} \Lambda_+ + \Lambda_+^\dagger = A + A^* + \sqrt{2}(\theta\psi + \bar{\theta}\bar{\psi}) + \theta\theta F + \bar{\theta}\bar{\theta}F^* \\ - i\theta\sigma_\mu\bar{\theta}\partial_\mu(A - A^*) + \frac{i}{\sqrt{2}}\theta\theta\bar{\theta}\sigma_\mu\partial_\mu\psi \\ + \frac{i}{\sqrt{2}}\bar{\theta}^2\theta\sigma_\mu\partial_\mu\bar{\psi} + \frac{1}{16}\theta^2\bar{\theta}^2\Box(A + A^*), \end{aligned}$$

we find the components transforming as follows:

$$\begin{aligned} C &\to C + A + A^*, \\ \chi &\to \chi - i\sqrt{2}\psi, \\ M + iN &\to M + iN - 2iF, \\ V_\mu &\to V_\mu - i\partial_\mu(A - A^*), \\ \lambda &\to \lambda, \\ D &\to D, \end{aligned} \tag{9.5.15}$$

It is thus clear that the transformation given in eqn. (9.5.14) is the supersymmetric generalization of the ordinary gauge transformations. We can now choose

$$\text{Re } A = -C,$$

$$\psi = -\frac{i}{\sqrt{2}}\chi,$$

and

$$F = \frac{1}{2i}(M + iN), \tag{9.5.16}$$

so as to write the vector multiplet in the form

$$V = (0, 0, 0, 0, V_\mu, \lambda, D). \tag{9.5.17}$$

This gauge is known as the Wess–Zumino gauge which contains, along with the gauge field V_μ, a Majorana spinor partner. Again, we see that off-shell, the number of bosonic and fermionic components match (since V_μ, due to gauge invariance, has off-shell only three degrees of freedom). On-shell D and one more component of V_μ is removed so that we have two real bosonic degrees of freedom. Since the coefficient of V_μ is quadratic in the Grassmann variable, for V_μ and λ to be dynamical fields with canonical dimension 1 and 3/2, the superfield must have canonical dimension 0.

Having given the gauge transformations we can write down the supersymmetric generalization of the gauge covariant (invariant in the abelian case) tensor $F_{\mu\nu}$ as follows: let us consider the abelian group first. The quantity

$$W_a = -\tfrac{1}{4}\bar{D}\bar{D}D_a V \tag{9.5.18}$$

can be shown to be invariant under the transformation (9.5.13). *Proof*:

$$\begin{aligned}\bar{D}\bar{D}D_a(\Lambda_+ + \Lambda_+^\dagger) &= \bar{D}\bar{D}D_a\Lambda_+ \\ &= \bar{D}_{\dot{b}}[i(\sigma P)_a^{\dot{b}}\Lambda_+ - D_a\bar{D}^{\dot{b}}\Lambda_+] \\ &= i(\sigma P)_a^{\dot{b}}\bar{D}_{\dot{b}}\Lambda_+ = 0.\end{aligned} \tag{9.5.19}$$

It is also clear from eqn. (9.5.18) that W_a is a chiral field. Similarly, we can define a gauge invariant antichiral field as follows:

$$\overline{W}_{\dot{a}} = -\tfrac{1}{4}DD\bar{D}_{\dot{a}}V. \tag{9.5.20}$$

We will now show that the components of W contain the gauge covariant field tensor $F_{\mu\nu}$. Remembering that

$$\frac{\partial}{\partial\bar{\theta}^{\dot{c}}}\frac{\partial}{\partial\bar{\theta}_{\dot{c}}}\bar{\theta}^2 = 4 \tag{9.5.21}$$

we get

$$W_a = -i\lambda_a + \theta_b[\delta_a^b D - i(\sigma_\mu\bar{\sigma}_\nu)_a^b F_{\mu\nu}] + \theta^2(\sigma\partial)_{\dot{a}}\bar{\lambda}^{\dot{a}} + \text{other terms}. \tag{9.5.22}$$

Since we have already shown that W_a is a chiral field the other terms are dictated by this to arise from the exponential $e^{i\theta\sigma_\mu\bar{\theta}\partial_\mu}$ operating on W with the three terms shown above.

To complete this section we give the gauge invariant coupling of the matter fields to the gauge fields. For this purpose, we note that under a gauge transformation, a matter field Φ transforms as follows:

$$\Phi \to e^{-g\Lambda}\Phi,$$
$$\Phi^\dagger \to \Phi^\dagger e^{-g\Lambda^\dagger}. \tag{9.5.23}$$

It then follows that the gauge invariant coupling of Φ and V is

$$\mathscr{L}_\Phi = \Phi^\dagger e^{gV}\Phi. \tag{9.5.24}$$

We will see, in the following chapter, that this gives rise to the gauge couplings of the matter fields after we expand the exponential and note that in the Wess–Zumino gauge, $V^n = 0$ for $n \geq 3$, i.e.,

$$\mathscr{L}_\Phi = \Phi^\dagger\Phi + g\Phi^\dagger V\Phi + \tfrac{1}{2}g^2\Phi^\dagger V^2\Phi. \tag{9.5.25}$$

References

[1] For excellent reviews see
A. Salam and J. Strathdee, *Fortschr Phys*, **26**, 57 (1978);
P. van Niuwenhuizen, *Phys. Rep.* **68**, 189 (1981);
P. Fayet and S. Ferrara, *Phys. Rep.* **32**, 249 (1977);
S. J. Gates, M. T. Grisaru, M. Rocek, and W. Siegel, *Superspace*, Benjamin Cummings, New York 1983;
J. Wess and J. Bagger, *Introduction to Supersymmetry*, Princeton University Press, Princeton, NJ, 1983;
Some more recent reviews are
H. P. Nilles, *Phys. Rep.* **110**, 1 (1984);
H. Haber and G. Kane, *Phys. Rep.* **117**, 76 (1984);
A. Chamseddin, P. Nath, and R. Arnowitt, *Applied N = 1 Supergravity*, World Scientific, Singapore, 1984;
B. Ovrut, Lecture Notes by S. Kalara and M. Yamawaki, 1982.

[2] J. Wess and B. Zumino, *Nucl. Phys.* **B70**, 39 (1974); *Phys. Lett.* **49B**, 52 (1974).

[3] D. Volkov and V. P. Akulov, *JETP Lett.* **16**, 438 (1972).

[4] A. Salam and J. Strathdee, *Nucl. Phys.* **B76**, 477 (1974); *Phys. Lett.* **51B**, 353 (1974).

CHAPTER 10

Field Theories with Global Supersymmetry

§10.1. Supersymmetry Action

To apply supersymmetry to describe particle interaction we have to construct field theories that are invariant under supersymmetry transformations. We will then obtain certain constraints among the parameters of the bosonic and fermionic sectors of the theory and compare them with observations. The kind of field theories we are interested in will involve matter fields, which will be given by the chiral superfields and gauge fields, which in turn will be given by the real gauge superfield V. We will always work in the Wess–Zumino gauge for V. These matter and gauge superfields may (and, in general, will) belong to some irreducible representations of compact internal symmetry groups (local or global). Before going on to the discussion of the most general case, we first consider the simple case of a matter field Φ and illustrate how we can write a general interacting field theory for this.

A field theory consists of two parts: the kinetic energy and the potential energy, and both must be invariant under the supersymmetry transformations described in the previous chapter. As noted in Chapter 9, supersymmetry transformation corresponds to translations in a superspace with coordinates $(x_\mu, \theta, \bar{\theta})$. The volume element in superspace is $d^8z = d^4x\, d^2\theta\, d^2\bar{\theta}$; this is translation invariant. Therefore, the supervolume integrals of products of superfields will lead to a supersymmetric action as follows:

$$\int d^8z\, f(\Phi, \Phi^\dagger) \xrightarrow{\text{SUSY}} \int d^8z\, f(\Phi(x_\mu + a_\mu, \theta + \varepsilon, \bar{\theta} + \bar{\varepsilon}), \Phi^\dagger(x + a_\mu, \theta + \varepsilon, \bar{\theta} + \bar{\varepsilon})). \tag{10.1.1}$$

If we now redefine the coordinates $z \to z' = z + z_0$ where $z_0 \equiv (a_\mu, \varepsilon, \bar{\varepsilon})$ then

$d^8z = d^8z'$ and the action is invariant. Furthermore, if we have the additional property for a particular product of fields that either θ or $\bar{\theta}$ multiplies only terms which are space derivatives, we can define a six-dimensional volume integral $d^6z \equiv d^4x\, d^2\theta$ or $d^6\bar{z} \equiv d^4x\, d^2\bar{\theta}$ which can also lead to a supersymmetric action. As an example of why this is so, consider an arbitrary product of chiral $(\phi_{1+}\phi_{2+}\phi_{3+}\ldots)$ or antichiral fields $(\phi_{1+}^*\phi_{2+}^*\ldots)$. We remind the reader of the form of a chiral field Φ (we drop the subscript $\pm$; instead field $\Phi^\dagger$ will be used to denote the antichiral field whereas a field without † will denote chiral fields)

$$\begin{aligned}\Phi(x, \theta, \bar{\theta}) &= \exp(i\theta\sigma_\mu\bar{\theta}\partial_\mu)[A(x) + \sqrt{2}\theta\psi + \theta^2 F] \\ &= A(x) + \sqrt{2}\theta\psi + \theta^2 F + i\theta\sigma_\mu\bar{\theta}\partial_\mu A \\ &\quad - \frac{i}{\sqrt{2}}\theta^2\partial_\mu\psi\sigma_\mu\bar{\theta} + \tfrac{1}{4}\theta^2\bar{\theta}^2\partial^2 A. \end{aligned} \tag{10.1.2}$$

Any product of all chiral fields also has this expression where we see that all terms involving $\bar{\theta}$ have space derivatives in them so that they will vanish after integration over d^4x. For this case, the superspace becomes effectively six dimensional. So we can write a supersymmetric Lagrangian as follows:

$$S_2 = \int d^6z\, W(\Phi) + \int d^6\bar{z}\, [W(\Phi)]^\dagger. \tag{10.1.3}$$

The type of action in eqn. (10.1.1) is called D-type action whereas the one in (10.1.3) is called F-type action. The reasons for the names will become obvious soon.

Since we have given volume integrals in the space of Grassman coordinates, we must give the rules for integration and precise definition of measure. The rules of integration are as follows:

$$\int \theta_a\, d\theta_a = \delta_{ab}, \qquad \int d\theta_a = 0, \tag{10.1.4}$$

and similarly for $\bar{\theta}$.

$$\begin{aligned} d^2\theta &= -\tfrac{1}{4} d\theta^a\, d\theta_a \\ &= -\tfrac{1}{4}\varepsilon_{ab}\, d\theta^a\, d\theta^b = +\tfrac{1}{2} d\theta^1\, d\theta^2, \end{aligned} \tag{10.1.5}$$

$$\begin{aligned} \int d^2\theta\, \theta^2 &= \tfrac{1}{2}\int d\theta^1\, d\theta^2\, \varepsilon_{ab}\theta^a\theta^b \\ &= \tfrac{1}{2}\int d\theta^1\, d\theta^2\, (-2\theta^1\theta^2) \\ &= +\int d\theta^1\, \theta^1 \int d\theta^2\, \theta^2 = 1. \end{aligned} \tag{10.1.6}$$

Similarly, we defines

$$d^2\bar{\theta} = -\tfrac{1}{4}d\bar{\theta}_{\dot{a}}\, d\bar{\theta}^{\dot{a}}$$
$$= -\tfrac{1}{4}\varepsilon_{\dot{a}\dot{b}}\, d\bar{\theta}^{\dot{b}}\, d\bar{\theta}^{\dot{a}} = -\tfrac{1}{2}d\bar{\theta}^{\dot{1}}\, d\bar{\theta}^{\dot{2}},$$

$$\int d^2\bar{\theta}\,\bar{\theta}^2 = -\tfrac{1}{2}\int d\bar{\theta}^{\dot{1}}\, d\bar{\theta}^{\dot{2}}\, \varepsilon_{\dot{a}\dot{b}}\bar{\theta}^{\dot{b}}\bar{\theta}^{\dot{a}}$$
$$= \int d\bar{\theta}^{\dot{1}}\, d\bar{\theta}^{\dot{2}}\, \bar{\theta}^{\dot{2}}\bar{\theta}^{\dot{1}} = 1. \tag{10.1.7}$$

Let us also note some other properties of θ-integration. Suppose we have a function f of θ, $\bar{\theta}$, and x. Then the following identity holds:

$$\int d^8z\, f(x, \theta, \bar{\theta}) = -\tfrac{1}{4}\int d^6z\, \bar{D}^2 f$$
$$= -\tfrac{1}{4}\int d^6\bar{z}\, D^2 f. \tag{10.1.8}$$

This follows because

$$\bar{D} = \frac{\partial}{\partial\bar{\theta}} + i\theta(\sigma\cdot\partial)$$

and since the second term in $\bar{D}$ is a total space divergence its volume integral is zero by the Gauss theorem. So, inside a volume integral, D and $\bar{D}$ behave as if they only have the first term. Then we note that integrating and differentiating twice with respect to θ or $\bar{\theta}$ amounts to the same thing, i.e., picking up the coefficient of θ^2 or $\bar{\theta}^2$. Hence the proof.

Another property of θ-space that follows from the integration rules is that

$$\theta^2 = \delta^2(\theta) \quad \text{and} \quad \bar{\theta}^2 = \delta^2(\bar{\theta}). \tag{10.1.9}$$

Also the mass dimension of $\int d\theta$ is $+\tfrac{1}{2}$.

Now we can give the action that describes the interacting field theory of a chiral superfield

$$S = \int d^8z\, \Phi^\dagger\Phi + \int d^6z\, W(\Phi) + \int d^6\bar{z}\, W(\Phi)^\dagger. \tag{10.1.10}$$

The first point we note is that, in units where $\hbar = c = 1$, S must be dimensionless. Since the mass dimension of $(d\theta) = +\tfrac{1}{2}$ and that of Φ is $+1$ the first term of eqn. (10.1.10) is clearly dimensionless. As far as the second and third terms go the cubic terms in Φ will have dimensionless coupling and any lower power of Φ will have the powers of mass in the coupling, and any higher power will be suppressed by inverse powers of mass.

Now let us verify that the first term indeed yields the correct form for the kinetic energy term. We have to evaluate

$$\int d^2\theta\, d^2\bar{\theta}\, \Phi^\dagger\Phi = ?.$$

Since, by the integration rules given earlier, $\int d^4\theta$ projects out only the coefficient of $\theta^2\bar{\theta}^2$ it is like the D-term in the expansion of a vector superfield. Another way to see that this term is supersymmetric is to note that under supersymmetric variation

$$\delta D = \varepsilon\sigma_\mu \cdot \partial_\mu \bar{\lambda} + \bar{\varepsilon}\bar{\sigma}_\mu \cdot \partial_\mu \lambda, \tag{10.1.11}$$

this being a four divergence which vanishes on integration. From eqn. (10.1.2) we can pick out the D-term from the product $\Phi^\dagger\Phi$, it is

$$\mathscr{L}_{\text{K.E.}} = \tfrac{1}{2}A^* \,\square\, A - \tfrac{1}{2}\partial_\mu A^* \partial_\mu A - \frac{i}{2}\partial_\mu \psi \sigma_\mu \bar{\psi} + \frac{i}{2}\psi\sigma_\mu \partial_\mu \bar{\psi} + F^*F. \tag{10.1.12}$$

Thus we get precisely the familiar kinetic energy term. Let us now look at the second term in eqn. (10.1.10). As discussed earlier, in the expansion of the chiral field $W(\Phi)$ we simply have to pick up the coefficient of θ^2, i.e., the F-term. To illustrate how it works we choose $W(\Phi) = \lambda\Phi^3 + m\Phi^2$. The F-term of this is easily evaluated to be

$$\int d^2\theta\, W(\Phi) = m(FA - \psi\psi) + \lambda(FA^2 - \psi\psi A). \tag{10.1.13}$$

Thus, in terms of component fields, the action can be written as

$$\begin{aligned} \mathscr{L} = &-\partial_\mu A^* \partial_\mu A - i\partial_\mu \psi \sigma_\mu \bar{\psi} + F^*F + (FA - \psi\psi) + \lambda(FA^* - \psi\psi A) \\ &+ (F^*A^* - \bar{\psi}\bar{\psi}) + \lambda(F^*A^{*2} - \bar{\psi}\bar{\psi}A^*). \end{aligned} \tag{10.1.14}$$

In this Lagrangian F is an auxiliary field which has no kinetic energy term associated with it. So we can eliminate it by writing down the field equation for F obtained by varying the Lagrangian with respect to F:

$$-F = mA^* + \lambda A^{*2}. \tag{10.1.15}$$

On substituting it in eqn. (10.1.14) we get

$$\begin{aligned} \mathscr{L} = &-(\partial_\mu A^*)(\partial_\mu A) - i\partial_\mu \psi \sigma_\mu \bar{\psi} - m(\psi\psi + \bar{\psi}\bar{\psi}) \\ &- \lambda\psi\psi A - \lambda\bar{\psi}\bar{\psi}A^* - |mA + \lambda A^2|^2. \end{aligned} \tag{10.1.16}$$

From this Lagrangian we can easily see the constraints imposed on the parameters of the theory by supersymmetry. For example, it implies

$$m_\psi = m_A$$

and the coupling constant relation

$$g_{A^4} = g^2_{\bar{\psi}\psi A}. \tag{10.1.17}$$

Thus, we already note that, in order for supersymmetry to be useful for the description of particle interactions it must be broken since we do not observe any fermion boson pair degenerate in mass.

For future use we also note that the scalar potential in the Lagrangian is obtained as follows (for theories without gauge fields):

$$V = |F|^2 = \left|\frac{\partial W}{\partial A}\right|^2, \tag{10.1.18}$$

where $W = W(\theta = 0)$. For any arbitrary theory, not involving gauge fields, the field is generalized to

$$V = \sum_i \left|\frac{\partial W}{\partial A_i}\right|^2. \tag{10.1.19}$$

§10.2. Supersymmetric Gauge Invariant Lagrangian

In this section we study the gauge invariant supersymmetric Lagrangian [1]. For simplicity we will consider abelian gauge invariance and gauge coupling of a chiral scalar field with U(1) charge +1. The gauge and supersymmetrically invariant action consists of the following two pieces:

$$S = \tfrac{1}{4}\int d^6z\, W^a W_a + \int d^8z\, \Phi^\dagger e^{gV}\Phi + \text{h.c.} \tag{10.2.1}$$

We now show that the first term consists of the kinetic energy term for the gauge field, and the second term is the gauge invariant kinetic energy term for the matter field Φ. To see this, recall that

$$W^a = -i\lambda^a + \theta^b\left[\delta^a_b D - \frac{i}{2}(\sigma_\mu\bar{\sigma}_\nu)^a_b F_{\mu\nu}\right] + \theta^2(\sigma\cdot\partial)^a_{\dot{a}}\bar{\lambda}^{\dot{a}}. \tag{10.2.2}$$

To obtain the first term we simply pick out the coefficient of θ^2 in the product of $W^a W_a$ and we find, on adding the Hermitian conjugate piece, that

$$S_1 = \int d^4x\,[-\tfrac{1}{4}F_{\mu\nu}F_{\mu\nu} - i\lambda\sigma_\mu\partial_\mu\bar{\lambda} + \tfrac{1}{2}D^2]. \tag{10.2.3a}$$

Now let us look at the second term in eqn. (10.2.1) and project out the coefficient of $\theta^2\bar{\theta}^2$ (the D-component)

$$S_2 = \int d^8z\,[\Phi^\dagger\Phi + g\Phi^\dagger V\Phi + \tfrac{1}{2}g^2\Phi^\dagger V^2\Phi]. \tag{10.2.3b}$$

The D-component of these terms have already been calculated:

$$g\Phi^\dagger V\Phi|_D = \frac{ig}{2}V_\mu(A^*\partial_\mu A - A\partial_\mu A^*) + \frac{ig}{\sqrt{2}}(A^*\psi\lambda - A\bar{\psi}\bar{\lambda}) + \frac{g}{2}DA^*A \tag{10.2.4}$$

$$g^2\Phi^\dagger V^2\Phi|_D = \frac{g^2}{4}V_\mu^2 A^*A. \tag{10.2.5}$$

Combining all these we find

$$S_2 = \int d^4x \left[(D_\mu A)^*(D_\mu A) + \frac{ig}{\sqrt{2}}(A^*\psi\lambda - A\bar{\psi}\bar{\lambda}) + \frac{g}{2}DA^*A \right]. \tag{10.2.6}$$

We now point out several important consequences of the Lagrangian in eqn. (10.2.6) which give the supersymmetric coupling of matter to gauge fields.

(a) There is a gauge fermion λ which is the fermionic partner of the gauge boson. This transforms in the same way under the gauge group as the gauge fields and is a feature common to all supersymmetric gauge theories. This particle will be called gaugino.
(b) In addition to the couplings expected from gauge invariance there is an additional interaction between the fermionic matter field ψ, its scalar partner A, and the gaugino λ. The strength of this interaction is also given by the gauge coupling g and this is also a general feature of supersymmetric gauge theories.
(c) Gauge invariance implies the masslessness of the gaugino field as well as the gauge fields. Thus, for these theories to be realistic, supersymmetry will have to be broken.

Using the field equation for the D-term we can isolate the scalar potential (i.e., that part of the Lagrangian not involving any derivatives or any fermions) as

$$V = \tfrac{1}{4}D^2. \tag{10.2.7}$$

Combining this with eqn. (10.1.18) we find that the scalar potential in a theory with matter and gauge fields coupled to each other can be written as

$$V = |F|^2 + \tfrac{1}{4}D^2. \tag{10.2.8}$$

This is the expression we have to analyze in order to study the symmetry breaking in these theories.

These considerations can be generalized to the non-abelian groups [2]. The W^a is then defined as follows:

$$W_a = \tfrac{1}{4}\bar{D}\bar{D}e^{-gV}D_a e^{gV}. \tag{10.2.9}$$

Other definitions remain the same. This leads to the Yang–Mills action for the gauge field with the gaugino belonging to the adjoint representation of the gauge group.

`10.3. Feynman Rules for Supersymmetric Theories [3]

In this section we will derive the Feynman rules for superfields and describe some of their applications. We start by writing down the general form for the supersymmetric action for a chiral field Φ coupled to a gauge field V

$$S = \tfrac{1}{2}\int d^4x\, d^4\theta\, \Phi^\dagger e^{gV}\Phi + \frac{1}{64g^2}\int d^4x\, d^2\theta\, W^a W_a$$
$$+ \int d^4x\, d^2\theta\, W(\Phi) + \text{Gauge fixing terms}$$
$$+ \int d^4x\, d^2\theta\, J\Phi + \tfrac{1}{2}\int d^4x\, d^4\theta\, KV + \text{h.c.} \tag{10.3.1}$$

We will now use the following identities to convert S into a form, in which we can easily invert the kinetic term for matter fields, to obtain the propagator:

$$\frac{\delta}{\delta J(z_1)} J(z_2) = -\tfrac{1}{4}\bar{D}_1^2 \delta^8(z_{12}), \tag{10.3.2a}$$

$$\int d^4x\, d^2\theta\, (-\tfrac{1}{4}\bar{D}^2 f) = \int d^8z\, f, \tag{10.3.2b}$$

where f is a function of $(x, \theta, \bar{\theta})$. In particular, if f is a chiral field (i.e., $\bar{D}f = 0$) then

$$\begin{aligned}\bar{D}^2 D^2 f &= \bar{D}_{\dot{a}}\bar{D}^{\dot{a}} D^b D_b f \\ &= [\bar{D}_{\dot{a}} D_b 2i(\bar{\sigma}\cdot\partial)^{\dot{a}b} - \bar{D}_{\dot{a}} D^b (2i\sigma\cdot\partial)^{\dot{a}}_{\dot{b}}] f \\ &= 8(\sigma\cdot\partial)_{b\dot{a}}(\bar{\sigma}\cdot\partial)^{\dot{a}b} f \\ &= 16\Box f,\end{aligned} \tag{10.3.3}$$

where we have used $\{D_a, \bar{D}_{\dot{b}}\} = -2i(\sigma\cdot\partial)_{a\dot{b}}$ and $\{\bar{D}^{\dot{a}}, D^b\} = 2i(\sigma\cdot\partial)^{\dot{a}b}$. This leads to the identity

$$\begin{aligned}\int d^6z\, f &= \int d^6z \cdot \frac{\bar{D}^2 D^2 f}{16\Box} \\ &= -\int d^8z \frac{D^2}{4\Box} f.\end{aligned} \tag{10.3.4}$$

To derive the Feynman rules let us first consider the chiral fields and ignore the gauge fields which can be treated in a similar manner. Also let us assume that the superpotential has the following simple form:

$$-W(\Phi) = \tfrac{1}{2}m\Phi^2 + \frac{1}{3!}\lambda\Phi^3. \tag{10.3.5}$$

Let us now try to write $\tfrac{1}{2}m\Phi^2$ in a useful form

$$\tfrac{1}{2}m\int d^4x\, d^2\theta\, \Phi^2 = \tfrac{1}{2}m\int d^4x\, d^2\theta\, \Phi \frac{\bar{D}^2 D^2 \Phi}{16\Box}. \tag{10.3.6a}$$

But

$$\Phi\bar{D}^2\frac{D^2}{16\Box}\Phi = +\frac{\bar{D}^2}{4}\left(\Phi\frac{D^2}{4\Box}\Phi\right), \tag{10.3.6b}$$

since Φ is a chiral field and hence obeys the condition $\bar{D}\Phi = 0$. This implies

$$\tfrac{1}{2}m\int d^4x\, d^2\theta\, \Phi^2 = +\tfrac{1}{2}m\int d^8z\, \Phi\left(-\frac{D^2}{4\Box}\Phi\right). \tag{10.3.6c}$$

The bilinear part of the chiral field action can now be written as

$$\begin{aligned} S_0 &= \int d^8z\left[\bar{\Phi}\Phi - \tfrac{1}{2}m\Phi\left(-\frac{D^2}{4\Box}\Phi\right) - \tfrac{1}{2}m\bar{\Phi}\left(-\frac{\bar{D}^2}{4\Box}\bar{\Phi}\right)\right.\\ &\qquad\left. + J\left(-\frac{D^2}{4\Box}\Phi\right) + \bar{J}\left(-\frac{\bar{D}^2}{4\Box}\bar{\Phi}\right)\right] \\ &= \int d^8z\,(\tfrac{1}{2}\psi^T A\psi + \psi^T B), \end{aligned} \tag{10.3.7}$$

where

$$\psi^T = (\Phi \quad \bar{\Phi}),$$

$$A = \begin{pmatrix} +\dfrac{mD^2}{4\Box} & 1 \\ 1 & \dfrac{m\bar{D}^2}{4\Box} \end{pmatrix}, \tag{10.3.8a}$$

$$B = \begin{pmatrix} -\dfrac{1}{4}\dfrac{D^2}{\Box}J \\ -\dfrac{1}{4}\dfrac{\bar{D}^2}{\Box}\bar{J} \end{pmatrix}, \tag{10.3.8b}$$

To calculate the propagator we follow the procedure employed in the functional approach to conventional nonsupersymmetric field theories, i.e., we write the generating functional

$$Z(J) = \int d\Phi\, d\bar{\Phi}\, \exp\left[i\int d^8z\, S(\Phi, \bar{\Phi}) + i\int d^6z\, J\Phi + i\int d^6\bar{z}\, \bar{J}\bar{\Phi}\right]. \tag{10.3.9}$$

Using eqn. (10.3.7) we can write

$$Z(J) = \exp\left[i\int d^6z\frac{\lambda}{3!}\frac{\delta^3}{\delta J(z)^3}\right]Z_0(J), \tag{10.3.10}$$

where

$$Z_0(J) = \int d\Phi\, d\bar{\Phi}\, \exp\left\{i\int d^8z[\tfrac{1}{2}\psi^T A\psi + \psi^T B]\right\}. \tag{10.3.11}$$

Redefining $\psi' = A^{1/2}\psi$ we can integrate the exponential in $Z_0(J)$ to obtain

$$Z_0(J) = \exp\left(-\frac{i}{2}\int d^8z\, B^T A^{-1} B\right). \tag{10.3.11a}$$

To obtain the various propagators such as $\langle\Phi\Phi\rangle$, $\langle\bar{\Phi}\Phi\rangle$, and $\langle\bar{\Phi}\bar{\Phi}\rangle$ we rewrite $Z_0(J)$ as follows:

$$Z_0(J) = \exp\left\{-\frac{i}{2}\int d^8z\, d^8z'\, [\tfrac{1}{2}J(z)\Delta_{\phi\phi}(z, z')J(z')\right\}$$
$$+ J(z)\Delta_{\phi\bar{\phi}}(z, z')\bar{J}(z') + \tfrac{1}{2}\bar{J}(z)\Delta_{\bar{\phi}\bar{\phi}}(z, z')\bar{J}(z')], \tag{10.3.12}$$

where the Δ's represent the correspond propagators. To give their explicit form we have to evaluate $B^T A^{-1} B$

$$B^T A^{-1} B = \left(-\frac{1}{4}\frac{D^2}{\Box}J,\ -\frac{1}{4}\frac{\bar{D}^2}{\Box}\bar{J}\right) A^{-1} \begin{pmatrix} -\frac{1}{4}\frac{D^2}{\Box}J \\ -\frac{1}{4}\frac{\bar{D}^2}{\Box}\bar{J} \end{pmatrix}, \tag{10.3.13}$$

where

$$A^{-1}A = 1.$$

Remembering the identity $D^2\bar{D}^2D^2 = 16\Box D^2$ and $\bar{D}^2D^2\bar{D}^2 = 16\Box\bar{D}^2$ we can easily find A^{-1} to be

$$A^{-1} = \begin{pmatrix} -\frac{1}{4}\frac{m\bar{D}^2}{\Box - m^2} & 1 + \frac{m^2\bar{D}^2D^2}{16\Box(\Box - m^2)} \\ 1 + \frac{m^2D^2\bar{D}^2}{16\Box(\Box - m^2)} & -\frac{1}{4}\frac{mD^2}{\Box - m^2} \end{pmatrix}. \tag{10.3.14}$$

This leads to

$$\Delta_{\bar{\phi}\phi}(z, z') = \frac{1}{(\Box - m^2)}, \tag{10.3.15}$$

$$\Delta_{\phi\phi}(z, z') = \frac{1}{4}\frac{mD^2}{\Box(\Box - m^2)}. \tag{10.3.16}$$

In momentum space we can write the propagates as

$$\Delta_{\phi\bar{\phi}}(p, \theta) = \frac{-i}{p^2 + m^2}\delta^4(\theta_1 - \theta_2), \tag{10.3.17}$$

$$\Delta_{\phi\phi}(p, \theta) = \frac{-\frac{i}{4}mD^2(p, \theta)}{p^2(p^2 + m^2)}. \tag{10.3.18}$$

To write the complete set of Feynman rules we have to look at the chiral vertex arising from the $\lambda\phi^3$ term. For this purpose let us write down the generating functional

$$Z(J) = \exp\left[-i\int d^6z'' \frac{\lambda}{3!}\left(\frac{1}{i}\frac{\delta}{\delta J(z'')}\right)^3\right] + \text{h.c.}$$
$$\times \exp\left\{-\frac{i}{2}\int d^8z\, d^8z'\, [J(z)\Delta_{\phi\phi}(z, z')J(z') + \cdots]\right\}. \qquad (10.3.19)$$

Let us look at the effect of the lowest order term in λ coming from the first exponent and we find

$$Z_1 = +\frac{i\lambda}{3!}\int d^6z''\left[\int d^8z\, d^8z'\, \delta^6(z - z'')\Delta_{\phi\phi}(z, z')J(z')\right]^3 + \cdots. \qquad (10.3.20)$$

Using the identity $\delta^6(z - z'') = -\frac{1}{4}\bar{D}^2_{z''}\, \delta^8(z - z'')$ and doing partial integration we get

$$Z_1 = \frac{i\lambda}{3!}\int d^6z''\left[\int d^8z' - \tfrac{1}{4}\bar{D}^2_{z''}\Delta_{\phi\phi}(z'', z')J(z')\right]^3 + \cdots.$$

Now one of the $-\frac{1}{4}\bar{D}^2_{z''}$ from the integrand can be removed by converting $\int d^6z''$ to $\int d^8z''$. Then, at the vertex, we are left with two $-\frac{1}{4}\bar{D}^2_{z''}$ factors for three legs. Furthermore, since the S-matrix is obtained by the following operation

$$S = \exp\left[-i\int \Phi_{in}\Delta^{-1}_{\phi\phi}\frac{\delta}{\delta J}\, d^6z\, \delta^2(\bar{\theta})\right] Z(J) \qquad (10.3.21)$$

for each external leg, we must remove $-\frac{1}{4}\bar{D}^2\Delta_{\Phi\Phi}$ (or $-\frac{1}{4}D^2\Delta_{\bar{\phi}\bar{\phi}}$). We can now state the Feynman rules for supersymmetric field theories in momentum space

$$\Delta_{\phi\bar{\phi}}(p, \theta_1 - \theta_2) = \frac{-i\delta^4(\theta_1 - \theta_2)}{p^2 + m^2}, \qquad (10.3.17)$$

$$\Delta_{\phi\phi}(p, \theta_1 - \theta_2) = \frac{-(i/4)mD^2(p, \dot{\theta}_1 - \theta_2)\delta^4(\theta_1 - \theta_2)}{p^2(p^2 + m^2)}. \qquad (10.3.18)$$

As far as the vertices are concerned each chiral (antichiral) vertex will have a $-\frac{1}{4}\bar{D}^2$ $(-\frac{1}{4}D^2)$ factor for each chiral superfield, but omitting one for converting d^6z to d^8z and omitting one for external legs. Each vertex has a $d^4\theta$ integration. Using these rules for the simple theory described let us evaluate a couple of Feynman diagrams.

The first diagram

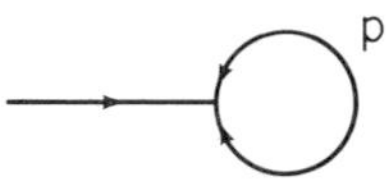

Figure 10.1

$$A_1 = -i\lambda m \tfrac{1}{16} \int D^2\bar{D}^2\delta(\theta_1 - \theta_2)|_{\theta_1=\theta_2}\phi\, d^4\theta_1\, I(p), \tag{10.3.22}$$

where $I(p)$ is the divergent momentum integral. We now note that

$$\tfrac{1}{16}D^2\bar{D}^2\delta^4(\theta_1 - \theta_2)|_{\theta_1=\theta_2} = 1, \tag{10.3.23}$$

this gives $\int \phi\, d^4\theta = 0$.

The One-loop Correction to the Propagator

Using the same Feynman rules we can obtain

$$\frac{1}{(2\pi)\phi}\int d^4\theta_1\, d^4\theta_2\, \tfrac{1}{16}D_1^2\delta^4(\theta_{12})\bar{D}_2^2\delta^4(\theta_{21})\int \frac{d^4k}{k^2(-k+p)^2}$$

$$= \int d^4\theta_1\, d^4\theta_2\, \tfrac{1}{16}\delta^4(\theta_{12})D_1^2\bar{D}_2^2\delta^4(\theta_{21})I(p). \tag{10.3.24}$$

We note that the $\sigma\cdot\partial$ term in D vanishes since it is inside space integration, which helps us to write each D or $\bar{D}$ inside an integral as $\partial/\partial\theta$ or $\partial/\partial\bar{\theta}$, respectively. We then use the identity that

$$\tfrac{1}{16}D_1^2\bar{D}_2^2\delta^4(\theta_{21}) = 1.$$

Integrating over $d^4\theta_2$ we find the effective action is $\int d^4\theta\, \Phi^\dagger\Phi$. This procedure can be repeated for arbitrary loops to show that all loop effects are of the form $\int d^4\theta\, f(\phi^\dagger, \phi)$. This, therefore, implies that the superpotential is completely unaffected by loop corrections.

§10.4. Allowed Soft Breaking Terms

We will now discuss the allowed soft-supersymmetry breaking terms that do not disturb the renormalizality of the theory [4]. To study these we first give the rules of power counting in supersymmetric field theories. These rules are different from the conventional field theories and are as follows:

(a) each vertex has a factor D^4 which is $\sim p^2$;
(b) propagators: $\Delta_{\phi\bar{\phi}} \sim 1/p^2$ and $\Delta_{\phi\phi} \sim 1/p^3$;
(c) the external line has $1/D^2 \sim 1/p$;
(d) the loop integral $\sim d^4p/p^2$ due to the fact that there is a $d^4\theta$ integration or due to the fact that four D's are needed to cancel the final $\delta^4(\theta)$ in the loop, thus leaving $\int d^4\theta$ in the end.

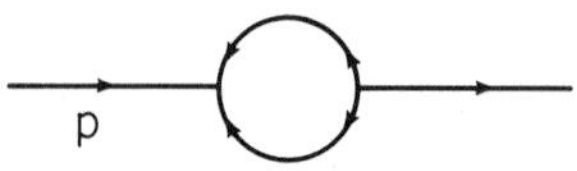

Figure 10.2

For a graph like that in Fig. 10.2 we can easily count that

$$d = \underset{\text{external leg}}{-2} \quad \underset{\text{two vertices}}{+4} \quad \underset{\text{two propagators}}{-4} \quad \underset{\text{loop integration}}{+2} \quad = 0.$$

It is therefore clear that if we have additional D factors at the vertices it will worsen the divergence structure of the theory. This has two implications

(a) $\lambda\Phi^n$, $n > 3$ would imply more than one D^4 at each vertex. Thus, the maximum allowed n is three for the theory to be renormalizable.
(b) To study allowed soft-supersymmetry breaking terms [5], we first note that by introducing constant spurion superfields we can write them in a manifestly supersymmetric form. For instance, if we introduce a spurion $U = \theta^2\bar{\theta}^2$, we can write a term such as $A^\dagger A = \int d^4\theta\, U\Phi^\dagger\Phi$. It is then possible to do power counting to see which kind of soft-breaking terms introduce new divergences into the theory.

Allowed soft-breaking terms should not involve D's in their vertices when expressed in manifestly supersymmetric form. This has important implications. For instance, if we want to add a fermion mass term to explicitly break supersymmetry, we can write it as

$$\int d^4\theta\, UD^a\Phi|D_a\Phi_2, \tag{10.4.1}$$

where $U = \theta^2\bar{\theta}^2$. Note that $D\Phi$ is not a chiral field (i.e., $\bar{D}D\phi \neq 0$), we cannot have a $d^2\theta$ integral and have manifest supersymmetry for the Lagrangian. So we must make it $d^4\theta$. This will add a term $\mu\psi_1\psi_2$ to the Lagrangian without its corresponding superpartner term $F_1^\dagger A_2 + F_2^\dagger A_1$. But this will not be allowed since, at each vertex, it will introduce D^6 (two powers of D from each ϕ and two explicit D's) making the theory nonrenormalizable. However, soft-breaking terms such as $\int d^4\theta\, U'\Phi^\dagger\Phi$ which give $\mu^2 A^\dagger A$ (choosing $U^1 = \mu^2\theta^2\bar{\theta}^2$) are allowed. The gaugino mass $\lambda\lambda$ is also allowed since it is of the form $\int U\, d^4\theta\, W^a W_a$.

References

[1] J. Wess and B. Zumino, *Nucl. Phys.* **B78**, 1 (1974).
[2] S. Ferrara and B. Zumino, *Nucl. Phys.* **B79**, 413 (1974);
A. Salam and J. Strathdee, *Phys. Lett.* **51B**, 353 (1974).
[3] M. T. Grisara, M. Rocek, and W. Siegel, *Nucl. Phys.* **B159**, 429 (1979).
[4] J. Wess and B. Zumino (Ref. [1]);
A. Slavnov, *Nucl. Phys.* **B97**, 155 (1975);
B. DeWit, *Phys. Rev.* **D12**, 1628 (1975);
S. Ferrara and O. Piguet, *Nucl. Phys.* **B93**, 261 (1975);
R. Delbourgo, M. Ramon Medrano, *Nucl. Phys.* **B110**, 473 (1976);
R. Delbourgo, *J. Phys.* **G1**, 800 (1975).
[5] L. Girardello and M. T. Grisaru, *Nucl. Phys.* **B194**, 65 (1982).

CHAPTER 11

Broken Supersymmetry and Application to Particle Physics

§11.1. Spontaneous Breaking of Supersymmetry

We pointed out in the previous chapter that in the exact supersymmetric limit fermions and bosons are degenerate in mass, a situation for which there appears to be no evidence in nature. Therefore, in order to apply supersymmetry to particle physics, we must consider models where supersymmetry is broken. There are two ways to break symmetries of Lagrangian field theories (see Chapter 2): first, where extra terms are added to the Lagrangian that are not invariant under the symmetry; and second, the Lagrangian is kept invariant whereas the vacuum is allowed to be noninvariant under the symmetry. The first method introduces an arbitrariness into the theory thereby reducing its predictive power. The condition that the divergence structure should not be altered very much reduces this arbitrariness somewhat; yet it is not a very satisfactory approach. On the other hand, the second method, the Nambu–Goldstone realization of the symmetry provides a unique, appealing, and more predictive way to study the consequences of symmetry noninvariances. We will, therefore, study this approach in this chapter.

In contrast with ordinary bosonic symmetries supersymmetry breaking requires more careful consideration for the following reason. We have

$$\{Q_a, \bar{Q}_{\dot{b}}\} = 2(\sigma_\mu P_\mu)_{a\dot{b}}. \tag{11.1.1}$$

Taking vacuum expection values of both sides we find [1]

$$\langle 0|Q_a\bar{Q}_{\dot{b}} + \bar{Q}_{\dot{b}}Q_a|0\rangle = 2\langle 0|H|0\rangle\delta_{a\dot{b}}$$

or

$$|Q_a|0\rangle|^2 = \langle 0|H|0\rangle. \tag{11.1.2}$$

This equation implies the following.

As long as the vacuum state has zero energy supersymmetry is unbroken. Thus, to break supersymmetry, we find from eqn. (10.2.8) that we must have either $\langle F \rangle \neq 0$ or $\langle D \rangle \neq 0$ or both. The first possibility is called the F-type or O'Raifeartaigh [2] mechanism for supersymmetry breaking whereas the second mechanism is called the D-type or Fayet–Illiopoulos [3] mechanism. In the subsequent sections we will give examples of both these mechanisms. Right now we show that, in an exactly analogous manner to the case of bosonic symmetries, the spontaneous breaking of supersymmetry leads to the existence of massless fermionic states which will be called Goldstino.

We note that, for a chiral field, supersymmetric transformation gives

$$\delta\psi_a = \varepsilon_a F - \sigma_{\mu,a\dot{a}}\bar{\varepsilon}^{\dot{a}}\partial_\mu A. \tag{11.1.3}$$

Taking vacuum expectation values of both sides we find

$$\langle \delta\psi_a \rangle = \varepsilon_a \langle F \rangle \neq 0, \tag{11.1.4}$$

which is the condition for supersymmetry breaking. But

$$\delta\psi_a = \varepsilon^b \{Q_b, \psi_a\}. \tag{11.1.5}$$

Equation (11.1.4) implies that

$$\langle 0 | \{Q_b, \psi_a\} | 0 \rangle \neq 0. \tag{11.1.6}$$

Using the supersymmetry current and its conservation condition we can rewrite eqn. (11.1.6) as

$$\int d^4x\, \partial_\mu I_\mu(x) \equiv \int d^4x \frac{\partial}{\partial x_\mu} \langle 0 | T\{S_{\mu b}(x), \psi(0)\} | 0 \rangle \neq 0. \tag{11.1.7}$$

We can convert the above integral into a surface integral and take the surface at infinity. If all fermionic states of the theory are massive then the T-product falls off as $e^{-m|x|}/x^3$ for large x and leads to vanishing of the surface integral. On the other hand, if there is at least one massless spin 1/2 fermionic state $|X\rangle$ in the theory, two possibilities occur.

(a) We may have

$$\langle 0 | S_{\mu,b} | X(P) \rangle = f P_\mu \delta_{ab}. \tag{11.1.8}$$

Then we have

$$\partial_\mu I_\mu(x) = \partial^2 \left(\frac{\bar{\sigma}_\mu \cdot x_\mu}{x^4} \right). \tag{11.1.9}$$

The integral in this case vanishes.

(b) On the other hand, if we have

$$\langle 0 | S_{\mu,b} | X_a(P) \rangle = (\sigma_\mu)_{b\dot{a}} f_X, \tag{11.1.10}$$

we get

$$\partial_\mu I_\mu(x) = \partial_\mu^{Tr} \frac{(\sigma_\mu \cdot \bar{\sigma}_\nu) x_\nu}{x^4}. \tag{11.1.11}$$

Clearly, for this case, the integral in eqn. (11.1.7) goes like $\int d^4x/x^4$ giving rise to a nonzero value of the integral. In this crude manner we see that spontaneous breaking of supersymmetry leads to massless fermionic states—the Goldstinos.

Similar arguments can be given for the D-terms by noting that under supersymmetry transformation

$$\delta\lambda_a = \varepsilon_a D + \varepsilon^b(\sigma_\mu\bar{\sigma}_\nu)_{ba}F_{\mu\nu}. \tag{11.1.12}$$

Thus, since $\langle 0|F_{\mu\nu}|0\rangle = 0$, $\langle D\rangle \neq 0$ implies $\langle\delta\lambda\rangle \neq 0$. We will see that, in the F-type breaking case, the field ψ is the Goldstino field where, in the D-type case, the corresponding field is λ.

§11.2. Supersymmetric Analog of the Goldberger–Treiman Relation

It is well known in hadronic weak interactions that spontaneous breaking of axial SU(2) symmetry leads to a relation between pion (the Goldstone boson for $SU(2)_A$ symmetry) coupling to nucleons, the pion decay constant, and the value of the current matrix element between nucleons at zero momentum transfer. This is the celebrated Goldberg–Treiman (GT) relation. For the case of spontaneously broken supersymmetry an analogous relation exists. Before deriving this we remind the reader about the derivation of the Goldberg–Treiman relation in hadron physics. Consider the matrix element

$$K_\mu \equiv \langle N(P_1)|A_\mu|N(P_2)\rangle = \bar{u}[\gamma_\mu\gamma_5 G_A(q^2) + \gamma_5 q_\mu F_A(q^2)]u. \tag{11.2.1}$$

Using dispersion relations for the left-hand side we see that K_μ receives contributions from the pion (the massless pole and the continuum are shown in Fig. 11.1). The first term implies that

$$F_A(q^2) = \frac{g_{NN\pi}f_\pi}{q^2} + \text{continuum}, \tag{11.2.2}$$

where

$$\langle 0|A_\mu|\pi\rangle = q_\mu f_\pi. \tag{11.2.3}$$

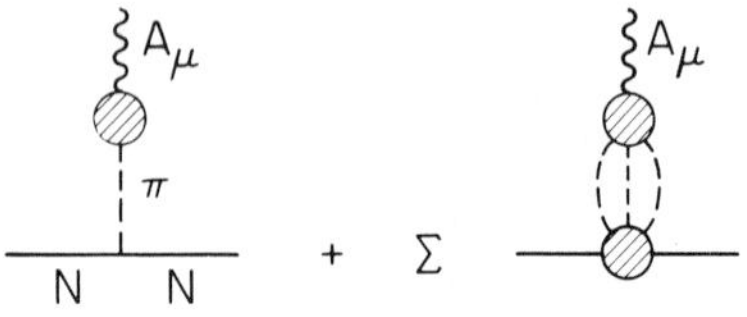

Figure 11.1

Now, taking the divergence of the axial current in K_μ and setting it equal to zero (for massless pions), we find

$$2G_A(0)M_N + g_{NN\pi}f_\pi = 0, \tag{11.2.4}$$

which is the Goldberg–Treiman relation.

Coming now to supersymmetry we take the matrix element of the supersymmetry current $S_{\mu,a}$ (see Chapter 9) between a fermionic and bosonic state

$$\langle B(P_1)|S_{\mu,a}|F(P_2)\rangle = f_1(q^2)(\bar{\sigma}_\mu\psi)_a + (P_1 + P_2)_\mu\psi_a f_2(q^2) + q_\mu\psi_a f_3(q^2). \tag{11.2.5}$$

If we assume the supersymmetry to be spontaneously broken, with χ being the associated Goldstino, then eqn. (11.1.10) tells us that the matrix element has a fermion pole at $q = 0$ and the pole actually occurs in the function $f_1(q^2)$ and we have

$$f_1(q^2) = \frac{g_{FB\chi}f_\chi}{\not{q}} + \text{continuum}. \tag{11.2.6}$$

Now using current conservation, as before, we conclude that

$$g_{FB\chi}f_\chi + q\cdot(p_1 + p_2)f_2 + q^2 f_3 = 0. \tag{11.2.7}$$

(i) If F and B are single-particle states being members of the same supermultiplet, then $q(p_1 + p_2) = M_F^2 - M_B^2$ and taking $q^2 \to 0$ as the limit, we find

$$M_F^2 - M_B^2 \simeq \frac{g_{FB\chi}f_\chi}{f_2(0)}. \tag{11.2.8}$$

This implies that spontaneous breaking of supersymmetry actually lifts the boson–fermion mass degeneracy within a multiplet and as symmetry breaking disappears, i.e., $f_\chi \to 0$, mass degeneracy is restored.

(ii) If F and B are not single-particle states but multiparticle states then $g_{FB\chi} \equiv A_{FB\chi}$, i.e., a scattering amplitude involving Goldstino in the initial state (or final state). Then eqn. (11.2.7) is the constraint on that scattering amplitude analogous to the Adler zero for the pion scattering amplitude. To see the usefulness of this formula let us assume that the neutrino is a Goldstino. Equation (11.2.7) becomes, then, a constraint on all weak decay amplitudes involving the neutrino. It implies that the β decay amplitudes must vanish as neutrino momentum goes to zero; but we know that it does not behave like that, which means that we cannot interpret neutrino as a Goldstino [4].

§11.3. *D*-Type Breaking of Supersymmetry

Breaking supersymmetry by *D*-terms was first suggested by Fayet–Illiopoulos [3]. The basic ides is that, if the theory has an abelian U(1) gauge invariance, the Lagrangian can include (apart from the terms described earlier) a term linear in the gauge superfield V. As an example consider

$$S = \tfrac{1}{4}\int d^2\theta\, W^a W_a + \tfrac{1}{4}\int d^2\bar{\theta}\, \overline{W_{\dot{a}}}\,\overline{W^{\dot{a}}} + \int [\Phi_+^\dagger e^{gV}\Phi_+ - 2kV]\, d^4\theta$$

$$+ \int [\Phi_-^\dagger e^{-gV}\Phi_-]\, d^4\theta + m\int \Phi_+\Phi_-\, d^2\theta + m\int \Phi_+^\dagger\Phi_-^\dagger\, d^2\bar{\theta}, \qquad (11.3.1)$$

where we have coupled two chiral matter fields $\Phi_\pm$ with equal and opposite U(1) quantum numbers to the U(1) gauge field V.

The scalar potential in this model arises from the following term:

$$V = -(\tfrac{1}{2}D^2 + |F_+|^2 + |F_-|^2) - kD - \frac{g}{2}D(A_+^* A_+ - A_-^* A_-)$$

$$- (F_+^* A_+^* - F_-^* A_-^* + F_+ A_+ + F_- A_-). \qquad (11.3.2)$$

The field equations for D and $F_\pm$ are given by

$$D + \frac{g}{2}(A_+^* A_+ - A_-^* A_-) + k = 0,$$

$$F_+ + mA_+^* = 0,$$

$$F_- + mA_-^* = 0. \qquad (11.3.3)$$

Using this we can rewrite the potential V as follows:

$$V = \tfrac{1}{2}\left\{k + \frac{g}{2}(A_+^* A_+ - A_-^* A_-)\right\}^2 + m^2(A_+^* A_+ + A_-^* A_-) \quad 1$$

$$\equiv \tfrac{1}{2}k^2 + \left(m^2 + \frac{kg}{2}\right)A_+^* A_+^* + \left(m^2 - \frac{kg}{2}\right)A_-^* A_-^*$$

$$+ \tfrac{1}{8}g^2(A_+^* A_+ - A_-^* A_-)^2. \qquad (11.3.4)$$

Now, we note that, if $m^2 \pm kg/2 \geq 0$, the minimum of this potential corresponds to $\langle 0|H|0\rangle > 0$ which means supersymmetry is spontaneously broken. Since $\langle D\rangle \neq 0$ for $\langle F_\pm\rangle = 0$ this is a D-type breaking with $\langle D\rangle = -k$. Since, in this situation, the gaugino field $\lambda \to \lambda + \delta\lambda$ with $\delta\lambda = \varepsilon k$ it must remain massless and correspond to the Goldstino. To see the manifestation of supersymmetry breaking in the particle spectrum, note that, the fermionic fields of the superfield $\Phi_\pm$ have equal mass $M_{\psi_\pm} = m$ whereas their superpartners $M_{A_\pm}$ have the following masses:

$$m_{A_\pm} = \sqrt{m^2 \pm \frac{kg}{2}}. \qquad (11.3.5)$$

In this case $f_\chi = k$ and $|g_{FB\chi}| = g/2$, i.e., the gauge coupling. It is worth noting that, since the minimum corresponds to $A_\pm = 0$, the gauge invariance is unbroken even though supersymmetry is broken.

§11.4. O'Raifeartaigh Mechanism or F-Type Breaking of Supersymmetry

We first study this case without any reference to gauge symmetries and subsequently we will include the gauge fields. The most trivial example of this type is to consider a singlet chiral superfield X with the superpotential

$$W = m^2 X. \tag{11.4.1}$$

The potential for this case is given by

$$V = m^4 > 0 \quad \text{for } m \neq 0 \tag{11.4.2}$$

and $F = m^2$; thus supersymmetry is broken. The Goldstino in this case is ψ_x—the fermion field in X.

To discuss a somewhat nontrivial example we consider the model of the previous section but augment it with the inclusion of two singlets X and Y. Let us consider the following superpotential:

$$W = \lambda_1 X(\Phi_+ \Phi_- - m^2) + \lambda_2 Y \Phi_+ \Phi_-. \tag{11.4.3}$$

This superpotential has a U(1) symmetry under which Φ_+ and Φ_- have equal and opposite charges and X and Y are neutral. The various F-terms in this case are the following:

$$F_X^* = \lambda_1(A_+ A_- - m^2), \tag{11.4.4a}$$

$$F_Y^* = \lambda_2 A_+ A_-, \tag{11.4.4b}$$

$$F_+^* = (\lambda_1 A_X + \lambda_2 A_Y) A_-, \tag{11.4.4c}$$

$$F_-^* = (\lambda_1 A_X + \lambda_2 A_Y) A_+. \tag{11.4.4d}$$

The value of the potential energy of the ground state is determined by $V = \sum_i |F_i|^2$ where $i = X, Y, +, -$. The fields will choose that value for which V is minimum. It is clear that, at the minimum, $F_+ = F_- = 0$ which has two solutions

$$\lambda_1 A_X + \lambda_2 A_Y = 0 \tag{11.4.5a}$$

or

$$A_+ = A_- = 0. \tag{11.4.5b}$$

For the second choice $F_Y = 0$ but $F_X \neq 0$. If both $A_+, A_- \neq 0$ then (11.4.5a) must hold and the fields $A_\pm$ must be such so as to minimize V, i.e.,

$$\frac{\partial V}{\partial A_+} = \frac{\partial V}{\partial A_-} = 0. \tag{11.4.6}$$

This implies

$$\lambda_1(A_+ A_- - m^2) + \lambda_2 A_+ A_- = 0, \tag{11.4.7}$$

since W is symmetric in A_+ and A_-, at the ground state $\langle A_+\rangle = \langle A_-\rangle = k$, where

$$k^2 = \frac{\lambda_1 m^2}{(\lambda_1 + \lambda_2)} \tag{11.4.8}$$

and $\langle V\rangle_{\text{vac}} \neq 0$ and both supersymmetry and the U(1) symmetry of W are broken.

To study the mass spectrum of the model, and isolate the Goldstino and Goldstone bosons, we have to write down the Yukawa coupling and V, and the resulting fermion and boson masses at $\langle A_\pm\rangle_{\text{vac}} = k$ and $\langle A_X\rangle_{\text{vac}} = \langle A_Y\rangle_{\text{vac}} = 0$

$$\begin{aligned}\mathscr{L}_Y = {} & \lambda_1(\psi_X\psi_+ A_- + \psi_-\psi_X A_+ + \psi_+\psi_- A_X) \\ & + \lambda_2(\psi_Y\psi_+ A_- + \psi_-\psi_Y A_+ + \psi_+\psi_- A_Y).\end{aligned} \tag{11.4.9}$$

At the minimum, we find that $[1/(\lambda_1^2 + \lambda_2^2)](\lambda_1\psi_X + \lambda_2\psi_Y)$ and $(1/\sqrt{2})(\psi_+ + \psi_-)$ form a Dirac spinor with mass $(\lambda_1^2 + \lambda_2^2)k$. The remaining fermions $\chi \equiv (\lambda_2\psi_X - \lambda_1\psi_Y)/\lambda_1^2 + \lambda_2^2$ and $(\psi_+ - \psi_-)/\sqrt{2} \equiv \theta$ remain massless. Since, under supersymmetry $\delta\psi_X \to \langle F_X\rangle\varepsilon$ and $\delta\psi_Y \to \varepsilon\langle F_Y\rangle$ where ε is the constant spinor, the massless fermion χ is the Goldstino and θ is an additional massless fermion which we will soon see is the fermionic partner of the Goldstone boson corresponding to spontaneous breakdown of the U(1) symmetry.

From the bosonic mass spectrum we infer that

$$G \equiv \frac{\text{Im } A_+ - \text{Im } A_-}{\sqrt{2}}$$

and

$$R = \frac{\text{Re } A_+ - \text{Re } A_-}{\sqrt{2}} \tag{11.4.10}$$

has zero mass. Clearly, G is the Nambu–Goldstone boson corresponding to U(1) symmetry breakdown.

The field R is also a massless particle which exists to complete the supersymmetric multiplet corresponding to G and θ, i.e.,

$$(R + iG, \theta, F') \tag{11.4.11}$$

form a massless supermultiplet. The particle R, henceforth, will be called a quasi-Nambu–Goldstone (QNG) boson, and θ the QNG-fermion.

To study this model further we make U(1) a local symmetry, so that there is a Higgs phenomenon along with supersymmetry breaking. Let us denote the components of the gauge multiplet by (V_μ, λ, D). As is clear, from Chapter 2 on the Higgs mechanism, the massless boson G becomes the longitudinal mode of the gauge boson V_μ. To see the impact of supersymmetry let us see what happens to the gaugino. As discussed in eqn. (10.2.4) the gaugino couples to the matter chiral multiplet as follows:

$$\mathscr{L}_\lambda = \frac{ig}{\sqrt{2}}(\lambda\psi_+ A_+^* - \lambda\psi_- A_-^*) + \text{h.c.} \tag{11.4.12}$$

On substituting the vacuum expectation values for $A_\pm$, this leads to a fermion bilinear as follows:

$$\mathscr{L}_\lambda = igk\lambda\theta + \text{h.c.} \tag{11.4.13}$$

Thus, the gaugino acquires a Dirac mass by "eating" up the superpartner of the Goldstone boson field and its mass is the same as the mass of the gauge boson after the Higgs mechanism. This will be a general feature of supersymmetric gauge theories with spontaneous symmetry breaking. The gaugino picking up mass is independent of the question of supersymmetry breaking. We also note that the quasi-Nambu–Goldstone boson field R now acquires mass gk from the D-term of the Lagrangian. Thus, the three components of the massive gauge field V_μ and massive R-field make up the four real massive superpartners of the massive four-component Dirac spinor $(\lambda, \bar{\theta})$.

§11.5. A Mass Formula for Supersymmetric Theories and the Need for Soft Breaking

As we have argued at the beginning of this chapter, if supersymmetry is to have a role in understanding the world of quarks and leptons, it must be a broken symmetry since no fermions and bosons are observed with degenerate mass. It is for this reason that we started the study of spontaneous breakdown of supersymmetry. In this section we show that the spontaneous breaking of supersymmetry by F- or D-terms does not lead to an acceptable particle spectrum. To show this, we derive a mass formula relating the bosons and fermions in a supersymmetric theory with or without spontaneous symmetry breaking. We will do this in two steps: first, we consider a theory without gauge fields; and second, we will generalize the formula by adding the gauge fields.

Let us consider a set of chiral fields Φ_a, $a = 1, \ldots, N$, and consider a superpotential $W(\Phi_a)$ which is an arbitrary polynomial (usually of degree ≤ 3), which is an analytic function of Φ_a. The Lagrangian can then be written as (where (A, ψ) denote the scalar and fermion fields)

$$\mathscr{L} = \mathscr{L}_{\text{K.E.}} + \mathscr{L}_Y - V, \tag{11.5.1}$$

where

$$\mathscr{L}_Y = \sum_{a,b} \frac{\partial W}{\partial A_a\, \partial A_b} \psi_a \psi_b \tag{11.5.2}$$

and

$$V = \sum_a \left(\frac{\partial W}{\partial A_a}\right)\left(\frac{\partial W^*}{\partial A_a^*}\right). \tag{11.5.3}$$

Taking the second derivative of V with respect to the bosonic fields, we get the general boson mass matrix as a function of fields

$$M_0^2 = \begin{matrix} & A_b^* & A_b \\ \begin{matrix} A_a \\ A_a^* \end{matrix} & \begin{pmatrix} \frac{1}{2}W''_{ac}W''^{cb*} & W'''_{abc}W'^{c*} \\ W'''^{abc*}W'_c & \frac{1}{2}W''_{ac}W''^{cb*} \end{pmatrix} \end{matrix}. \tag{11.5.4}$$

From eqn. (11.5.2) we get the fermion mass matrix

$$(M_{1/2})_{ab} = W''_{ab}. \tag{11.5.5}$$

It follows from the above two equations that (choosing all parameters in the Lagrangian to be real)

$$T_r(M_0^2 - M_{1/2}^2) = 0. \tag{11.5.6}$$

Let us now couple the chiral fields to a general non-abelian gauge field denoted by (V_μ, λ, D). The gauge kinetic term contributes to M_0^2, as well as to $M_{1/2}$, via the gaugino matter field coupling. For simplicity of discussion we see the gauge coupling to one (i.e., $g/2 = 1$). We can then write

$$\mathscr{L}_Y = \sum_{a,b} W''_{ab}\psi_a\psi_b + i\sqrt{2}\lambda_i\psi_a T_i A_a^* + \text{h.c.} \tag{11.5.7}$$

and

$$V = \sum_a W'_a W'^{a*} + \tfrac{1}{2}(A^* T_i A)^2. \tag{11.5.8}$$

The diagonal elements of M_0^2 then get modified to:

$$(M_0^2)_{ab^*} = \tfrac{1}{2}W''_{ac}W''^{cb*} + \tfrac{1}{2}(T_i)_a^b A^* T_i A + \tfrac{1}{2}(T_i A)^b (A^* T_i)_a, \tag{11.5.9}$$

$$M_{1/2} = \begin{matrix} & \psi & \lambda \\ \begin{matrix} \psi \\ \lambda \end{matrix} & \begin{pmatrix} W''_{ab} & \sqrt{2}(A^* T_i) \\ \sqrt{2}T_i A & 0 \end{pmatrix} \end{matrix}. \tag{11.5.10}$$

The mass matrix for the gauge bosons is

$$M_{ij}^2 = (A^* T_i, T_j A). \tag{11.5.11}$$

It is clear from this that, using trace condition $T_r T_i = 0$ for non-abelian groups, we obtain

$$T_r(M_0^2 - M_{1/2}^2 + 3M_1^2) = 0. \tag{11.5.12}$$

This mass formula was derived by Ferrara, Girardello, and Palumbo [5] in 1979 and has important implications for model building. The generality of the mass formula implies that, regardless of whether the supersymmetry is broken by F-terms or whether the internal symmetry is broken, all the boson masses cannot be heavier than the fermion masses as would be required for useful model building. It there is a linear D-term present in the theory the right-hand side of eqn. (11.5.12) is proportional (becomes proportional) to $D_i T_r(T_i)$ which may improve the situation somewhat except that, in order to cancel gauge anomalies, it is often desirable to have $T_r T_i = 0$ so that, again in realistic gauge models, eqn. (11.5.12) is the actual constraint.

Therefore, spontaneous breaking of supersymmetry would appear to be too restrictive a requirement for model building. We may argue that radiative corrections [6] may induce changes in the above mass formula but studies of semirealistic models in this regard do not appear very promising. Therefore, we will now proceed to a discussion of soft-breaking of supersymmetry.

As we saw in the previous chapter, requirements of renormalizability allow us to introduce only three classes of soft-breaking terms.

(i) Scalar mass terms, $\mu^2 A^* A$-type.
(ii) Trilinear scalar interactions of type $W(\Phi)|_{\theta=0}$.
(iii) Mass terms for the gaugino, i.e., $\lambda\lambda$.

We will see in the next chapter that these three kinds of terms are enough to lead to realistic models for particle physics; in fact, without additional constraints they lead to a large proliferation of parameters.

References

[1] J. Illiopoulos and B. Zumino, *Nucl. Phys.* **B76**, 310 (1974).
[2] L. O'Raifeartaigh, *Nucl. Phys.* **B96**, 331 (1975).
[3] P. Fayet and J. Illiopoulos, *Phys. Lett.* **51B**, 461 (1974).
[4] B. deWit and D. Freedman, *Phys. Rev. Lett.* **35**, 827 (1975);
W. Bardeen, unpublished.
[5] S. Ferrara, Ll Girardello, and F. Palumbo, *Phys. Rev.* **D20**, 403 (1979).
[6] For radiative corrections to supersymmetry breaking see
B. Zumino, *Nucl. Phys.* **B89**, 535 (1975);
S. Weinberg, *Phys. Lett.* **62B**, 111 (1976);
C. Nappi and B. A. Ovrut, *Phys. Lett.* **113B**, 175 (1982);
M. Dine and W. Fischler, *Nucl. Phys.* **B204**, 346 (1982);
M. Huq, *Phys. Rev.* **D14**, 3548 (1976);
E. Witten, Trieste lectures, 1981; *Nucl. Phys.* **B195**, 481 (1982).

CHAPTER 12

Phenomenology of Supersymmetric Models

§12.1. Supersymmetric Extension of the Standard Model

In the previous three chapters we have laid the foundation for applying the ideas of supersymmetry to building models of particle physics. At present there exists a successful (at low energies) model of electro-weak and strong interactions—the standard $SU(2)_L \times U(1)_Y \times SU(3)_c$ model. The recent discovery of W- and Z-bosons at the CERN $Sp\bar{p}S$ machine has proved the correctness of this theory. Also, everybody believes that there is more physics beyond the standard model. In Chapters 6, 7, and 8 we have discussed some interesting classes of models that provide examples of possible new physics. In this chapter we consider the possibility that new physics may be related to supersymmetry. Specifically, if elementary scalar bosons are to be part of the unified gauge theory framework, existence of a hidden supersymmetry might not only make their field theory better "behaved" but also establish a connection between fermions and bosons. As already emphasized in such a case, supersymmetry must be broken by soft terms since we do not observe any degenerate multiplets containing bosons and fermions. In this first section we will present a supersymmetric extension of the standard model. All fermions and bosons of the standard model must be accompanied by their supersymmetric partners which are bosons and fermions, respectively. Moreover, since supersymmetry commutes with the electro-weak symmetry, the transformation properties of the known particles and their superpartners must be the same, and the part of the Lagrangian that breaks supersymmetry must respect the electro-weak symmetry. These are the general guidelines that we will follow for constructing supersymmetric models. In the rest of this chapter we will call the supersymmetric partners (squark, slepton, ..., with a prefix s) and denote them with a tilde over the symbol representing the corresponding particle. In Table 12.1 we give the particle spectrum along with their electro-weak quantum numbers for one generation. We will also choose all particles to be left-handed (or chiral) particles so that a right-handed field (i.e., u_R) will be denoted as a left-handed antiparticle field (i.e., u_L^c).

We see from Table 12.1 that we have introduced two Higgs doublets. We will see that the supersymmetry actually requires it in order to give mass to the fermions. We are now ready to write down the Lagrangian to study the properties of these new particles [1]. We will write the action in supersymmetric notation so as to denote the explicit soft-breaking terms and we will introduce the following constant superfield $\eta = \theta^2$. We can then write

$$S = S_0 + S_1, \tag{12.1.1}$$

$$S_0 = S_g + \int W\, d^2\theta + \text{h.c.}, \tag{12.1.2}$$

where S_g represents the gauge part of the action which can be written down following the rules in Chapter 10 and we will give this in component notation soon. (Again, we carry out our discussion for one generation for simplicity.)

$$W = h_u Q^T \tau_2 H_u U + h_d Q^T \tau_2 H_d D + h_e L^T \tau_2 H_d E + \lambda X(H_u^T \tau_2 H_d - \mu^2). \tag{12.1.3}$$

We have introduced a singlet superfield X in order to have a realistic model of spontaneous symmetry breaking [2]. We also note that if we did not have two Higgs doublets either the U or D superfield will be absent from the superpotential and will therefore remain massless. It is important to emphasize at this stage that we have not included in the superpotential all possible terms allowed by the electro-weak symmetry, i.e., LH_u, $\varepsilon_{ijk} U^i D_a^j D_b^k$, a and b denoting generations, etc., since they lead to lepton and baryon number violation which we do not wish to discuss at this stage.

We will choose S_1 to include all allowed terms that break supersymmetry softly

$$\begin{aligned} S_1 = \int & d^2\theta\, \eta[W(\mu^2 = 0) + m_0 \mathbf{W}^a \mathbf{W}_a + m_1 W_B^a W_{Ba}] + \text{h.c.} \\ & + d^4\theta\, \eta^\dagger \eta [m_{\tilde{Q}}^2 Q^* Q + m_{\tilde{U}}^2 U^* U + m_{\tilde{D}}^2 D^* D + m_{\tilde{L}}^2 L^* L + m_{\tilde{E}}^2 E^* E], \end{aligned} \tag{12.1.4}$$

where $\eta = \theta^2$.

As we will see below, all but the first term within the bracket are needed to give arbitrary masses to the superpartners. Using the component field notation of Table 12.1 we can write down the Lagrangian for this model

$$\mathscr{L} = \mathscr{L}_{\text{gauge}} + \mathscr{L}_{\text{matter}} + \mathscr{L}_Y - V + \mathscr{L}_1. \tag{12.1.5}$$

Table 12.1

Superfield	Component fields	$SU(2)_L \times U(1) \times SU(3)$ quantum number	Name
	Matter fields		
Q	$\begin{pmatrix} u_L \\ d_L \end{pmatrix} \equiv Q_L$	$(2, \frac{1}{3}, 3)$	Quark
	$\begin{pmatrix} \tilde{u}_L \\ \tilde{d}_L \end{pmatrix} \equiv \tilde{Q}$		Squark
U	u_L^c	$(1, -\frac{4}{3}, 3^*)$	Quark
	$\tilde{u}_L^c$		(denotes right-handed up-quark)
D	d_L^c	$(1, +\frac{2}{3}, 3^*)$	Squark
	$\tilde{d}_L^c$		(denotes right-handed down-quark)
L	$\begin{pmatrix} \nu_L \\ e_L^- \end{pmatrix}$	$(2, -1, 1)$	Lepton
	$\begin{pmatrix} \tilde{\nu}_L \\ \tilde{e}_L \end{pmatrix}$		Slepton
E	e_L^c	$(1, +2, 1)$	Antilepton
	$\tilde{e}^c$		Antislepton
	Gauge fields		
V	$\begin{pmatrix} W^{\pm} \\ W^3 \end{pmatrix}$	$(3, 0, 1)$	Gauge bosons
	$\begin{pmatrix} \tilde{W}^{\pm} \\ \tilde{W}^3 \end{pmatrix}$		Gaugino
B	B	$(1, 0, 1)$	Gauge bosons
	$\tilde{B}$		Gaugino
	Higgs fields		
H_u	$\begin{pmatrix} \phi_u^+ \\ \phi_u^0 \end{pmatrix}$	$(2, +1, 1)$	Higgs field
	$\begin{pmatrix} \tilde{\phi}_u^+ \\ \tilde{\phi}_u^0 \end{pmatrix}$		Higgsino
H_d	$\begin{pmatrix} \phi_d^0 \\ \phi_d^- \end{pmatrix}$	$(2, -1, 1)$	Higgs field
	$\begin{pmatrix} \tilde{\phi}_d^0 \\ \tilde{\phi}_d^- \end{pmatrix}$		Higgsino

We will have the familiar four-component notation

$$\mathscr{L}_{\text{gauge}} = -\tfrac{1}{4}\mathbf{W}_{\mu\nu}\cdot\mathbf{W}_{\mu\nu} - \tfrac{1}{4}B_{\mu\nu}B_{\mu\nu} - \bar{\tilde{W}}\gamma\cdot\nabla\tilde{W} - \bar{\tilde{B}}\gamma\cdot\partial\tilde{B}, \tag{12.1.6a}$$

$$\mathscr{L}_{\text{matter}} = \sum - \bar{\psi}\gamma_\mu(D_\mu)\psi - \sum(D_\mu A_\psi)^+(D_\mu A_\psi)$$
$$+ i\frac{g}{\sqrt{2}}\sum\bar{\psi}_L\boldsymbol{\tau}\cdot\tilde{\mathbf{W}}A_\psi + \text{h.c.} + i\frac{g'}{\sqrt{2}}\sum\bar{\psi}_L\tilde{B}YA_\psi + \text{h.c.} \tag{12.1.6b}$$

Summation goes over $\psi = Q, U, D, L, E, \tilde{H}_u, \tilde{H}_d, \psi_x$, and

$$A_\psi = \tilde{Q}, \tilde{U}, \tilde{D}, \tilde{L}, \tilde{E}, H_u, H_d, \tilde{\psi}_x,$$

$$D_\mu = \partial_\mu - \frac{ig}{2}\boldsymbol{\tau}\cdot\mathbf{W}_\mu - \frac{ig'}{2}YB_\mu, \tag{12.1.6c}$$

Y being the appropriate $\mathrm{U}(1)_Y$ quantum number. The $\boldsymbol{\tau}\cdot\mathbf{W}$ term will be absent when a particle is $\mathrm{SU}(2)_L$ singlet. ∇ denotes the covariant derivative for the appropriate gauge fields

$$\mathscr{L}_Y = h_u(Q_L^T C^{-1}\tau_2 H_u U_L^c + Q_L^T C^{-1}\tau_2\tilde{H}_u + \tilde{U}^c\tilde{H}_c^T C^{-1}\tau_2\tilde{Q}U_L^c) + (u\to d)$$
$$+ h_e(L^T C^{-1}\tau_2 H_d e^c + L^T C^{-1}\tau_2\tilde{H}_d\tilde{e}_c + \tilde{H}_d^T C^{-1}\tau_2\tilde{L}e_L^c)$$
$$+ \lambda(\tilde{H}_d^T C^{-1}\tau_2 H_d\psi_x + \tilde{H}_d^T C^{-1}\tau_2 H_u\psi_x + \tilde{H}_u C^{-1}\tau_2\tilde{H}_d\tilde{\psi}_x) + \text{h.c.} \tag{12.1.7}$$

Here C is the Dirac charge conjugation matrix and τ_2 is the second Pauli matrix. We identify the first terms in each of the first three bracketed expressions in eqn. (12.1.7) as the Yukawa couplings present in a two Higgs extension of the standard model (see Chapter 4). The remaining term involving fermions is the soft supersymmetry breaking Majorana mass term, i.e.,

$$\mathscr{L}_{\text{soft}} = m'\tilde{\mathbf{W}}^T C^{-1}\tilde{\mathbf{W}} + m''\tilde{B}^T C^{-1}\tilde{B} + \text{h.c.} \tag{12.1.8}$$

Let us now turn to the potential

$$V = |F|^2 + D^2 + V_{\text{soft}}, \tag{12.1.9}$$

$$|F|^2 = |h_u\tilde{Q}\tilde{U}^c + \lambda\tilde{\psi}_x H_d|^2 + |h_d\tilde{Q}\tilde{d}^c + h_e\tilde{L}\tilde{e}^c + \lambda\tilde{\psi}_c H_u|^2$$
$$+ |h_u H_u\tilde{u}^c + h_d H_d\tilde{d}^c|^2 + h_u^2|\tilde{Q}^T\tau_2 H_u|^2 + h_d^2|\tilde{Q}^T\tau_2 H_d|^2$$
$$+ h_e^2(H_d^+ H_d\tilde{e}^{c*}\tilde{e}^c + |\tilde{L}^T\tau_2 H_d|^2) + \lambda^2|H_u^T\tau_2 H_d - \mu^2|^2, \tag{12.1.10a}$$

$$V_{\text{soft}} = h_u\tilde{Q}^T\tau_2 H_u\tilde{u}^c + h_d\tilde{Q}^T\tau_2 H_d\tilde{d}^c + h_e\tilde{L}^T\tau_2 H_d\tilde{e}^c$$
$$+ \lambda\tilde{\psi}_x H_u^T\tau_2 H_d + \text{h.c.} + m_Q^2\tilde{Q}^+\tilde{Q} + m_L^2\tilde{L}^+\tilde{L} + m_u^2\tilde{u}^{c*}\tilde{u}^c$$
$$+ m_D^2\tilde{d}^{c*}\tilde{d}^c + m_E^2\tilde{e}^{c*}\tilde{e}^c, \tag{12.1.10b}$$

$$\tfrac{1}{4}D^2 = \tfrac{1}{4}g^2\sum_a\left|\sum_{A_\psi}A_\psi^+\tau_a A_\psi\right|^2 + \tfrac{1}{4}g'^2\left|\sum_{A_\psi}A_\psi^+ YA_\psi\right|^2. \tag{12.1.10c}$$

Let us now study the spontaneous breaking of the gauge symmetry. A look at eqns. (12.1.9) and (12.1.10) makes it clear that for $m_{\tilde{Q}}^2$, $m_{\tilde{L}}^2$, $m_{\tilde{U}}^2$, $m_{\tilde{D}}^2$, $m_{\tilde{E}}^2$ positive, the minimum of V corresponds to

$$\langle \tilde{Q} \rangle = \langle \tilde{L} \rangle = \langle \tilde{e}^c \rangle = \langle \tilde{u}^c \rangle = \langle \tilde{d}^c \rangle = 0 \tag{12.1.11}$$

and

$$\langle H_u^0 \rangle = \langle H_d^0 \rangle = \frac{v}{\sqrt{2}} \equiv \mu. \tag{12.1.12}$$

The above equations break the $SU(2)_L \times U(1)$ symmetry leaving $SU(3)_c \times U(1)_{em}$ symmetry intact. This also gives nonzero masses to the quarks and leptons. As far as the Higgs sector is concerned (decomposing $H_u^0 = \mu + \sigma_a + i\chi_a$), the three Higgs–Kibble bosons are

$$\frac{1}{\sqrt{2}}(\chi_u - \chi_d), \qquad \frac{1}{\sqrt{2}}(H_u^{\pm} - H_d^{\pm}),$$

which are absorbed to become longitudinal modes of the gauge bosons. The combination $(1/\sqrt{2})(\sigma_u - \sigma_d)$ acquires a mass $gv/2$ due to the D-terms in (12.1.10). The remaining physical Higgs fields $(1/\sqrt{2})(\sigma_u + \sigma_d)$, $(1/\sqrt{2})(\chi_u + \chi_d)$, and $(1/\sqrt{2})(H_u^{\pm} + H_d^{\pm})$ all pick up masses λv. The masses of W and Z are given by

$$m_W = \frac{gv}{\sqrt{2}} \tag{12.1.13}$$

and

$$m_Z = \sqrt{g^2 + g'^2}\,\frac{v}{\sqrt{2}}. \tag{12.1.14}$$

Fermion Spectrum

Let us now turn to the masses of those fermions which arise as supersymmetric partners of known particles. By our choice of soft-breaking terms, the winos ($\tilde{W}$), bino ($\tilde{B}$) already have a Majorana mass. The Higgsino masses, on the other hand, cannot be put in as part of the soft-breaking Lagrangian since they spoil renormalizability of the theory. To study their mass spectra, subsequent to spontaneous breakdown, we look at the last terms in eqn. (12.1.6b) and find the following fermion bilinears in the Lagrangian:

$$\begin{aligned}\mathscr{L}_{\text{mass}} = \frac{iv}{2}\{&\bar{\tilde{H}}_u^0(g\tilde{W}^3 + g'\tilde{B}) - \bar{\tilde{H}}_d^0(g\tilde{W}_3 + g'\tilde{B}) \\ &+ \sqrt{2}gv(\bar{\tilde{H}}_u^- \tilde{W}^- + \bar{\tilde{H}}_d^- \tilde{W}^+)\} + \text{h.c.},\end{aligned} \tag{12.1.15}$$

where

$$\tilde{W}^{\pm} = \frac{1}{\sqrt{2}}(\tilde{W}^1 \mp i\tilde{W}^2).$$

Thus, we see that the gauginos acquire Dirac mass through their mixing with Higgsinos and

$$m^D_{\tilde{W}_\pm} = m_W. \tag{12.1.16}$$

Similarly, the combination $[1/(\sqrt{g^2 + g'^2})](g\tilde{W}^3 + g'\tilde{B})$ (to be called zino, $\tilde{Z}$) combines with $(\tilde{H}^0_u - \tilde{H}^0_d)/\sqrt{2}$ to form a Dirac fermion whose mass is the same as m_Z. This degeneracy between the gauge boson and gaugino mass is to be expected since the mass written down in eqn. (12.1.15) does not "know" about supersymmetry breaking. In fact, in terms of the two-component notation, the various Dirac particles can be written as follows:

$$W^+ = \begin{pmatrix} \tilde{W}^+ \\ \bar{\tilde{H}}^-_d \end{pmatrix}$$

and

$$\omega^- = \begin{pmatrix} \tilde{W}^- \\ \bar{\tilde{H}}^+_d \end{pmatrix},$$

$$\tilde{Z} = \begin{pmatrix} \dfrac{g\tilde{W}^3 + g'\tilde{B}}{\sqrt{g^2 + g'^2}} \\ \dfrac{\tilde{H}^0_u - \tilde{H}^0_d}{\sqrt{2}} \end{pmatrix}. \tag{12.1.17}$$

Note that, as to be expected, the combination $(g'\tilde{W}^3 - g\tilde{B})/(\sqrt{g^2 + g'^2})$ (denoted by $\tilde{\gamma}$, the photino) remains massless in the absence of supersymmetry breaking.

As mentioned in eqn. (12.1.9) supersymmetry breaking can bring in Majorana mass terms for the gauginos which can be rewritten as follows:

$$\mathscr{L}_{\text{soft}} = m'\tilde{W}^{+^T}C^{-1}\tilde{W}^- + m'\tilde{W}^{3^T}C^{-1}\tilde{W}^3 + m''\tilde{B}^T C^{-1}\tilde{B} + \text{h.c.} \tag{12.1.18}$$

In terms of the Dirac particles defined in eqn. (12.1.17) the first term is easily expressed as a mixing term as follows:

$$m'\tilde{W}^{+^T}C^{-1}\tilde{W} = m'\tilde{\omega}^{+^T}C^{-1}\tilde{\omega}^-. \tag{12.1.19}$$

As far as the neutral bosons are concerned, in the presence of the mixing terms in eqn. (12.1.19), the situation looks more complicated. The general mass matrix is expressed as follows (using $\tan\theta_W = g'/g$):

$$\begin{aligned}\mathscr{L}^{\text{neutral}}_{\text{soft}} &= (m'\cos^2\theta_W + m''\sin^2\theta_W)\tilde{Z}^T_L C^{-1}\tilde{Z}_L \\ &\quad + (m'\sin^2\theta_W + m''\cos^2\theta_W)\tilde{\gamma}^T C^{-1}\tilde{\gamma} \\ &\quad + 2(m' - m'')\cos\theta\sin\theta\tilde{Z}^T_L C^{-1}\tilde{\gamma}_L + \text{h.c.}\end{aligned} \tag{12.1.20}$$

This gives a nonzero mass to the photino and leads to mixing between the zino and photino. Also, if we choose $m' = m''$, the $\tilde{\gamma} - \tilde{Z}_L$ mixing disappears. The mass matrix for $\tilde{Z} - (1/\sqrt{2})(\tilde{H}_u^0 - \tilde{H}_d^0)$ system takes the following form:

$$\begin{array}{cc} & \tilde{Z} \quad \frac{1}{\sqrt{2}}(\tilde{H}_u^0 - \tilde{H}_d^0) \\ \begin{array}{c}\tilde{Z}\\ \frac{1}{\sqrt{2}}(\tilde{H}_u^0 - \tilde{H}_d^0)\end{array} & \begin{pmatrix} m'' & m_Z \\ m_Z & 0 \end{pmatrix}. \end{array} \tag{12.1.21}$$

The two eigenstates of this matrix represent Majorana fermions $\tilde{Z}_1$ and $\tilde{H}^0$

$$\tilde{Z}_1 = \cos\alpha \tilde{Z} + \frac{\sin\alpha}{\sqrt{2}}(\tilde{H}_u^0 - \tilde{H}_d^0),$$

$$\tilde{H}^0 = -\sin\alpha \tilde{Z} + \frac{\cos\lambda}{\sqrt{2}}(\tilde{H}_u^0 - \tilde{H}_d^0), \tag{12.1.22}$$

where

$$\tan 2\alpha = \frac{2m_Z}{m''}$$

and eigenvalues

$$m_{\tilde{Z}_1} = \frac{m'' + \sqrt{m''^2 + 4m_Z^2}}{2}$$

and

$$m_{\tilde{H}^0} = \frac{m'' - \sqrt{m''^2 + 4m_Z^2}}{2}. \tag{12.1.23}$$

$m_{\tilde{H}^0}$ is negative and can be made positive by a γ_5-transformation on the spinors. Furthermore, $m_{\tilde{H}^0}$ is lighter than m_Z and can be produced in Z-decay. This discussion will be important in the phenomenological considerations that will follow.

To complete the discussion of fermion masses in the model presented, we note that the singlet fermion ψ_X acquires a Dirac mass after spontaneous breakdown through its mixing with $(\tilde{H}_u^0 + \tilde{H}_d^0)(1/\sqrt{2})$. Calling the new Dirac particle ψ we have

$$\psi = \begin{pmatrix} \psi_x \\ \frac{1}{\sqrt{2}}(\tilde{H}_u^0 + \tilde{H}_d^0) \end{pmatrix} \tag{12.1.24}$$

and

$$m_\psi = \lambda v = m_{(\sigma_u + \sigma_v)}. \tag{12.1.25}$$

Finally, we note that we have not discussed the $SU(3)_c$ sector. This will involve the interactions of the quarks and squarks with gluons, G, and gluinos, $\tilde{G}$. The general form of their interaction is similar to the electro-weak gauge interactions with appropriate modifications, i.e.,

$$\begin{aligned}\mathscr{L}^{\text{color}}_{\text{gauge+matter}} = &-\tfrac{1}{4}G^{i}_{\mu\nu_j}G^{j}_{\mu\nu_i} - \bar{\tilde{G}}^{i}_{j}\gamma\cdot(\nabla_G\tilde{G})^{j}_{i} - \bar{Q}^{i}\gamma_\mu D^{j}_{\mu,i}Q_j - (D_\mu\tilde{Q})^{+i}(D_\mu\tilde{Q})_i \\ &+ \frac{ig_s}{\sqrt{2}}\bar{\tilde{Q}}^{i}(\tilde{G})^{j}_{i}Q_j + \text{h.c.} + \tfrac{1}{4}(\tilde{Q}^{+i}\tilde{Q}_j)(\tilde{Q}^{+j}\tilde{Q}_i) \\ &+ \text{similar terms replacing } Q \text{ by } U, D. \end{aligned} \tag{12.1.26}$$

Without supersymmetry breaking the gluino fields have no mass. As in the previous case, the supersymmetry breaking introduces Majorana masses for the gluinos, $\tilde{G}$, which we choose as $m_{\tilde{G}}$. By $SU(3)_c$ invariance all eight gluinos have the same Majorana mass

$$\mathscr{L}_{\tilde{G}} = m_{\tilde{G}}(\tilde{G}^{T\,i}_{j}C^{-1}\tilde{G}^{j}_{i} + \text{h.c.}). \tag{12.1.27}$$

§12.2. Constraints on the Masses of Superparticles

In this section we will discuss the phenomenological constraints on the masses of the superpartners of the "known" particles. (We emphasize that, by "known," we do not necessarily mean that a particle has been observed but simply that it is familiar to us from our study of nonsupersymmetric models.) We will use the Lagrangian discussed in the previous section as representing the broad features of the interactions between particles and superparticles, although we will let the parameters be arbitrary. An important point to remember is that the interaction between particles and their superpartners obeys a selection rule, even in the presence of the most general supersymmetry breaking terms. Let us call it R-parity [3]. We can assign R-parity $+1$ to "normal" particles and -1 to their superpartners. Then all interaction vertices must respect R-parity invariance, i.e., they must contain an even number of super- (tilded) particles. One way to break R-parity is to give a vacuum expectation value to the scalar neutrino [4] $\langle\tilde{\nu}\rangle \neq 0$. But this also gives rise to lepton number violation implying that $\langle\tilde{\nu}\rangle \leq 10^{-6}\, m_W$. Therefore, in general, its effects will be very small.

To get an idea about phenomenologically allowed masses for various s-particles we look at their effects on known processes, such as e^+e^- annihilation and low-energy weak processes. We consider them one by one and we will also assume that the lightest s-particles are $\tilde{\nu}$, $\tilde{G}$, and $\tilde{\gamma}$. Before we proceed further it is also important to point out that the interactions of $\tilde{\gamma}$, $\tilde{\nu}$, and $\tilde{G}$ are all extremely weak. For instance, $\tilde{\gamma}$-interaction with matter proceeds through the diagrams in Fig. 12.1 and is given by [1]

$$\sigma_{\tilde{\gamma}\text{-matt}} \approx 2\times 10^{-37}\,(E_{\tilde{\gamma}} \text{ in G3V})\left(\frac{m_W}{M_{\tilde{q}}}\right)^4 F\ \text{cm}^2, \tag{12.2.1}$$

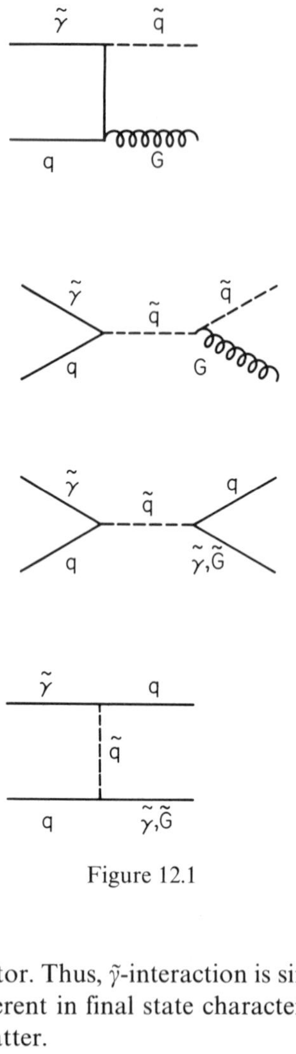

Figure 12.1

where F is a structure factor. Thus, $\tilde{\gamma}$-interaction is similar to ν-interaction in cross section though different in final state characteristics. Thus, $\tilde{\gamma}$ interacts only very weakly with matter.

(a) Scalar Leptons

There are two kinds of charged scalar leptons: $\tilde{l}_L$ and $\tilde{l}_R$ ($l = e, \mu, \tau$). In general, supersymmetry breaking will mix these states. These states will be produced in e^+e^- annihilation: $e^+e^- \to \tilde{l}^+\tilde{l}^-$. Since these are spin 0 bosons, their cross section will be $\sim(\beta^3/4)\sigma(e^+e^- \to l^+l^-)$ where $\beta = v/c$, the P-ware threshold factor. Because of the β^3-factor, the rise in the ratio $R = \sigma(e^+e^- \to \tilde{l}^+\tilde{l}^-)/\sigma_{\text{tot}}$ will only be observable far above the threshold. However, in general

$$\tilde{l}^{\pm} \to l^{\pm}\tilde{\gamma}. \tag{12.2.2}$$

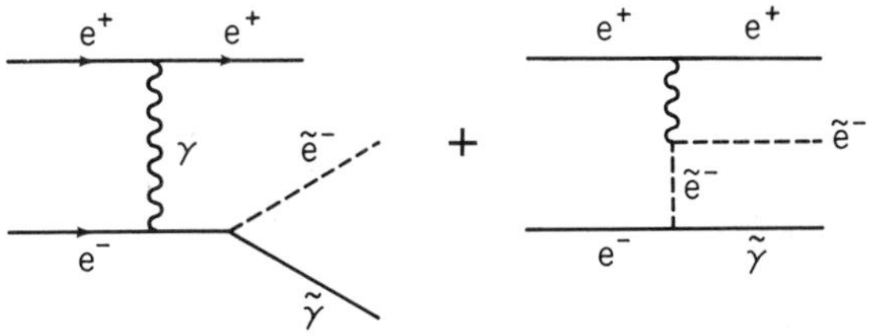

Figure 12.2

Therefore, above the threshold for the production of $\tilde{l}^{\pm}$, we would expect the product of $l^{\pm}$ with a lot of missing energy similar to the product of $\tau^{\pm}$. Since $\tau^{\pm}$ is a calculable background this can be calculated and subtracted. So far, no such events have been observed in e^+e^- colliders [5] leading to bounds of 18 GeV on $\tilde{e}$ and $\tilde{\mu}$. A more sensitive method to search for $\tilde{l}^{\pm}$ is to look for [6]

$$e^+e^- \to e^+\tilde{e}^-\tilde{\gamma},$$

which proceeds via the Feynman diagrams in Fig. 12.2. For light photinos, $m_{\tilde{\gamma}} < 1$ GeV. In the Weizsacker–Williams equivalent photon approximation the cross section [6] turns out to be

$$\frac{\sigma(e^+e^- \to e^+\tilde{e}_L^-\tilde{\gamma})}{\sigma(e^+e^- \to \mu^+\mu^-)} = \frac{\alpha}{12\pi}\log\left(\frac{E}{m_e}\right)$$
$$\times\, [2/x + 18 - 54x + 34x^2 + 3(3 - 3x - 4x^2)\log x - 9x\log^2 x], \tag{12.2.3}$$

where $x = M_{\tilde{e}}^2/4E_e^2$. This method leads to a lower bound on the $\tilde{e}$ mass of about 22 GeV. We can also look at processes where a virtual scalar lepton is involved, such as $e^+e^- \to \gamma\tilde{\gamma}\tilde{\gamma}$, which proceed through the Feynman diagrams of Fig. 12.3.

The difficulty of experimental isolation of the two photino final states, in inclusive $e^+e^- \to \gamma + x$, has been the reason that no useful lower bound on $\tilde{e}^{\pm}$ results from the consideration [7].

Finally, if $m_{\tilde{e}} < m_Z/2$, we will expect the Z-decaying to $\tilde{e}_L^+\tilde{e}_L^-$ and $\tilde{e}_R^+\tilde{e}_R^-$ final states. The coupling strengths are the same as that for the electron, except it is a p-wave leading to k^3-dependence of the decay width where k is the

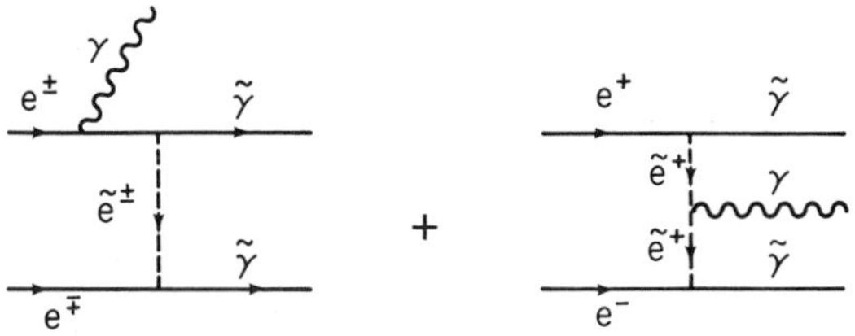

Figure 12.3

$\tilde{e}$-momentum, i.e.,

$$\frac{\Gamma(Z^0 \to \mathbf{e}_L^+ \mathbf{e}_L^- + \tilde{e}_R^+ \tilde{e}_R^-)}{\Gamma(Z_0 \to e^+ e^-)} = \frac{1}{2}\left(1 - \frac{4M_{\tilde{e}}^2}{M_Z^2}\right)^{3/2}. \tag{12.2.4}$$

For $M_Z > M_{\tilde{e}} \geq M_Z/2$ there will be supersymmetric decay modes of Z^0 of type $Z^0 \to e^\pm \tilde{e}^\pm \tilde{\gamma}$, which should provide e^+e^- with missing energy in Z^0-decays.

(b) Scalar Neutrino

In contrast to the superpartner of the charged lepton there is only one superpartner of ν_L to be denoted by $\tilde{\nu}_L$. This, being a neutral particle, will be produced in e^+e^- collision only via the Z_0 intermediate state. Its cross section will therefore be extremely small ($\sim G_F^2$) and e^+e^- scattering does not lead to any bound on its mass. One way to obtain a lower bound on its mass is to study τ-decays. If $m_{\tilde{\nu}_e} + m_{\tilde{\nu}_\tau} < (m_\tau - m_e)$ then there will be a supersymmetric contribution to the τ-decay arising from the diagram shown in Fig. 12.4. For expected values of $\tilde{W}^\pm$-mass this amplitude will be comparable to that of three-lepton and other decay modes. Present observations on τ-lepton decays would imply (for $m_{W^+} \simeq m_{\tilde{W}^+}$) that $m_{\tilde{\nu}_e} + m_{\tilde{\nu}_\tau} > (m_\tau - m_e)$.

Since we have assumed $m_{\tilde{\gamma}} \approx 1$ GeV, properties of $\tilde{\nu}$ will depend on whether $m_{\tilde{\nu}} < m_{\tilde{\gamma}}$ or $> m_{\tilde{\gamma}}$. We will assume that the gluino mass is much bigger than 1 GeV as there appear to be indications from collider experiments. If $m_{\tilde{\nu}} < m_{\tilde{\gamma}}$, it will be an absolutely stable particle except in the model of Ref. [4], where $\tilde{\nu}_e \to e^+e^-, \nu\tilde{\nu}, \ldots$, etc. through the diagrams typically of the type in Fig. 12.5.

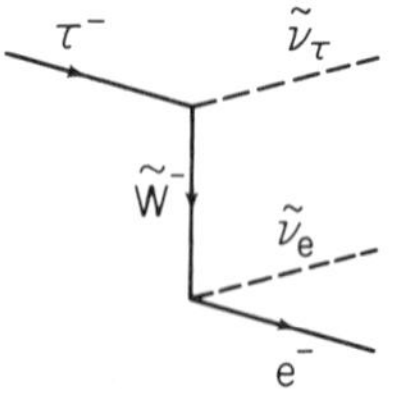

Figure 12.4

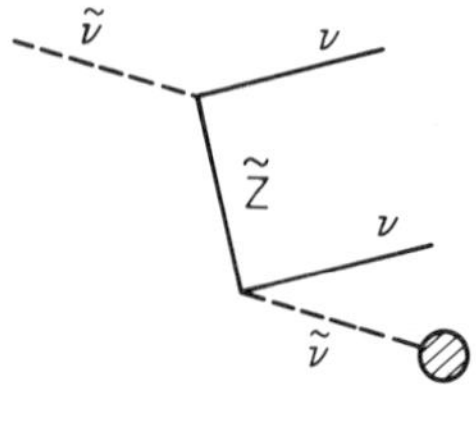

Figure 12.5

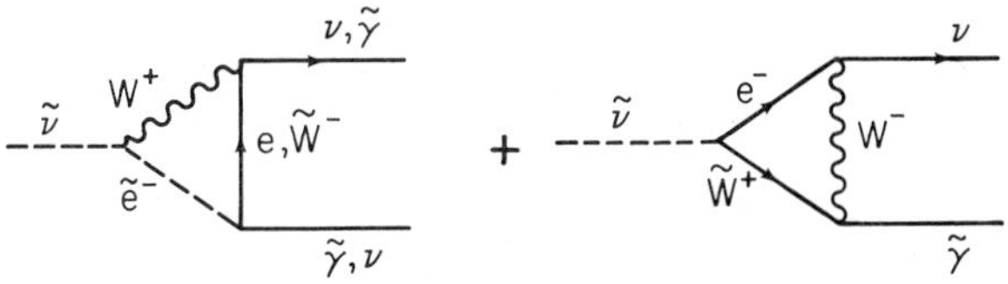

Figure 12.6

These amplitudes are proportional to

$$A(\tilde{\nu} \to \nu\nu) \simeq G_F m_{\tilde{\nu}} \langle \tilde{\nu} \rangle. \tag{12.2.5}$$

Since $\langle \tilde{\nu} \rangle \leq 1$ KeV, this decay mode will imply a lifetime of $\tilde{\nu}$ of about 10^{-2} s from $m_{\tilde{\nu}} \simeq 1$ GeV. However, the model of Ref. [4] also leads naturally to $m_{\tilde{\nu}}$ in the KeV range, in which case its lifetime would be 10^3 s. On the other hand, if $m_{\tilde{\gamma}} < m_{\tilde{\nu}}$ the dominant decay mode turns out to be [8] $\tilde{\nu} \to \nu + \tilde{\gamma}$ arising from the diagrams shown in Fig. 12.6. The strength of this amplitude is given by [8]

$$A(\tilde{\nu} \to \nu\tilde{\gamma}) \simeq \frac{ie^3}{32\pi^2 \sqrt{2} \sin^2 \theta_W} \tag{12.2.6}$$

leading to a lifetime of order 10^{-16} s for $M_{\tilde{\nu}} \simeq 1$ GeV. In this case there exist weaker decay modes ($\sim G_F$) which involve charge final states such as $\tilde{\nu} \to \nu e^+ e^- \tilde{\gamma}$, $\nu q \bar{q} \tilde{\gamma}$, etc.

As far as the production of scalar neutrino is concerned it could appear in Z^0-decay, singly if $M_Z/2 < M_{\tilde{\nu}} < M_Z$, and in pairs if $M_{\tilde{\nu}} < M_Z/2$. The ratio of decay widths of $\Gamma(Z^0 \to \tilde{\nu}\bar{\tilde{\nu}})$ to $\Gamma(Z^0 \to \nu\bar{\nu})$ is given by the same formula as in eqn. (12.2.4).

(c) Scalar Quarks

The properties of scalar quarks ($\tilde{q}_L$, $\tilde{q}_R$) are similar to those of scalar leptons, with two important exceptions, i.e., they have fractional charges and strong interactions. Again, in this discussion, we will assume photino mass to be less than $m_{\tilde{q}}$. This immediately implies that $\tilde{q} \to q\tilde{\gamma}$. If $m_{\tilde{G}} < m_{\tilde{q}}$ there exists the additional decay mode $\tilde{q} \to q\tilde{g}$. In this case the production of $\tilde{q}$ will be indicated by a quark jet with missing energy.

As in the case of sleptons we would expect the production of squarks in e^+e^- annihilation, with differential cross section to be given by

$$\left(\frac{d\sigma}{d\Omega}\right)_{e^+e^- \to \tilde{q}\bar{\tilde{q}}} = \frac{3}{8}\frac{\alpha^2}{s} e_q^2 \beta^3 \sin^2 \theta, \tag{12.2.7}$$

where e_q is the quark charge, $\beta = \sqrt{1 - (4M_{\tilde{q}}^2/S)}$, and θ is the center of mass production angle.

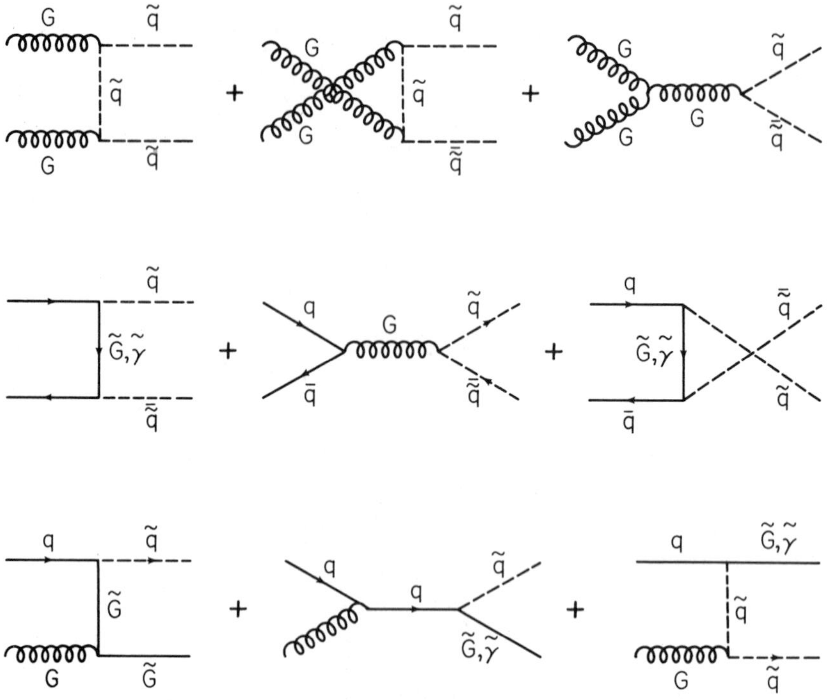

Figure 12.7

Present experimental limits from e^+e^- collisions give 3.1 GeV $< M_{\tilde{q}} <$ 17.8 GeV and 7.4 GeV $< M_{\tilde{q}} <$ 16.0 GeV, assuming $\tilde{q} \to q\tilde{\gamma}$ to be the dominant decay mode.

Because the squarks carry strong interaction quantum numbers it can be produced in $p\bar{p}$ collision (through the Drell–Yan mechanism) through collisions of quarks, antiquarks, gluons, or quarks and gluons through the diagrams [9] shown in Fig. 12.7. The quark produced in the final state will decay to $q + \tilde{\gamma}$ which would appear as a monojet in $p\bar{p}$ collision.

Finally, as in the case of $\tilde{l}$ and $\tilde{\nu}$, $\tilde{q}$ can also be produced in pairs, or singly, with $\tilde{\gamma}$ in Z^0-decay.

(d) Gluino Physics

Gluino ($\tilde{G}$) is the spin 1/2 superpartner of the gluon and is similar to the photino except that there are eight of them corresponding to the eight generators of $SU(3)_c$ and they have only strong interactions and no weak and electromagnetic interactions. We will again assume that it is more massive than $\tilde{\gamma}$. A dominant decay mode of the gluino is $\tilde{G} \to q\bar{q}\tilde{\gamma}$ arising from the

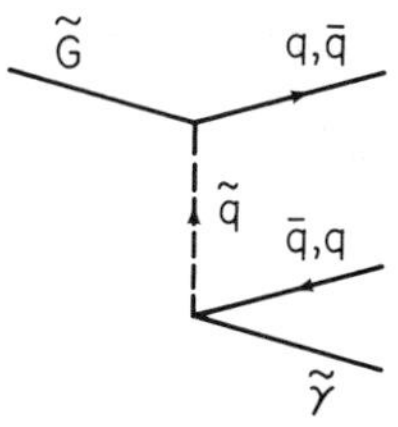

Figure 12.8

Feynman diagram in Fig. 12.8. For $M_{\tilde{q}_{L,R}} \gg M_{\tilde{G}}$ the decay width of $\tilde{G}$ is given by [10] (with $x = E_{\tilde{\gamma}}/M_{\tilde{G}}$; $y = M_{\tilde{\gamma}}/M_{\tilde{G}}$)

$$\frac{d\Gamma}{dx} = \frac{\alpha\alpha_s}{2\pi} e_q^2 M_{\tilde{q}}^5 \left(\frac{1}{M_{\tilde{q}_L}^4} + \frac{1}{M_{\tilde{q}_R}^4}\right)(x^2 - y^2)^{1/2}$$
$$\times [x(1 - \tfrac{4}{3}x) + y(1 - 2x) - (\tfrac{2}{3} - x)y^2 + y^3)]. \tag{12.2.8}$$

This is similar to a weak decay since $\Gamma \sim (\alpha\alpha_s M_{\tilde{G}}^2/M_{\tilde{q}}^4)$ as in the case for typical weak decays.

An interesting decay mode of the gluino is $\tilde{G} \to G\tilde{\gamma}$. This violates both parity and charge conjugation. To see why it violates parity, let us note that both $\tilde{G}$ and $\tilde{\gamma}$ are Majorana particles which satisfy the condition

$$\psi = C\bar{\psi}^T,$$

where

$$\psi = \tilde{G}, \tilde{\gamma}. \tag{12.2.9}$$

Let

$$\psi \xrightarrow{P} \eta\gamma_4\psi. \tag{12.2.10}$$

Substituting eqn. (12.2.10) into eqn. (12.2.9) we find that consistency requires [11] $\eta = \pm i$. Thus, unlike a Dirac fermion, Majorana fermions of all varieties have a unique parity transformation. Since G has odd parity the decay, $\tilde{G} \to G\tilde{\gamma}$ is parity forbidden. However, if the masses of left- ($\tilde{q}_L$) and right- ($\tilde{q}_R$) squarks are different, then through the higher order graph in Fig. 12.9, this

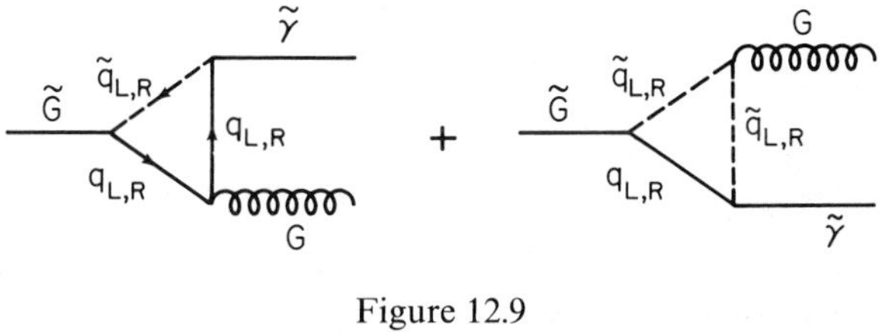

Figure 12.9

process can arise. The amplitude for this process is of order

$$A(\tilde{G} \to G\tilde{\gamma}) \approx e\alpha_s M_{\tilde{G}}^2 \left(\frac{1}{M_{\tilde{q}_L}^4} - \frac{1}{M_{\tilde{q}_R}^4} \right). \tag{12.2.11}$$

The ratio of these two decar rates is given by

$$\frac{\Gamma(\tilde{G} \to G\tilde{\gamma})}{\Gamma(\tilde{G} \to q\bar{q}\tilde{\gamma})} = \frac{3\alpha_s}{4\pi} \frac{(M_{\tilde{q}_L}^2 - M_{\tilde{q}_R}^2)^2}{(M_{\tilde{q}_L}^4 + M_{\tilde{q}_R}^4)}. \tag{12.2.12}$$

This ratio can therefore be 10% or more in general supersymmetric models. This would be important in experimental searches for the gluino in high-energy hadron experiments.

Interesting spectroscopy implications can arise if the gluino is light since it can form bound states [12] with quarks and antiquarks: $q\bar{q}\tilde{G}$. These particles will be fermions. Some of these particles (both charged and neutral) will be stable with respect to strong interactions and will be long lived and can leave tracks or gaps in detectors. There can also exist two gluino bound states, the gluinonium, whose spectroscopy has been studied [13]. The dominant decay mode of the $\eta_{\tilde{G}}$ is to hadrons. Because of the color octet nature of $\tilde{G}$, production of $\eta_{\tilde{G}}$ is enhanced by a factor of $\frac{27}{2}$ with respect to η_c and would thus be copiously produced in $p\bar{p}$ collision.

It has recently been suggested that several exotic events with jets, but with large missing transverse momentum, found in the CERN $p\bar{p}$ experiment [13a] can be explained in terms of gluino production [13b] $q\bar{q} \to \tilde{G}\tilde{G}$, $GG \to \tilde{G}\tilde{G}$, and subsequent gluino decay to $q\bar{q}\tilde{\gamma}$ provided 15–25 GeV $<$ $m_{\tilde{G}} < 45$ GeV. Further tests of this hypothesis are clearly important for supersymmetry.

(e) Wino

As we saw in the previous section supersymmetric models of weak interactions lead to new fermions $\tilde{W}^{\pm}$, which are Dirac particles consisting of the weak gaugino and the Higgsino as its components. The Winos will couple to $q\bar{\tilde{q}}$, $l\bar{\tilde{l}}$, etc. If we assume that they are heavier than $\tilde{q}$ their decay models will be of type

$$\begin{aligned} \tilde{W}^{\pm} &\to q\bar{q}\tilde{G} \\ &\to q\bar{q}\tilde{\gamma} \\ &\to q\bar{q}\tilde{Z}, \end{aligned}$$

if

$$m_{\tilde{Z}} \to m_{\tilde{W}^+}.$$

Their lifetimes are of order $\sim (M_{\tilde{q}}^4/g^4 M_{\tilde{W}}^5)$ and we therefore expected then to be too short to be directly observable. Their production in lepton and hadron collisions has been extensively studied [14]. In e^+e^- collision no evidence for $\tilde{W}^{\pm}$ production exists yet, leading to a lower bound on their mass of about

20 GeV. Production of $\tilde{W}^+\tilde{W}^-$ would lead to an increase in R by one unit and should therefore be observable. The final state characteristics can then be used to distinguish them from heavier leptons or quarks.

(f) Zino and Neutral Higgsino

The existence of fermionic partners of Z-boson and netural Higgs bosons are also characteristic of supersymmetric extensions of electro-weak models. The general mass matrix can be written as

$$(\tilde{\gamma}\quad \tilde{Z}\quad \tilde{\chi}_1^0\quad \tilde{\chi}_2^0)\begin{pmatrix} m_{\tilde{\gamma}} & m_{\tilde{\gamma}\tilde{Z}} & 0 & 0 \\ m_{\tilde{\gamma}\tilde{Z}} & m_{\tilde{Z}} & M_Z^D & 0 \\ 0 & M_Z^D & M_1 & 0 \\ 0 & 0 & 0 & M_2 \end{pmatrix}\begin{matrix} \tilde{\gamma} \\ \tilde{Z} \\ \tilde{\chi}_1^0 \\ \tilde{\chi}_2^0 \end{matrix}. \tag{12.2.13}$$

As we saw before $M_Z^D = M_Z$ and $m_{\tilde{\gamma}\tilde{Z}}$ can be small or zero. This matrix has to be diagonalized to find the mass eigenstates as well as the eigenvalues. This has been analyzed in great detail in Ref. [1]. These particles will have typical decay modes of the following type:

$$\tilde{\chi}_{1,2}^0 \to \begin{cases} \bar{u}_{L,R} u_{L,R} \tilde{\gamma} \\ \bar{d}_{L,R} d_{L,R} \tilde{\gamma} \end{cases}.$$

They will be produced in e^+e^- collision through Z- and slepton-exchange graphs as shown in Fig. 12.10. Similarly, in $p\bar{p}$ collision, they can be produced by the constituent quark–antiquark annihilation diagrams shown in Fig. 12.11.

At present there does not exist any useful bound on these neutral fermions.

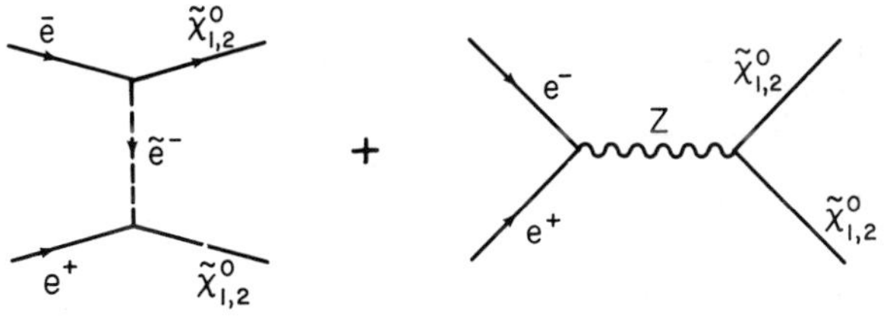

Figure 12.10

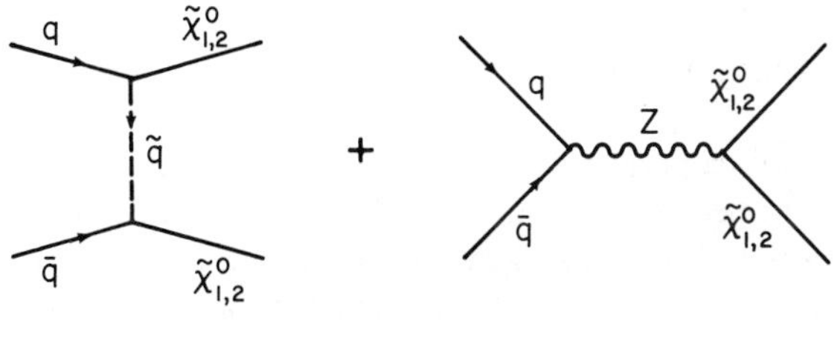

Figure 12.11

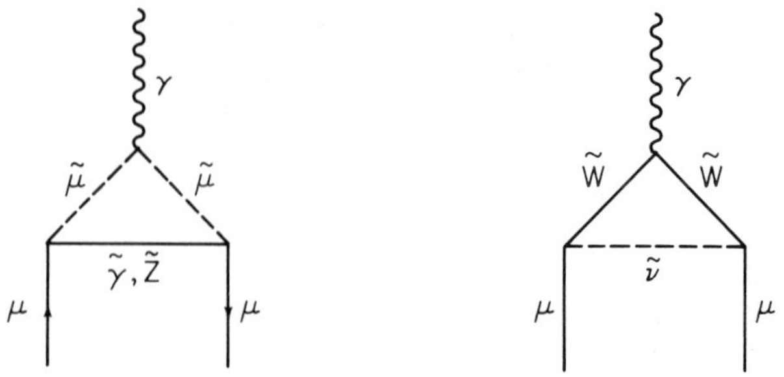

Figure 12.12

§12.3. Other Effects of Superparticles

The existence of superparticles in the 1–100 GeV mass range can give rise to new effects in processes where they are exchanged as virtual particles. In this section we enumerate a few of them.

(a) $g - 2$ of the Electron and the Muon

Two of the most precisely measured parameters of elementary particle physics are the $g - 2$ of the electron and the muon. Therefore, they provide useful constraints on the parameters of any theoretical extension of the QED. The quantity $a_\mu \equiv (g - 2)_\mu/2$ is so precisely measured and calculated that any new contribution to a_μ (called Δa_μ) must be bounded as follows [15]:

$$-2 \times 10^{-8} < \Delta a_\mu < 2.6 \times 10^{-8} \quad (95\% \text{ confidence limit}). \tag{12.3.1}$$

In supersymmetric models new contributions to a_μ from the diagrams shown in Fig. 12.12 are [16]

$$\Delta a_\mu \simeq \frac{\alpha}{12\pi} \frac{\cos\theta \sin\theta m_\mu (M_{\tilde{\mu}_L}^2 - M_{\tilde{\mu}_R}^2}{M_{\tilde{\chi}}^3}, \tag{12.3.2}$$

where θ is a mixing angle between $\tilde{\mu}_L$ and $\tilde{\mu}_R$, and $M_{\tilde{\chi}}$ is a typical superparticle mass. These constraints are typically of the form $\Delta a_\mu \sim (\alpha/12\pi)(m_\mu/m_{\tilde{\chi}})^2\theta$ which leads to bounds on masses $M_{\tilde{\mu}} \geq 18$ GeV.

(b) Flavor Changing Neutral Currents

The existence of superparticles of arbitrary masses upsets the delicate cancellations implied by the Glashow–Illiopoulos–Mariani mechanism for understanding the observed suppression of flavor changing neutral currents (FCNC) as well as the $\Delta S = 2$ effective Hamiltonians. The new diagrams to

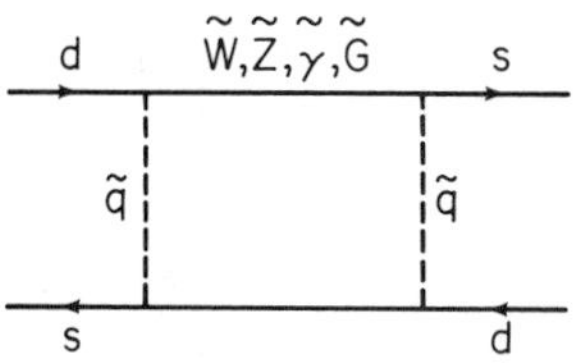

Figure 12.13

$K^0 - \bar{K}^0$ mixing are shown in Fig. 12.13. These and other FCNC effects have been estimated by several groups [17]. Typically they contribute to $\Delta S = 2$ matrix elements an amount

$$M_{\text{SUSY}}^{\Delta S=2} \simeq \frac{G_F\alpha}{4\pi}\left(\frac{\Delta M_{\tilde{q}}^2}{M_{\tilde{q}}^2}\right) \leq 10^{-12}\ \text{GeV}^{-2}. \tag{12.3.3}$$

This implies

$$\frac{\Delta M_{\tilde{q}}^2}{M_{\tilde{q}}^2} \leq 10^{-4}. \tag{12.3.4}$$

This constraints the mass splitting between squarks of different flavor.

(c) Parity Violation in Atomic Physics

The existence of mass splitting between left- and right-handed sleptons can lead to parity violating effects in atomic physics arising from graphs of the type shown in Fig. 12.14. The effective parity violating Hamiltonian is of the form

$$H_{\text{wk}} = \frac{\alpha^2}{q^2}(q^2\delta_{\mu\nu} - q_\mu q_\nu)\bar{e}\gamma_\mu\gamma_5 e\bar{q}\gamma_\mu q\left(\frac{1}{M_{\tilde{L}}^2} - \frac{1}{M_{\tilde{R}}^2}\right). \tag{12.3.5}$$

This implies that

$$\frac{\alpha^2\Delta M_{LR}^2}{M_{L,R}^4} \leq G_F/10. \tag{12.3.6}$$

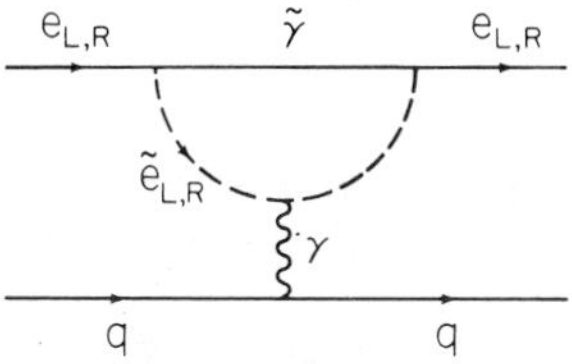

Figure 12.14

For strong interactions [18] the factor α is replaced by α_s which leads to a somewhat stronger bound on splitting between left- and right-handed squarks.

(d) Correction to the ρ_W-Parameter

As discussed in Chapter 3 $\rho_W = 1$ at the tree level in the standard model. In perturbation theory, breaking of weak isospin leads to deviations from one and has been calculated. In supersymmetric theories the radiative corrections involving superparticles also lead [19] to deviations from $\rho_W = 1$. The corrections $\delta\rho_W$ are sensitive to mass splitting between the left-squarks $\tilde{t}_L$ and $\tilde{b}_L$ and are given by

$$\delta\rho_W = \frac{3\alpha}{16\pi \sin^2\theta m_W^2}\left[m_{\tilde{t}_L}^2 + m_{\tilde{b}_L}^2 - \frac{2m_{\tilde{t}_L}^2 m_{\tilde{b}_L}^2}{m_{\tilde{t}_L}^2 - m_{\tilde{b}_L}^2}\ln\frac{m_{\tilde{t}_L}^2}{m_{\tilde{b}_L}^2}\right]. \tag{12.3.7}$$

As is clear in the limit of $m_{\tilde{t}_L} = m_{\tilde{b}_L}$, $\delta\rho_W = 0$. To obtain the constraints on $m_{\tilde{t}_L}$ and $m_{\tilde{b}_L}$ we note that, experimentally,

$$\rho_W^{\text{expt}} = 1.002 \pm 0.015. \tag{12.3.8}$$

If $m_{\tilde{b}_L} = 0$ this leads to a constraint on $m_{\tilde{t}_L} < 300$ GeV. This implies that splitting between members of a weak isospin multiplet must be less than 100 GeV and is not a very severe limit on model building.

(e) $K \to \pi\tilde{\gamma}\tilde{\gamma}$

The last effect related to superparticles which we discuss here is the rare decay mode $K^+ \to \pi^+\tilde{\gamma}\tilde{\gamma}$. The similar weak decay modes are $K^+ \to \pi^+\nu\bar{\nu}$, $K^+ \to \pi^+ +$ axion which arise in the context of other models. Thus observation of $K \to \pi +$ nothing could be interpreted in terms of either of the three processes described. However, $K^+ \to \pi^+\nu\bar{\nu}$ can be predicted in the standard model and is expected to give [20] $BR(K^+ \to \pi^+\nu\bar{\nu}) \simeq (1.4\text{–}11) \times 10^{-10}$ per neutrino. Thus, any excess in branching ratio over this value can be interpreted as new physics. The Feynman diagrams shown in Fig. 12.15, contributing to $K^+ \to \pi^+\tilde{\gamma}\tilde{\gamma}$, have been calculated and lead to [21]

$$BR(K^+ \to \pi^+\tilde{\gamma}\tilde{\gamma}) \simeq 7 \times 10^{-11}\left(\frac{20 \text{ GeV}}{M_{\tilde{c}}}\right)^4. \tag{12.3.9}$$

Since $M_{\tilde{c}}$ is expected to be larger than 20 GeV the box graph contributions are negligible. It has, however, been observed that $K^+ \to \pi^+\pi^0$ and π^0 may subsequently decay to $\tilde{\gamma}\tilde{\gamma}$. This is also expected to be of order 10^{-11} and is again unlikely to be observed.

The existence of supersymmetry will also have effects on considerations of CP-violation as has recently been speculated [22]. But again, due to the

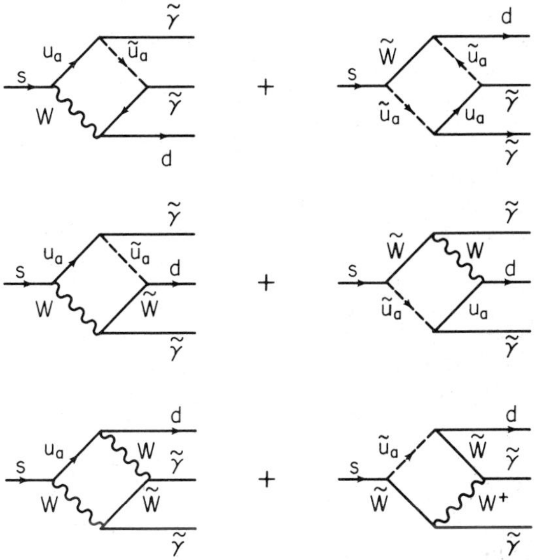

Figure 12.15

presence of mnay unknown parameters, definite conclusions are not easily drawn.

To summarize this chapter the existence of low-energy supersymmetry touches practically all areas of elementary particle physics. We have tried to discuss only some of the prominent places where these effects will be significant. Hopefully, we have given enough diagrams and illustrations to aid the imagination of the enthusiastic reader in looking for effects of supersymmetry in his area of interest. Finally, a comment is in order concerning the model presented in this chapter. We have added arbitrary soft-breaking terms to the Lagrangian to make it look realistic. A more appealing approach would be to obtain realistic supersymmetry breaking terms via spontaneous supersymmetry breaking. Several such attempts [23], which use global supersymmetry, exist in the literature but considerable sacrifice in economy, and elegance appears to be required. Using supergravity appears to be much more suitable for this purpose as we discuss in Chapter 15.

References

[1] There exist several excellent recent reviews of the subject:
H. Haber and G. Kane, *Phys. Rep.* **117**, 76 (1984).

[2] H. P. Nilles, *Phys. Rep.* **110**, 1 (1984);
R. Arnowitt, A. Chamseddine, and P. Nath, *N* = 1 *Supergravity*, World Scientific, Singapore, 1984.

[3] P. Fayet, *Nucl. Phys.* **B90**, 104 (1975);
R. K. Kaul and P. Majumdar, *Nucl. Phys.* **B199**, 36 (1982).

[4] C. S. Aulakh and R. N. Mohapatra, *Phys. Lett.* **121B**, 147 (1983);
L. Hall and M. Suzuki, *Nucl. Phys.* **B231**, 419 (1984).
For subsequent work, see
G. G. Ross and J. W. F. Valle, *Phys. Lett.* **151B**, 375 (1985).
[5] CELLO: H. Behread *et al.*, *Phys. Lett.* **114B**, 287 (1982);
JADE: W. Bartel *et al.*, *Phys. Lett.* **114B**, 211 (1982);
MARK J: D. Barber *et al.*, *Phys. Rev. Lett.* **45**, 1904 (1981);
TASSO: R. Brandelik *et al.*, *Phys. Lett.* **117B**, 365 (1982).
[6] M. K. Gaillard, L. Hall, and I. Hinchliffe, *Phys. Lett.* **116B**, 279 (1982);
M. Kuroda, K. Ishikawa, T. Kobayashi, and S. Yamada, *Phys. Lett.* **127B**, 467 (1983).
[7] L. Gladney *et al.*, *Phys. Rev. Lett.* **51**, 2253 (1983);
E. Fernandez *et al.*, *Phys. Rev. Lett.* **52**, 22 (1984).
[8] R. M. Barnett, H. E. Haber, and K. Lackner, *Phys. Lett.* **126B**, 64 (1983).
[9] For an exhaustive study see
E. Eichten, I. Hinchliffe, K. Lane, and C. Quigg, Fermilab preprint, 1984.
[10] H. Haber and G. Kane, *Nucl. Phys.* **B232**, 333 (1984).
[11] B. Kayser, Private communication, 1983.
[12] M. Chanowitz and S. Sharpe, *Phys. Lett.* **126B**, 225 (1983);
A. Mitra and S. Ono, CERN preprint, 1983.
[13] W. Y. Keung and A. Khare, *Phys. Rev.* **D** (1984).
[13a] G. Arnison *et al.*, *Phys. Lett.* **139B**, 115 (1984).
[13b] G. Kane and J. Leveille, *Phys. Lett.* **112B**, 227 (1982);
E. Reya and D. P. Roy, *Phys. Rev. Lett.* **53**, 881 (1984);
J. Ellis and H. Kowalski, CERN preprint, 1984.
[14] S. Weinberg, *Phys. Rev. Lett.* **50**, 387 (1983);
V. Barger, R. W. Robinett, W. Y. Keung, and R. J. N. Phillips, *Phys. Lett.* **131B**, 372 (1983);
A. Chamseddine, R. Arnowitt, and P. Nath, *Phys. Rev. Lett.* **49**, 970 (1972);
D. A. Dicus, S. Nandi, W. Repko, and X. Tata, *Phys. Rev. Lett.* **51**, 1030 (1983); *Phys. Rev.* **D29**, 67 (1984);
J. Ellis, J. Hagelin, D. V. Nanopoulos, and M. Srednicki, *Phys. Lett.* **127B**, 233 (1983).
[15] J. Baily *et al.*, *Nucl. Phys.* **B150** (1979).
[16] P. Fayet, in *Unification of the Fundamental Particle Interactions*, (edited by S. Ferrara *et al.*) Plenum, New York, 1980, p. 587;
J. Ellis, J. Hagelin, and D. V. Nanopoulos, *Phys. Lett.* **116B**, 283 (1982);
R. Barbieri and L. Maiani, *Phys. Lett.* **117B**, 203 (1982).
[17] J. Ellis an D. V. Nanopoulos, *Phys. Lett.* **110B**, 44 (1982);
R. Barbieri and R. Gatto, *Phys. Lett.* **110B**, 211 (1982);
T. Inami and C. S. Lim, *Nucl. Phys.* **B207**, 533 (1982);
B. A. Cambell, *Phys. Rev.* **D28**, 209 (1983);
J. Donoghue, H. P. Nilles, and D. Wyler, *Phys. Lett.* **128B**, 55 (1983);
M. Suzuki, *Phys. Lett.* **115B**, 40 (1982);
E. Franco and M. Mangano, *Phys. Lett.* **135B**, 40 (1982).
[18] M. Suzuki, *Phys. Lett.* **115B**, 40 (1982);
M. Duncan, *Nucl. Phys.* **B214**, 21 (1983).
[19] T. K. Kuo and N. Nakagawa, *Nuovo Cim. Lett.* **36**, 560 (1983);
R. Barbieri and L. Maiani, *Nucl. Phys.* **B224**, 32 (1983);
C. S. Lim, T. Inami, and N. Sakai, *Phys. Rev.* **D29**, 1488 (1984).
[20] J. Ellis and J. Hagelin, *Nucl. Phys.* **B217**, 189 (1983).
[21] M. K. Gaillard, Y. C. Kao, I. H. Lee, and M. Suzuki, *Phys. Lett.* **123B**, 241 (1983).

[22] A. Raichoudhury *et al.*, CERN preprint, 1984;
P. Langacker and B. Sathiapalan, University of Pennsylvania preprint, 1984.
[23] C. Nappi and B. Ovrut, *Phys. Lett.* **113B**, 1751 (1982);
M. Dine and W. Fishler, *Phys. Lett.* **110B**, 227 (1982);
L. Alvarez-Gaume, M. Claudson and M. Wise, *Nucl. Phys.* **B207**, 96 (1982);
J. Ellis, L. Ibanez, and G. Ross, *Phys. Lett.* **113B**, 283 (1982);
R. Barbieri, S. Ferrara, and D. Nanopoulos, *Z. Phys.* **C13**, 267 (1982).

CHAPTER 13

Supersymmetric Grand Unification

§13.1. The Supersymmetric SU(5)

One of the original motivations for the application of supersymmetry to particle physics was to solve the gauge hierarchy problem that arises in the grand unification program. As has been emphasized in Chapter 5, the tree level parameters must be fine tuned to an accuracy of 10^{-26} or so, to generate the mass ratio $M_x/m_W \simeq 10^{12}$ in the SU(5) model. In other models, due to the presence to intermediate mass scales, the problem of fine tuning is not as severe but a lesser degree of fine tuning is always required. Since a non-supersymmetric theory with scalar bosons is plagued with quadratic divergences, such tree level fine tunings are upset in higher orders. This need not happen in supersymmetric theories due to the nonrenormalization theorem of Grisaru, Rocek, and Siegel described in Chapter 10. According to this theorem, the parameters of the superpotentials do not only receive infinite renormalization but they also do not receive finite renormalization in higher orders. Supersymmetry can, therefore, be used to solve one aspect of the gauge hierarchy problem, i.e., once we fine tune parameters at the tree level the radiative corrections do not disturb the hierarchy. This point was utilized by Dimopoulos and Georgi [1] and Sakai [2] to construct supersymmetric SU(5) models with partial solutions to the gauge hierarchy problem. We illustrate their procedure with a simple but realistic supersymmetric SU(5) model.

Chiral superfields ψ (belonging to the $\{\bar{5}\}$-dimensional representation) and T (belonging to the $\{10\}$-dimensional representation) are chosen to denote the matter fields. As in the supersymmetric generalization of the $SU(2)_L \times U(1)$ model we will need two sets of Higgs superfields denoted by H_u and H_d

belonging to the {5}- and {$\bar{5}$}-dimensional representations of SU(5) to generate fermion masses. We also choose a {24}-dimensional Higgs superfield Φ to break the SU(5) group down to $SU(3)_c \times SU(2)_L \times U(1)_Y$. Except for an additional Higgs superfield (H_u or H_d) all other fields are supersymmetric generalizations of fields present in the SU(5) model of Chapter 5.

The first stage in the construction of the model is to give the superpotential $W(H_u, H_d, \phi, \psi, T)$. The superpotential has to fulfil two functions:

(i) to give realistic breaking for the SU(5) symmetry down to the $SU(3)_c \times SU(2)_L \times U(1)_Y$ group and subsequently to the $SU(3)_c \times U(1)_{em}$ group; and
(ii) to give masses to the quarks and leptons.

We choose W to be the following:

$$W = h_u \varepsilon^{ijklm} T_{ij} T_{kl} H_{u,m} + h_d T_{ij} \psi^i H_d^j + z \operatorname{Tr} \Phi + x \operatorname{Tr} \Phi^2 + y \operatorname{Tr} \Phi^3 + \lambda_1 (H_{ui} H_d^j \phi_j^i + m' H_{u,i} H_d^i). \tag{13.1.1}$$

Since our aim is to make the fine tuning of parameters natural we would like supersymmetry to stay unbroken down to the weak interaction scale of $\sim m_W/g$. This requires that $F_\phi = 0 + O(m_W^2/g^2)$. From eqn. (13.1.1) we get

$$\langle F_{\phi j}^i \rangle = z\delta_j^i + 2x\langle \phi_j^i \rangle + 3y\langle \phi_k^i \phi_j^k \rangle. \tag{13.1.2}$$

Since $\operatorname{Tr}\langle F_\phi \rangle = 0$ this implies that

$$z = -\tfrac{3}{5} y \operatorname{Tr}\langle \Phi^2 \rangle. \tag{13.1.3}$$

Now, $F_{\phi j}^i = 0$ has the following three solutions:

(a)
$$\langle \Phi_j^i \rangle = 0, \tag{13.1.4a}$$

(b)
$$\langle \Phi_j^i \rangle = a \begin{pmatrix} 1 & & & & \\ & 1 & & & \\ & & 1 & & \\ & & & -3/2 & \\ & & & & -3/2 \end{pmatrix},$$

with

$$a = \frac{4}{3}\frac{x}{y}, \tag{13.1.4b}$$

(c)
$$\langle \Phi_j^i \rangle = b \begin{pmatrix} 1 & & & & \\ & 1 & & & \\ & & 1 & & \\ & & & 1 & \\ & & & & -4 \end{pmatrix},$$

with

$$b = \frac{10}{33}\frac{x}{y}. \tag{13.1.4c}$$

This implies that the superpotential in eqn. (13.1.1) leads to a threefold vacuum degeneracy, one corresponding to unbroken SU(5), $SU(3)_c \times SU(2)_L \times U(1)$ and $SU(4) \times U(1)$. This creates cosmological problems since, as the universe cools below temperature $T \sim M_X/g$, it finds three degenerate minima including the one it is in, i.e., the SU(5) symmetric one. Thus we fail to understand why, below the SU(5) scale, $SU(3)_c \times SU(2) \times U(1)$ is the symmetry of electro-weak physics in this model. Again, radiative corrections cannot change this picture due to the nonrenormalization theorem. This is the first price we pay for the introduction of supersymmetry. The first minimum is easily removed by the introduction of a singlet field X and rewriting the part of the superpotential with ϕ's only as

$$W(\phi) = \lambda_2 X(\mathrm{Tr}\,\Phi^2 - \mu^2). \tag{13.1.5}$$

However, there is no way to split the $SU(3)_c \times SU(2)_L \times U(1)_Y$ and $SU(4) \times U(1)$ vacuum degeneracy. As we will see in Chapter 15 introduction of supergravitational interactions does help to remove this degeneracy. We will therefore assume, for now, that somehow the vacuum symmetry is uniquely chosen to be $SU(3)_c \times SU(2)_L \times U(1)_Y$ below the grand unification scale.

Let us now look at the breaking of $SU(2)_L \times U(1)_Y$ down to $U(1)_{em}$ symmetry. In the most general SU(5) invariant superpotential given in eqn. (13.1.1) we now focus on the terms involving $H_{u,d}$ and Φ. We see that, subsequent to SU(5) breaking, the last term leads to an effective low-energy Higgs potential of the form

$$|F^2| = \lambda_1^2\left(m' - \frac{3a}{2}\right)^2(\phi_u^+\phi_u + \phi_d^+\phi_d) + (\text{triplet terms}), \tag{13.1.6}$$

where $\phi_{u,d}$ denote the $SU(2)_L$ doublet piece of the Higgs fields $H_{u,d}$. Note that if the fine tune m' is such that

$$m' - \frac{3a}{2} = \mathrm{O}(m_W/g), \tag{13.1.7}$$

then, weak interaction breaking scales will emerge. (The sign of the mass term in eqn. (13.1.6) is, however, positive; but soft supersymmetry breaking terms with negative $(\text{mass})^2$ of the same order can fix the sign.) Since m' and a are determined in terms of parameters present in the superpotential they are not effected by radiative corrections at all. This partially resolves the gauge hierarchy problem.

Before we leave this section we wish to point out that as in the supersymmetry extension of the $SU(3)_c \times SU(2)_L \times U(1)$ model, we will add appropriate soft-breaking terms that do not disturb renormalizability to make the model phenomenologically acceptable. The supersymmetry breaking terms must, of course, respect SU(5) symmetry.

§13.2. Proton Decay in the Supersymmetry SU(5) Model

In the supersymmetric grand unified models there are several sources for proton decay: (i) the conventional lepto–quark (X, Y) vector boson exchange; and (ii) the new supersymmetric contributions [3]. From our discussion in Chapter 5 we learn that the conventional X, Y gauge boson exchange contributions depend on the magnitude of the unification mass. Since supersymmetric models involve new fields there will be new contributions to the β-function that will effect the evolution of the various gauge coupling constants, and this will change the value of the unification mass M_U. This has been studied by various authors [4]. To study these effects we note that for the supersymmetric SU(N) gauge group the β-function is given by the following formula:

$$\beta_i(g_i) = b_{0i}\frac{g_i^3}{16\pi^2}, \tag{13.2.1}$$

where

$$b_{0i} = -3N + T_F + T_H, \tag{13.2.2}$$

where T_F and T_H represent the contributions of the fermionic and Higgs fields. The gauge contribution changes from $-11N/3$ to $-3N$ due to the contribution of the gauginos and, similarly, the changes in fermionic and Higgs contributions reflect the new contributions from supersymmetric partners. We must, however, be careful to use this formula only in the domain where supersymmetry is exact. In the case under discussion in this chapter, this formula will be applicable between $m_W < \mu < M_X$. We note that this reduces the slope in the evolution of the α_i^{-1} in its approach to the grand unified value α_i^{-1} and, as a result, the unification is "delayed," i.e.,

$$M_{x,\mathrm{SUSY}} \simeq 6 \times 10^{16}\, \Lambda_{\overline{MS}} \approx 10^{16}\ \mathrm{GeV} \tag{13.2.3}$$

and we get

$$\sin^2 \theta_W(m_W) = 0.236 \pm 0.003. \tag{13.2.4}$$

Because of the larger M_X the proton lifetime is predicted to be $\tau_p \simeq 10^{35}$ yr. Thus, gauge boson induced baryon nonconservation in the supersymmetric SU(5) model is not measurable. This is not, however, the whole story of baryon nonconservation in supersymmetric models and there exist other potentially large sources of baryon number violation in these theories.

To understand these new class of phenomena we note that, in supersymmetric theories, there exist two classes of possible operators for baryon

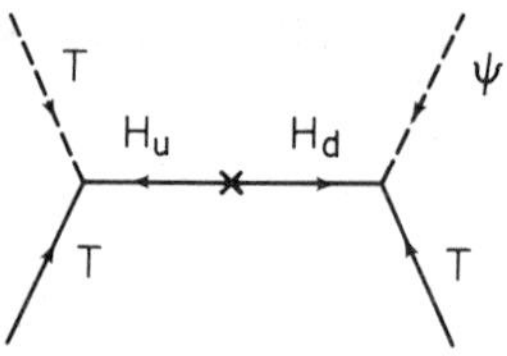

Figure 13.1. Typical Feynman diagram that induces the dimension 5 operator with $\Delta B \neq 0$.

number violating processes:

(a) D-type $$\frac{1}{M_U^2}\int d^4\theta\, \Phi^+\Phi\Phi^+\Phi, \tag{13.2.5}$$

where Φ denotes the matter superfields, ψ, T, etc.

(b) F-type $$\frac{1}{M_U}\int d^2\theta\, \Phi^4. \tag{13.2.6}$$

The first class is of the type that arises from gauge boson exchange and has already been discussed and shown to be small. The second class of terms has lower dimension and *a priori* could be very large, leading to rapid proton disintegration. They must, therefore, be carefully examined in order to uncover any additional suppression. In the component language the F-type terms give rise to the baryon number violating operator of dimension 5 typically of the following type:

$$O_{\Delta B\neq 0} = \frac{1}{M_U} Q_i^{\mathrm{T}} C^{-1}\tau_2 Q_j \tilde{Q}_k^{\mathrm{T}}\tau_2 \tilde{L}\varepsilon^{ijk}. \tag{13.2.7}$$

This operator could arise from the Feynman diagram shown in Fig. 13.1. Note that, as it stands, the above operator involves squark and slepton fields and, therefore, cannot induce proton decay in the lowest order. The scalar quark and the scalar lepton fields could, however, turn into quarks and lepton fields on emitting weak gauginos as in Fig. 13.2.

The dominant contribution to proton decay [5] will arise from the graph

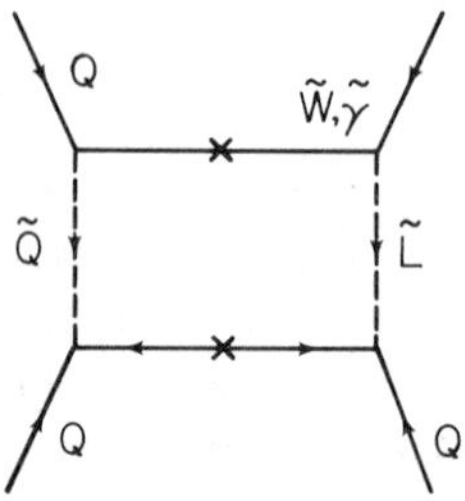

Figure 13.2. The box diagram involving gluino exchange that converts the dimension 5 operator of Fig. 13.1. into a four-Fermi operator that can induce proton decay.

where a gluino is exchanged to convert the squarks to quarks and has the following magnitude:

if

$$m_{\tilde{W}} > m_{\tilde{q}},$$

$$M_{qqql} \simeq \frac{g_2^2}{32\pi^2} \cdot \frac{Q^{\mathrm{T}}C^{-1}\tau_2 h_u Q \cdot Q^{\mathrm{T}}C^{-1}\tau_2 h_d L}{m_{\tilde{W}} m_H} \tag{13.2.8}$$

and

$$m_{\tilde{G}} \ll m_{\tilde{q}},$$

$$M_{qqql} \simeq \frac{8g_3^2}{48\pi^2} \cdot \frac{m_{\tilde{G}}}{m_{\tilde{q}}^2 M_H}. \tag{13.2.9}$$

The above operators involve purely left-handed quark-lepton fields. We can construct similar operators involving right-handed fields. To calculate the proton decay amplitude we have to find the dominant contributions out of (13.2.8) and (13.2.9). We note that $h_u = G_F^{1/2} M_u$ and $h_d = G_F^{1/2} M_d$. It is easy to find that the leading operator is of the form

$$\sim G_F m_u m_s \frac{2\alpha_s m_{\tilde{G}} \theta_c u_L^{\mathrm{T}} C^{-1} d_L s_L^{\mathrm{T}} C^{-1} \nu_{\mu L}}{3\pi m_{\tilde{q}}^2 M_H}. \tag{13.2.10}$$

This is of order

$$\sim 10^{-26} \frac{(m_{\tilde{G}} \text{ in GeV})}{m_{\tilde{q}}^2} \mathrm{GeV}^{-2}. \tag{13.2.11}$$

For $m_{\tilde{q}} \simeq 10^2$ GeV and $M_G \leq 1$ GeV, this amplitude $\simeq 10^{-30}$ GeV^2, which is compatible with presently known bounds on the proton lifetime. An important outcome of supersymmetry, however, is that the dominant decay mode is different from that of the SU(5) model and is

$$p \to K^+ \bar{\nu}_\mu,$$

$$n \to K^0 \bar{\nu}_\mu.$$

The discovery of the above decay modes will therefore be an indication of supersymmetric nature, or at least the important role played by the Higgs fields in proton decay.

Another important point to note is that the selection rule $\Delta(B - L) = 0$ holds for proton decay amplitudes as in the non-supersymmetric model.

§13.3. Inverse Hierarchy Alternative

The models of the type considered in the previous section adopt the "set it and forget it" type approach to the gauge hierarchy problem. An extremely interesting alternative mechanism for understanding the gauge hierarchy problem has been proposed by Witten [6]. This mechanism starts with a specific form of the superpotential, a typical example of which can be given using two {24}-dimensional Higgs superfields Φ and Z and a singlet field X

$$W = \lambda_1 \operatorname{Tr}(Z\Phi^2) + \lambda_2 X(\operatorname{Tr}\Phi^2 - M^2). \tag{13.3.1}$$

Let us analyze the F-terms resulting from this

$$F_X = \lambda_2(\operatorname{Tr}\Phi^2 - M^2), \tag{13.3.2a}$$

$$F^i_{Zj} = \lambda_1[(\Phi^2)^i_j - \tfrac{1}{5}\delta^i_j \operatorname{Tr}\Phi^2], \tag{13.3.2b}$$

$$F^i_{\Phi j} = \lambda_1[Z^i_k\Phi^k_j - \tfrac{1}{5}\delta^i_j \operatorname{Tr}(Z\Phi)] + 2\lambda_2 X[\Phi^i_j]. \tag{13.3.2c}$$

From eqn. (13.3.2) we note that both F_X and F_Z cannot vanish simultaneously. This implies that supersymmetry is spontaneously broken at the tree level. We can now write the Higgs potential as follows:

$$V = |F_x|^2 + \sum_{i,j}|F^i_{zj}|^2 + \sum_{i,j}|F^i_{\Phi j}|^2 + \tfrac{1}{2}\operatorname{Tr} D^2. \tag{13.3.3}$$

The minimum of this corresponds to

$$\langle\tilde{\Phi}\rangle = \Phi_0\begin{pmatrix}1 &&&& \\ & 1 &&& \\ && 1 && \\ &&& -3/2 & \\ &&&& -3/2\end{pmatrix},$$

$$\langle\tilde{Z}\rangle = Z_0\begin{pmatrix}1 &&&& \\ & 1 &&& \\ && 1 && \\ &&& -3/2 & \\ &&&& -3/2\end{pmatrix},$$

$$\langle\tilde{X}\rangle = X_0. \tag{13.3.4}$$

Substituting this in eqn. (13.3.2) we get

$$V = \lambda_2^2(\tfrac{15}{2}\phi_0^2 - M^2)^2 + \tfrac{15}{8}\lambda_1^2\phi_0^4 + \tfrac{15}{8}(\lambda_1 Z_0 - 2\lambda_2 X_0)^2\phi_0^2. \tag{13.3.5}$$

The minimum of this potential corresponds to

$$\phi_0^2 = \frac{8\lambda_2^2 M^2}{15(\lambda_1^2 + 30\lambda_2^2)} \tag{13.3.6}$$

and

$$Z_0 = \frac{2\lambda_2 X_0}{\lambda_1}. \tag{13.3.7}$$

Thus, the potential is flat for $\phi_0 = \text{const.}$ and $Z_0 = (2\lambda_2/\lambda_1)X_0$, i.e., Z_0 remains undetermined at the tree level. To find an absolute minimum of the potential we must therefore look at the one-loop corrections to this potential [7] which gives

$$V_{\text{one-loop}} = V_{\text{tree}} + \frac{1}{64\pi^2}\sum_s (-1)^{2s}(2s+1)M_s^4 \ln M_s^2, \tag{13.3.8}$$

where s is the spin of the state. This function has been computed by Witten who showed that

$$V_{\text{one-loop}} = V_{\text{tree}} + f(\lambda_1, \lambda_2)\frac{29\lambda_1^2 - 50g^2}{80\pi^2} \ln \frac{Z_0^2}{M^2}, \tag{13.3.9}$$

where $f(\lambda_1, \lambda_2) \geq 0$.

Let us now choose the theory so that g is asympotically free. Since λ_1 increases with momenta, if at $\mu \approx m_W$ we start with $\lambda_1^2 < \frac{50}{29}g^2$, the potential will decrease with increasing Z_0 and once $\lambda_1^2 > \frac{50}{29}g^2$ (which will happen at a very large Z_0), V will turn around and V will have a minimum along Z_0 at

$$Z_0^2 \approx M^2 e^{4\pi C/g^2}, \tag{13.3.10}$$

where C is some number and we obtain $Z_0^2 \gg M^2$. Thus, a large mass scale is generated starting from a lower mass scale. In principle, we could choose $M = m_W$; but that would require an extreme degree of fine tuning [8] of the coupling parameters. A clear way out [9] is to choose M at some intermediate scale, such as $M_I \simeq 10^{11}$ GeV or so. In this case, Dimopoulos and Raby [9] have argued that the weak interaction breaking scale can be induced in higher orders and is of order

$$M_M \simeq \frac{M_I^2}{M_U}. \tag{13.3.11}$$

If $M_U \simeq 10^{18}$ GeV this leads to the correct scale for weak interaction breaking. Thus, a very attractive picture emerges where both low and high scale are induced as quantum effects starting with an intermediate scale.

These ideas have been extended to the SO(10) model [10] as well as the SU(5) models with supergravity effects [11]. There are, however, some shortcomings to this approach: it does not lead to a unification scale below Planck mass without further "pollution" by additional Higgs multiplets. Thus, more research is called for before this attractive idea can find its rightful place in particle physics.

References

[1] S. Dimopoulos and H. Georgi, *Nucl. Phys.* **B193**, 150 (1981).

[2] N. Sakai, *Z. Phys.* **C11**, 153 (1981).

[3] S. Weinberg, *Phys. Rev.* **D25**, 287 (1982);
N. Sakai and T. Yanagida, *Nucl. Phys.* **B197**, 533 (1982).

[4] M. B. Einhorn and D. R. T. Jones, *Nucl. Phys.* **B196**, 475 (1982);
W. Marciano and G. Senjanovic, *Phys. Rev.* **D25**, 3092 (1982).

[5] J. Ellis, D. V. Nanopoulos, and S. Rudaz, *Nucl. Phys.* **B202**, 43 (1982);
For a recent analysis see
J. Milutinovich; P. Pal, and G. Senjanovic, ITP Santa Barbara preprint, 1984.
[6] E. Witten, *Phys. Lett.* **105B**, 267 (1981).
[7] S. Coleman and E. Weinberg, *Phys. Rev.* **D7**, 1888 (1973).
[8] H. Yamagishi, *Nucl. Phys.* **B216**, 508 (1983);
L. Hall and I. Hinchliffe, *Phys. Lett.* **119B**, 128 (1982).
[9] S. Dimopoulos and S. Raby, *Nucl. Phys.* **B219**, 479 (1983).
[10] S. Kalara and R. N. Mohapatra, *Phys. Rev.* **D28**, 2241 (1983).
[11] B. Ovrut and S. Raby, *Phys. Lett.* **125B**, 270 (1983).

CHAPTER 14

Local Supersymmetry ($N = 1$)

§14.1. Connection Between Local Supersymmetry and Gravity

In this chapter we will study the implications of the hypothesis that the parameters of supersymmetry transformation ε become a function of space–time, i.e., $\varepsilon = \varepsilon(x)$. We know, from Chapter 1, that invariance under local symmetry requires new fields in the theory which have spin 1, and have the same number of components as the number of independent parameters in the group. In analogy, local supersymmetry will require the introduction of the spin 3/2 field which is the Majorana type. This will bring us into a completely new domain of particle physics where new spin 3/2 elementary fields interact with ordinary matter fields. Furthermore, there will also be analogs of the Higgs mechanism once supersymmetry is spontaneously broken (the so-called super-Higgs effect). There is, however, a much more profound aspect to local supersymmetry. Once the spin 3/2 fields are introduced, to make the theory supersymmetric in the high spin sector, it will turn out that we will require a massless spin 2 field which can be identified with the graviton field $g_{\mu\nu}$, thus "unifying" gravitation with the other three forces of nature. This discovery was made independently by Freedman, Ferrara, and Van Niuenhuizen [1], and by Deser and Zumino [1], and opened up a whole new possibility, not only of unification of gravity with particle physics [2] but also of new consequences for particle physics with supersymmetry. We will call the spin 3/2 particle gravitino and denote it by a Majorana field ψ_μ.

To see the need for gravity in local supersymmetry, we first give some heuristic intuitive arguments which we follow up in the subsequent sections with more rigorous formulation. Let us look at the supersymmetry transformations for matter fields (ϕ, ψ, F)

$$\delta\phi = \varepsilon\psi,$$

$$\delta\psi = \varepsilon F - i\partial_\mu \phi\sigma_\mu\bar{\varepsilon},$$

$$\delta F = \frac{i}{2}\partial_\mu\psi\sigma_\mu\bar{\varepsilon}. \tag{14.1.1}$$

If ε is a function of x then we can look at the trivial free supersymmetric Lagrangian in (9.2.1) (with the addition of the F^*F-term) and see what modifications it will require. To see this, note that

$$\delta(\partial_\mu\phi) = \varepsilon\partial_\mu\psi + \partial_\mu\varepsilon\cdot\psi,$$

$$\delta(\partial_\mu\psi) = \varepsilon\partial_\mu F + \partial_\mu\varepsilon F - i\partial_\mu\partial_\nu\phi\sigma_\nu\bar{\varepsilon} - i\partial_\nu\phi\sigma_\nu\partial_\mu\bar{\varepsilon}. \tag{14.1.2}$$

Since global supersymmetry ($\varepsilon(x) = \text{const.}$) is a special case of this the change in the action can be written as

$$\delta S = \int d^4x\, \partial_\mu\varepsilon^a K_{\mu,a} + \text{h.c.}, \tag{14.1.3}$$

where

$$K_\mu^a \equiv -\partial_\mu\phi^*\cdot\psi^a - \frac{i}{2}\psi^b(\sigma_\mu\bar{\sigma}_\nu)_b^a\partial_\nu\phi^* \tag{14.1.4}$$

(a, b are spinor indices). To keep the action invariant we must add the corresponding gauge field (ψ_μ) coupling, i.e.,

$$S' = \kappa\int K_\mu^a\psi_a^\mu\, d^4x, \tag{14.1.5}$$

such that, under the local supersymmetry transformation,

$$\psi_a^\mu \to \psi_a^\mu + \kappa^{-1}\partial^\mu\varepsilon_a. \tag{14.1.6}$$

What is the parameter κ? Looking at (14.1.5) and (14.1.16) we can conclude that κ has dimension of M^{-1}. So, in contrast with local bosonic symmetries, local supersymmetry requires a dimensional coupling. Since the gravitational constant $G_N = M_P^{-2}$ provides such a dimensional parameter, we could define $\kappa = \sqrt{8\pi G_N}$ and use this as the universal gauge coupling associated with local supersymmetry. Furthermore, we will see that the spin 2 boson, which carries the gravitational force, provides the superpartner of the gravitino.

A second heuristic argument which implies a connection between local supersymmetry and gravity is the following: if we look at the commutator of two successive local supersymmetry transformations

$$[\delta(\varepsilon_1(x)), \delta(\varepsilon_2(x))]B \sim \bar{\varepsilon}_1(x)\gamma_\mu\varepsilon_2(x)\partial_\mu B. \tag{14.1.7}$$

(From this point on we will switch to four-component notation.) The right-hand side implies that the translation depends on the space–time point but this precisely represents the coordinate transformations which lead to general relativity. A more powerful argument is to consider the variation of $(S + S')$ when we find (from the variation of K^a_μ under supersymmetry transformation)

$$\delta(S + S') = \int d^4x \, \bar{\psi}_\mu \gamma_\nu \varepsilon T^{\mu\nu}, \tag{14.1.8}$$

where $T^{\mu\nu}$ is precisely the energy momentum tensor. The only way to cancel this is to add to the Lagrangian a new term $T^{\mu\nu} g_{\mu\nu}$ provided under supersymmetry transformation, $\delta g_{\mu\nu} = \bar{\psi}_\mu \gamma_\nu \varepsilon$; this is the usual way to couple gravity to matter fields. Before proceeding to this discussion we will give a brief introduction to free field theory of the spin 3/2 Rarita–Schwinger field and review the elements of general relativity.

§14.2. Rarita–Schwinger Formulation of the Massless Spin 3/2 Field

The massless Rarita–Schwinger field is described by a Majorana spinor with a Lorentz index, i.e., ψ_μ. The Lagrangian for this system is given by

$$\mathscr{L} = -\tfrac{1}{2}\varepsilon^{\mu\nu\rho\sigma}\bar{\psi}_\mu \gamma_5 \gamma_\nu \partial_\rho \psi_\sigma. \tag{14.2.1}$$

We see that it is invariant under the gauge transformation $\psi_\sigma \to \psi_\sigma + \kappa^{-1}\partial_\sigma \varepsilon$. The field equations are given by

$$-\varepsilon^{\mu\nu\rho\sigma}\gamma_5 \gamma_\nu \partial_\rho \psi_\sigma = 0. \tag{14.2.2}$$

We can evaluate the canonical momenta $\pi_\mu = \delta L/\delta(\partial_0 \psi_\mu)$ and we find $\pi_0 = 0$. Using the Dirac quantization procedure the gauge condition corresponding to this can be written as $\psi_0 = 0$; otherwise, $\pi_0 = 0$ cannot be maintained at all times. Then calculating π_k, we obtain $\pi_k^\dagger = \varepsilon^{ijk}\gamma_5\gamma_4\gamma_i\psi_j$ which implies that

$$\partial_\kappa \pi_k^\dagger - \varepsilon^{ijk}\gamma_5\gamma_4\gamma_i\partial_k\psi_j = 0. \tag{14.2.3}$$

To maintain this condition at all times we can choose a gauge condition

$$\partial_i \psi^i = 0 \tag{14.2.4}$$

After these conditions are implemented we are left with only two independent Majorana spinors. We, furthermore, note that there is a set of second class constraints

$$\pi_k^\dagger - \varepsilon^{ijk}\gamma_5\gamma_4\gamma_i\psi_j = 0. \tag{14.2.5}$$

We can now count the number of degrees of freedom by subtracting all the momentum and coordinate constraints from the total number of coordinates and momenta, i.e., 16 (coordinates) + 16 (momentum) – 4 ($\pi^0 = 0$) – 4 ($\psi^0 = 0$) – 4 (eqn. (14.2.3)) – 4 (eqn. (14.2.4)) – 12 (eqn. (14.2.5)) = 4 which corresponds to the two helicity states.

§14.3. Elementary General Relativity

The basic ideas of general relativity [3] can be described either in terms of a metric $g_{\mu\nu}$ which is a symmetric second rank tensor with $\mu, \nu = 0, 1, 2, 3$ or in terms of a vierbein e_μ^m. We start by introducing the properties of $g_{\mu\nu}$. The basic ingredient of the theory of relativity is the requirement that, under generalized coordinate transformations

$$x^\mu \to x^{\mu'}, \tag{14.3.1}$$

the physics remains invariant. The line element $ds^2 = g_{\mu\nu}\, dx^\mu\, dx^\nu$ is invariant under the coordinate transformations

$$dx'^\mu = \frac{\partial x'^\mu}{\partial x^\nu}\, dx^\nu, \tag{14.3.2}$$

which implies that

$$g'_{\mu\nu} = \frac{\partial x^\rho}{\partial x'^\mu} \frac{\partial x^\sigma}{\partial x'^\nu} g_{\rho\sigma}. \tag{14.3.3}$$

We can define a contravariant $g^{\lambda\rho}$ which has the property

$$g_{\mu\lambda} g^{\lambda\rho} = \delta_\mu^\rho \tag{14.3.4a}$$

and under coordinate transformation

$$g'^{\mu\nu} = \frac{\partial x'^\mu}{\partial x^\rho} \frac{\partial x'^\nu}{\partial x^\sigma} g^{\rho\sigma}. \tag{14.3.4b}$$

We can then define the Christoffel symbol (or the affine connection) $\Gamma_{\nu\rho}^\mu$

$$\Gamma_{\nu\rho}^\mu = \tfrac{1}{2} g^{\mu\lambda} \left[\frac{\partial g_{\lambda\nu}}{\partial x^\rho} + \frac{\partial g_{\lambda\rho}}{\partial x^\nu} - \frac{\partial g_{\rho\nu}}{\partial \chi^\lambda} \right]. \tag{14.3.5}$$

Using this we can define the covariant derivative of tensors

$$D_\sigma V^\mu \equiv V^\mu_{;\sigma} = \frac{\partial V^\mu}{\partial x^\sigma} + \Gamma_{\rho\sigma}^\mu V^\rho. \tag{14.3.6}$$

The curvature (Ricci) tensor is then given as follows:

$$R_{\mu\nu\sigma}^\lambda = \frac{\partial \Gamma_{\mu\nu}^\lambda}{\partial x^\sigma} - \frac{\partial \Gamma_{\mu\sigma}^\lambda}{\partial x^\nu} + \Gamma_{\mu\nu}^\rho \Gamma_{\sigma\rho}^\lambda - \Gamma_{\mu\sigma}^\rho \Gamma_{\nu\rho}^\lambda. \tag{14.3.7}$$

The gravitational Lagrangian can then be written as

$$\mathscr{L}_G = -\frac{1}{2\kappa^2}\sqrt{g} R, \tag{14.3.8}$$

where $R = g^{\mu\nu} R_{\mu\nu\lambda}^\lambda$. To this, we can add the matter piece.

An alternative formulation, more suitable for the discussion of local supersymmetry, is the vierbein (or tetrad) formalism, where we define the vierbein

e_μ^m where m transforms like a flat space coordinate while μ transforms like a curved space coordinate. This alternative formulation is possible because of the principle of equivalence, which states that at each space–time point we can choose a locally inertial coordinate system to describe physics in the presence of gravitational fields. Of course, this coordinate system will differ from point to point. We can then write the metric tensor $g_{\mu\nu}$ as follows:

$$g_{\mu\nu} = e_\mu^m(x) e_\nu^n(x) \eta_{mn}, \tag{14.3.9}$$

where η is metric in flat space–time. Under coordinate transformations

$$e_\mu^m \to e_\mu^{m'} = \frac{\partial x^\nu}{\partial x'^\mu} e_\nu^m. \tag{14.3.10}$$

We can use the vierbein to express any contravariant vector A^μ as a vector in the locally inertial coordinate system

$$A^m = e_\mu^m A^\mu. \tag{14.3.11}$$

We can raise and lower indices on e_μ^m by using $g^{\mu\nu}$ and η_{mn} to obtain e_m^μ which satisfies the following property:

$$e_m^\mu e_\mu^n = \delta_m^n \quad \text{and} \quad e_m^\mu e_\nu^m = \delta_\nu^\mu. \tag{14.3.12}$$

The principle of equivalence requires that special relativity should apply in locally inertial frames; therefore, the index m will transform as a flat space vector index under Lorentz transformation, i.e.,

$$e_\mu^m \to \Lambda_n^m e_\mu^n. \tag{14.3.13}$$

Let us proceed to the construction of the Lagrangian in terms of e_μ^m. We must require it to be invariant under both coordinate as well as *local* Lorentz transformations (the latter because, at each space–time, point, the inertial coordinate system is different). There is an excellent discussion of this point in the book by S. Weinberg [3]. We refer the reader to this book for details.

To write down the Lagrangian in this formalism we will have to express the curvature tensor in terms of the vierbeins. For this purpose we will need to express $\Gamma_{\nu\alpha}^\mu$ in terms of e_μ^m. This is done by using the spin connection defined as follows.

The vierbein formalism also requires the definition of a covariant derivative which must be such that it must not only transform appropriately under coordinate transformation, but also transform appropriately under local Lorentz transformation. As is obvious, this will require the introduction of a new quantity ω_μ^{mn}, called the spin connection, which will transform under local Lorentz transformations as

$$M_{mn}\omega_\mu^{mn} \to D(\Lambda) M_{mn} \omega_\mu^{mn} D^{-1}(\Lambda) - \partial_\mu D(\Lambda) D^{-1}(\Lambda). \tag{14.3.14}$$

The ω_μ^{mn} are the gauge fields corresponding to local Lorentz transformation. We now note that since the covariant derivative of the metric tensor is zero, that of the vierbein must also be zero, i.e.,

$$D_\rho e_\sigma^m = 0 \tag{14.3.15}$$

or

$$\partial_\rho e_\sigma^m + \omega_\rho^{mn} e_{n\sigma} - \Gamma_{\rho\sigma}^\alpha e_\alpha^m = 0. \tag{14.3.16}$$

Using this equation we can reexpress $\Gamma_{\rho\sigma}^\alpha$ in terms of ω and e and we can define the following curvature tensor:

$$R_{\mu\nu}^{mn} = \partial_\mu \omega_\nu^{mn} - \partial_\nu \omega_\mu^{mn} + \omega_\mu^{mp} \omega_{\nu p}^{\ \ n} - \omega_\nu^{mp} \omega_{\mu p}^{\ \ n}. \tag{14.3.17}$$

It is interesting to note that this is in the form of $F_{\mu\nu}$ for usual gauge fields. The usual curvature tensor is then defined as

$$R_{\tau\mu\nu}^\sigma e_{m\sigma} e_n^\tau = R_{\mu\nu,mn}. \tag{14.3.18}$$

The action for pure gravity is then given by

$$S^{(2)} = -\frac{1}{2\kappa^2} \int eR(e, \omega)\, d^4x. \tag{14.3.19}$$

Before closing this section we would like to cast the coordinate transformations, and the transformation of $g_{\mu\nu}$ under them, in a form that is reminiscent of the gauge theories. For this purpose define

$$x'_\mu = x_\mu + \zeta_\mu(x). \tag{14.3.20}$$

Note a very powerful implication of this equation which is: coordinate transformations can be thought of as a set of local translations. It would then be possible to think of gravitation as the gauge theory of translation and Lorentz transformations.

To obtain the transformation property of $g^{\mu\nu}$ we use eqn. (14.3.4b) and expand for infinitesimal ζ_μ

$$\begin{aligned} g^{\mu\nu'} &= \frac{\partial(x^\mu + \zeta^\mu)}{\partial x^\rho} \frac{\partial(x^\nu + \zeta^\nu)}{\partial x^\sigma} g^{\rho\sigma}(x + \zeta) \\ &= g^{\mu\nu} + g^{\mu\sigma}\partial_\sigma \zeta^\nu + g^{\rho\nu}\partial_\rho \zeta^\mu + \zeta^\rho \partial_\rho g^{\mu\nu} + O(\zeta^2). \end{aligned} \tag{14.3.21}$$

For the vierbein e_a^μ this implies

$$e_a^{\mu'} = e_a^\mu + e_a^\lambda \partial^\mu \zeta_\lambda + \zeta^\rho \partial_\rho e_a^\mu + O(\zeta^2). \tag{14.3.22}$$

§14.4. $N = 1$ Supergravity Lagrangian

At this stage we are ready to write down the supergravity Lagrangian

$$\mathscr{L} = \mathscr{L}^{(2)} + \mathscr{L}^{(3/2)}, \tag{14.4.1}$$

where

$$\mathscr{L}^{(2)} = -\frac{1}{2\kappa^2} eR(e, \omega)$$

and

$$\mathscr{L}^{(3/2)} = -\tfrac{1}{2}\varepsilon^{\mu\nu\rho\sigma}\bar{\psi}_\mu\gamma_5\gamma_\nu D_\rho\psi_\sigma, \tag{14.4.2}$$

where

$$D_\rho = \partial_\rho + \tfrac{1}{2}\sigma_{mn}\omega_\rho^{mn}. \tag{14.4.3}$$

Note that the absence of $\sqrt{g} \equiv e$ in eqn. (14.4.2) $\mathscr{L}$ is invariant under the following set of supersymmetry transformations:

$$\begin{aligned} \delta e_\mu^m &= \frac{\kappa}{2}\bar{\varepsilon}\gamma^m\psi_\mu, \\ \delta\psi_\mu &= \frac{1}{\kappa}(D_\mu\varepsilon), \\ \delta\omega_{\mu,ab} &= -\tfrac{1}{4}\bar{\varepsilon}\gamma_5\gamma_\mu\tilde{\psi}_{ab} + \tfrac{1}{8}\bar{\varepsilon}\gamma_5(\gamma^\lambda\tilde{\psi}_{\lambda b}e_{a\mu} - \gamma^\lambda\tilde{\psi}_{\lambda a}e_{b\mu}), \end{aligned} \tag{14.4.4}$$

where

$$\psi_{ab} = e_a^\mu e_b^\nu(D_\mu\psi_\nu - D_\nu\psi_\mu).$$

As in the case of global supersymmetry, if we check the commutator of two supersymmetry transformations, we find that they do not close. This is merely a reflection of the fact that the number of bosonic and fermionic degrees of freedom off-shell do not match, and auxiliary fields must be introduced to compensate for this mismatch. To see how many auxiliary fields are needed we can count the fermionic degrees of freedom: since, by local supersymmetry transformation, we can remove four fermionic degrees of freedom, we have, off-shell twelve fermionic degrees of freedom. On the other hand, from the *a priori* sixteen bosonic degrees of freedom in e_μ^a, by translation and Lorentz gauge transformation we can remove $4 + 6 = 10$ degrees of freedom. This leaves us with six bosonic degrees of freedom. So we need six auxiliary fields which we call (S, P, A_μ). For completeness, we give the supersymmetry transformation in the presence of auxiliary fields

$$\begin{aligned} \delta e_\mu^m &= \frac{\kappa}{2}\bar{\varepsilon}\gamma^m\psi_\mu, \\ \delta\psi_\mu &= \frac{1}{\kappa}\left(D_\mu + \frac{i\kappa}{2}A_\mu\gamma_5\right)\varepsilon - \tfrac{1}{2}\gamma_\mu\eta\varepsilon, \\ \delta S &= \tfrac{1}{4}\bar{\varepsilon}\gamma^\mu R_\mu^{\text{cov}}, \\ \delta P &= -\frac{i}{4}\bar{\varepsilon}\gamma_5\gamma^\mu R_\mu^{\text{cov}}, \\ \delta A_m &= \frac{3i}{4}\bar{\varepsilon}\gamma_5(R_m^{\text{cov}} - \tfrac{1}{8}\gamma_m\gamma\cdot R^{\text{cov}}), \end{aligned}$$

where

$$\eta = -\tfrac{1}{8}(S - i\gamma_5 P - iA\gamma_5),$$

$$R_\mu^{\text{cov}} = \varepsilon^{\mu\nu\rho\sigma}\gamma_5\gamma_\nu\left(D_\rho\psi\sigma - \frac{i}{2}A_\sigma\gamma_5\psi_\rho + \tfrac{1}{2}\gamma_\sigma\eta\psi_\rho\right) \tag{14.4.5}$$

and the supersymmetric Lagrangian is

$$\mathscr{L} = \mathscr{L}^{(2)}(e, \omega) + \mathscr{L}^{(3/2)}(e, \psi, \omega) - \frac{e}{3}(S^2 + P^2 - A_m^2). \qquad (14.4.6)$$

So far we have considered only pure supergravity. Now we can try to couple supergravity to matter field systems. One way is to look at the variation of various fields under local supersymmetry and construct covariant derivatives; for instance, writing ϕ in eqn. (14.1.1) as $\phi = A + iB$ we observe that local supersymmetry would require $D_\mu A = (\partial_\mu A - i\kappa\bar{\psi}\psi_\mu)$, etc. In the next section we will develop a systematic set of rules to couple matter fields and supergravity. This is known as tensor calculus.

§14.5. Group Theory of Gravity and Supergravity Theories

We will adopt a gauge-theoretic approach to supergravity where not only the gravitino field but also the gravitational field will arise as gauge potentials (or connections) corresponding to an underlying local symmetry. This will make it easier to obtain the coupling of the matter fields to supergravity simply by following the rules for the usual Yang–Mills theory. Important for this purpose is the discovery of the underlying symmetry group.

To get an idea about the underlying group we recall the algebra of supersymmetric generators Q in Chapter 9 and we find that they involve the momentum. Therefore, the gravitational field must be a result of gauging translational symmetry. As mentioned in this chapter, making the translation parameter local is equivalent to a coordinate transformation. This, therefore, appears to be an immensely satisfying framework for the study of gravity. Below we note some of the well-known facts about the algebraic structure of gravity theories [2].

(a) Poincaré Algebra

This is the algebra of angular momentum operators $M_{\mu\nu}$ and momentum P_μ and is given by ($g_{np} = +++-$)

$$[M_{mn}, M_{pq}] = g_{np}M_{mq} - g_{mp}M_{nq} - g_{nq}M_{mp} + g_{mq}M_{np}, \qquad (14.5.1)$$

$$[M_{mn}, P_q] = g_{nq}P_m - g_{mq}P_n, \qquad (14.5.2)$$

$$[P_m, P_n] = 0. \qquad (14.5.3)$$

A coordinate space representation for M_{mn} and P_n is

$$M_{mn} = x_m \frac{\partial}{\partial x_n} - x_n \frac{\partial}{\partial x_m},$$

$$P_m = \frac{\partial}{\partial x_m}. \qquad (14.5.4)$$

(b) de Sitter Algebra

We can define O(4, 1) algebra by considering a five-dimensional space and writing generalized angular momentum operators M_{AB} satisfying the algebra ($g_{AB} = +++--$)

$$[M_{AB}, M_{CD}] = g_{BC}M_{AD} - g_{AC}M_{BD} - g_{BD}M_{AC} + g_{AD}M_{BC}. \qquad (14.5.5)$$

We can identify $M_{\mu 5} = P'_\mu$; we then see that

$$[P'_\mu, P'_\nu] = M_{\mu\nu}. \qquad (14.5.6)$$

To obtain the Poincaré algebra we perform the Inonu–Wigner contraction by defining $P = \varepsilon P'$ and letting $\varepsilon \to 0$. The right-hand side of (15.5.6) then vanishes.

(c) Conformal Algebra

The Poincaré algebra can be extended by including the generators of scale (S) and conformal transformations (K_m) which have the following coordinate space representations:

$$S = x_m \frac{\partial}{\partial x_m}$$

and

$$K_m = 2x_m x_n \partial_n - x^2 \partial_m. \qquad (14.5.7)$$

The full conformal algebra is obtained by adding the following commutation relation to eqns. (14.5.1), (14.5.2), and (14.5.3):

$$\begin{aligned} [K_m, K_n] &= 0, \\ [K_m, P_n] &= -2(g_{mn}D + M_{mn}), \\ [K_m, D] &= -K_m, \\ [K_m, D] &= P_m. \end{aligned} \qquad (14.5.8)$$

This algebra has 15 elements and can also be obtained from the algebra of the O(4, 2) group by performing the Inonu–Wigner contraction. SU(2, 2) is also locally isomorphic to the O(4, 2) group.

(d) Super-Poincaré Algebra

In order to study the algebraic structure of supersymmetry we must extend the Poincaré algebra by including the spinorial (or odd) elements Q^a in the algebra, and which supplement the algebra in (14.5.1), (14.5.2), and (14.5.3) by addition of the following. In four-component notation

$$\{Q^a, \bar{Q}^b\} = (\gamma_m)^{ab} P_m, \tag{14.5.9}$$

$$[P_m, Q^a] = 0, \tag{14.5.10}$$

$$[M_{mn}, Q^a] = \tfrac{1}{2}(\sigma_{mn} Q)^a. \tag{14.5.11}$$

In terms of superfield coordinates (x, θ) we can write the generators as follows:

$$\begin{aligned} P_m &= \partial_m, \\ M_{mn} &= x_m \partial_n - x_n \partial_m + \bar{\theta}\sigma_{mn} \frac{\partial}{\partial \bar{\theta}}, \\ Q^a &= \frac{1}{2}\left(\frac{\partial}{\partial \bar{\theta}} - \gamma_m \partial_m \theta\right)^a. \end{aligned} \tag{14.5.12}$$

In fact, the de Sitter algebra can also be extended in a similar fashion to include supersymmetry. The graded Lie algebra (i.e., an algebra consisting of fermionic (odd) and bosonic (even) elements) associated with this is called 0 Sp(4/1) which leaves the following line element invariant:

$$ds^2 = x_M x_M + \theta_a (C^{-1})_{ab} \theta_b, \tag{14.5.13}$$

where $M = 1, \ldots, 5$ with $g_{MN} = (+++--)$. The group on bosonic coordinates is O(3, 2) and on fermionic coordinates is SP(4) which are isomorphic to each other. By group contraction we can obtain the super-Poincaré algebra from this.

(e) Superconformal Algebra

The super-Poincaré algebra can be further extended to include the entire conformal group in ordinary bosonic coordinates. To close the algebra we will need one new spinorial element, S^a, and another bosonic element, denoted by A, called the superchiral symmetry. The graded Lie algebra for the superconformal group can be written as

$$\begin{aligned} &[M_{mn}, M_{pq}] = g_{np} M_{mq} - g_{mp} M_{nq} - g_{nq} M_{mp} + g_{mq} M_{np}, \\ &[M_{mn}, P_q] = g_{nq} P_m - g_{mq} P_n, \\ &[P_m, P_n] = 0, \\ &[K_m, K_n] = 0, \qquad [K_m, P_n] = -2(g_{mn} D + M_{mn}), \\ &[K_m, D] = -K_m, \qquad [P_m, D] = P_m, \\ &\{Q^a, \bar{Q}^b\} = 2(\gamma \cdot P)^{ab}, \\ &[P_m, Q^a] = 0, \qquad [M_{mn}, Q^a] = -(\sigma_{mn} Q)^a, \end{aligned}$$

$$
\begin{aligned}
\{S^a, \bar{S}^b\} &= -2(\gamma \cdot K)^{ab}, \\
\{Q^a, \bar{S}^b\} &= 2[D\sigma^{ab} - (\sigma^{mn})^{ab}M_{mn} - i(\gamma_5)^{ab}A], \\
[M_{mn}, S^a] &= -(\sigma_{mn})^{ab}S^b, \\
[Q^a, A] &= -\frac{3i}{4}(\gamma_5)^a_b Q^b, \\
[S^a, A] &= \frac{3i}{4}(\gamma_5)^a_b S^b, \\
[Q^a, D] &= \tfrac{1}{2}Q^a, \qquad [S^a, D] = -\tfrac{1}{2}S^a, \\
[S^a, P_m] &= (\gamma_m)^a_b Q^b, \\
[Q^a, K_m] &= -(\gamma_m)^a_b S^b.
\end{aligned}
\tag{14.5.14}
$$

These elements also satisfy the Jacobi identities. We can now write down the super-coordinate representation for K, S, and A

$$
\begin{aligned}
K_m &= 2x_m x \cdot \partial - x^2\partial_m - \bar{\theta}\gamma \cdot x\gamma_m \frac{\partial}{\partial \bar{\theta}} - \tfrac{1}{2}(\bar{\theta}\theta)^2\partial_m - \bar{\theta}\theta(\bar{\theta}\gamma_m\partial/\partial\bar{\theta}), \\
S &= \tfrac{1}{2}\left(\bar{\theta}\theta\frac{\partial}{\partial\bar{\theta}} + \bar{\theta}\gamma_5\theta\gamma_5\frac{\partial}{\partial\bar{\theta}}\right) + \tfrac{1}{4}\bar{\theta}\gamma_5\gamma_m\theta\gamma_5\gamma_m\frac{\partial}{\partial\bar{\theta}} \\
&\quad + \left(-\tfrac{1}{2}\gamma \cdot x\frac{\partial}{\partial\bar{\theta}} + \tfrac{1}{2}\gamma \cdot x\gamma \cdot \partial\theta + \tfrac{1}{2}\bar{\theta}\theta\partial\!\!\!/\theta\right), \\
A &= -\tfrac{1}{4}i\bar{\theta}\gamma_5\frac{\partial}{\partial\bar{\theta}}.
\end{aligned}
\tag{14.5.15}
$$

Another representation for these operators can be given in terms of 4×4 matrices of the following type:

$$
\begin{aligned}
P_m &= -\tfrac{1}{2}\gamma_m(1 - \gamma_5), \\
K_m &= \tfrac{1}{2}\gamma_m(1 + \gamma_5), \\
D &= -\tfrac{1}{2}\gamma_5, \qquad M_{mn} = (\sigma_{mn}), \\
(Q^a)^\kappa_{4+i} &= \{-\tfrac{1}{2}(1 + \gamma_5)c^{-1}\}^{\kappa a}, \\
(S^a)^k_{4+i} &= \{-\tfrac{1}{2}(1 - \gamma_5)c^{-1}\}^{ka}, \\
(Q^a)^{4+i}_k &= \tfrac{1}{2}(1 - \gamma_5)^a_k, \\
(S^a)^{4+i}_k &= -\tfrac{1}{2}(1 - \gamma_5)^a_k.
\end{aligned}
\tag{14.5.16}
$$

The algebra of superconformal symmetry can be obtained from the graded Lie algebra SU(2, 2/1) which leads to the following metric invariant:

$$
ds^2 = |z_1|^2 + |z_2|^2 - |z_3|^2 - |z_\phi|^2 + \theta_1^*\theta^1 + \theta_2^*\theta^2 - \theta_3^*\theta^3 - \theta_4^*\theta^4.
\tag{14.5.17}
$$

§14.6. Local Conformal Symmetry and Gravity

Before we study supergravity, with the new algebraic approach developed, we would like to discuss how gravitational theory can emerge from the gauging of conformal symmetry. For this purpose we briefly present the general notation for constructing gauge covariant fields. The general procedure is to start with the Lie algebra of generators X_A of a group

$$[X_A, X_B] = f_{AB}^C X_C, \tag{14.6.1}$$

where f_{AB}^C are structure constants of the group. We can then introduce a gauge field connection h_μ^A as follows:

$$h_\mu \equiv h_\mu^A X_A. \tag{14.6.2}$$

Let us denote the parameter associated with X_A by ε^A. The gauge transformations on the fields h_μ^A are given as follows:

$$\delta h_\mu^A = \partial_\mu \varepsilon^A + h_\mu^B \varepsilon^C f_{CB}^A \equiv (D_\mu \varepsilon)^A. \tag{14.6.3}$$

We can then define a covariant curvature

$$R_{\mu\nu}^A = \partial_\nu h_\mu^A - \partial_\mu h_\nu^A + h_\nu^B h_\mu^C f_{CB}^A. \tag{14.6.4}$$

Under a gauge transformation

$$\delta_{\text{gauge}} R_{\mu\nu}^A = R_{\mu\nu}^B \varepsilon^C f_{CB}^A. \tag{14.6.5}$$

We can then write the general gauge invariant action as follows:

$$I = \int d^4x\, Q_{AB}^{\mu\nu\rho\sigma} R_{\mu\nu}^A R_{\rho\sigma}^B. \tag{14.6.6}$$

Let us now apply this formalism to conformal gravity. In this case

$$h_\mu = P_m e_\mu^m + M_{mn}\omega_\mu^{mn} + K_m f_\mu^m + D b_\mu. \tag{14.6.7}$$

The various $R_{\mu\nu}$ are

$$R_{\mu\nu}(P) = \partial_\nu e_\mu^m - \partial_\mu e_\nu^m + \omega_\mu^{mn} e_\nu^n - \omega_\nu^{mn} e_\mu^m - b_\mu e_\nu^m + b_\nu e_\mu^m, \tag{14.6.8}$$

$$R_{\mu\nu}(M) = \partial_\nu \omega_\mu^{mn} - \partial_\mu \omega_\nu^{mn} - \omega_\nu^{mp}\omega_{\mu,p}^n - \omega_\mu^{mp}\omega_{\nu,p}^n - 4(e_\mu^m f_\nu^n - e_\nu^m f_\mu^n), \tag{14.6.9}$$

$$R_{\mu\nu}(K) = \partial_\nu f_\mu^m - \partial_\mu f_\nu^m - b_\mu f_\nu^m + b_\nu f_\mu^m + \omega_\mu^{mn} f_\nu^n - \omega_\nu^{mn} f_\mu^n, \tag{14.6.10}$$

$$R_{\mu\nu}(D) = \partial_\nu b_\mu - \partial_\mu b_\nu + 2e_\mu^m f_\nu^m - 2e_\nu^m f_\mu^m. \tag{14.6.11}$$

The gauge invariant Lagrangian for the gravitational field can now be written down, using eqn. (14.6.6), as

$$S = \int d^4x\, \varepsilon_{mnrs}\varepsilon^{\mu\nu\rho\sigma} R_{\mu\nu}^{mn}(M) R_{\rho\sigma}^{rs}(M). \tag{14.6.12}$$

We also impose the constraint that

$$R_{\mu\nu}(P) = 0, \tag{14.6.13}$$

which expresses ω_μ^{mn} as a function of (e, b). The reason for imposing this constraint has to do with the fact that P_m transformations must be eventually identified with coordinate transformation. To see this point more explicitly let us consider the vierbein e_μ^m. Under coordinate transformations

$$\delta_{GC}(\xi^\nu)e_\mu^m = \partial_\mu \xi^\lambda e_\lambda^m + \xi^\lambda \partial_\lambda e_\mu^m. \tag{14.6.14}$$

Using eqn. (14.6.8) we can rewrite

$$\delta_{GC}(\xi^\nu)e_\mu^m = \delta_P(\xi e^n)e_\mu^m + \delta_M(\xi\omega^{mn})e_\mu^m + \delta_D(\xi b)e_\mu^m + \xi^\nu R_{\mu\nu}^m(P),$$

where

$$\delta_P(\xi^n)e_\mu^m = \partial_\mu \xi^m + \xi^n \omega_\mu^{mn} + \xi^m b_\mu. \tag{14.6.15}$$

If $R^{\mu\nu}(P) = 0$, the general coordinate transformation becomes related to a set of gauge transformations via eqn. (14.6.15).

At this point we also wish to point out how we can define the covariant derivative. In the case of internal symmetries $D_\mu = \partial_\mu - iX_A h_\mu^A$; now since momentum is treated as an internal symmetry we have to give a rule. This follows from eqn. (14.6.15) by writing a redefined translation generator $\tilde{P}$ such that

$$\delta_{\tilde{P}}(\xi) = \delta_{GC}(\xi^\nu) - \sum_{A'} \delta_{A'}(\xi^m h_m^A), \tag{14.6.16}$$

where A' goes over all gauge transformations excluding translation. The rule is

$$\delta_{\tilde{P}}(\xi^m)\phi = \xi^m D_m^C \phi. \tag{14.6.17}$$

We also wish to point out that for fields which carry spin or conformal charge, only the intrinsic parts contribute to D_m^C and the orbital parts do not play any rule.

Coming back to the constraints we can then vary the action with respect to f_μ^m to get an expression for it, i.e.,

$$e_\nu^m f_{\mu m} = -\tfrac{1}{4}[e_m^\lambda e_{n\nu} R_{\mu\lambda}^{mn} - \tfrac{1}{6} g_{\mu\nu} R], \tag{14.6.18}$$

where f_μ^m has been set to zero in R written in the right-hand side.

This eliminates (from the theory the degrees of freedom) ω_μ^{mn} and f_μ^m and we are left with e_μ^m and b_μ. Furthermore, these constraints will change the transformation laws for the dependent fields so that the constraints do not change.

Let us now look at the matter coupling to see how the familiar gravity theory emerges from this version. Consider a scalar field ϕ. It has conformal weight $\lambda = 1$. So we can write a convariant derivative for it, eqn. (14.6.17).

$$D_\mu^C \phi = \partial_\mu \phi - \phi b_\mu. \tag{14.6.19}$$

We note that the conformal charge of ϕ can be assumed to be zero since $K_m = x^2 \partial$ and is the dimension of inverse mass. In order to calculate $\Box^c \phi$ we start with the expression for d'Alambertian in general relativity

$$\frac{1}{e}\partial_\nu(g^{\mu\nu}eD_\mu^C\phi). \tag{14.6.20}$$

The only transformations we have to compensate for are the conformal transformations and the scale transformations. Since

$$\delta b_\mu = -2\xi_k^m e_{m\mu}, \qquad \delta(\phi b_\mu) = \phi\delta b_\mu = -2\phi f_\mu{}^m e_m^\mu = +\tfrac{2}{12}\phi R, \tag{14.6.21}$$

where, in the last step, we have used the constraint equation (14.6.18). Putting all these together we find

$$\square^C\phi = \frac{1}{e}\partial_\nu(g^{\mu\nu}eD_\mu^c\phi) + b_\mu D_\mu^c\phi + \tfrac{2}{12}\phi R. \tag{14.6.22}$$

Thus, the Lagrangian for conformal gravity coupled to matter fields can be written as

$$S = \int e\, d^4x\, \tfrac{1}{2}\phi\square^C\phi. \tag{14.6.23}$$

Now we can use conformal transformation to gauge $b_\mu = 0$ and local scale transformation to set $\phi = \kappa^{-1}$ leading to the usual Hilbert action for gravity. To summarize, we start with a Lagrangian invariant under full local conformal symmetry and fix conformal and scale gauge to obtain the usual action for gravity. We will adopt the same procedure for supergravity. An important technical point to remember is that, $\square^C$, the conformal d'Alambertian contains R, which for constant ϕ, leads to gravity. We may call ϕ the auxiliary field.

§14.7. Conformal Supergravity and Matter Couplings

Our primary goal in this chapter is to write a Lagrangian for the interaction of quarks and leptons that not only respects local electro-weak symmetries but also invariances under local supersymmetry. This provides a truly unified theory of all forces in nature. In the previous section the strategy was outlined for obtaining the Hilbert action for gravity. In this chapter we would like to follow the same procedure for supergravity, i.e., we will start with conformal supergravity so that a full conformal invariant Lagrangian can be written down for matter fields using the standard methods familiar from non-abelian gauge theories. Then we would reduce the gauge freedom by the use of auxiliary multiplets, thereby introducing the Planck constant into physics and obtaining the theory of supergravity coupled to matter. This approach was pioneered by Kaku, Townsend, and Van Nieuwenhuizen [4].

The Lie algebra valued connection and parameter for conformal supergravity are given by

$$h_\mu = P_m e_\mu^m + M_{mn}\omega_\mu^{mn} + Db_\mu + K_m f_\mu^m + \bar{Q}\psi_\mu + \bar{S}\phi_\mu + AA_\mu, \qquad (14.7.1)$$

$$\varepsilon = P_m \xi^m + \theta^{mn} M_{mn} + \bar{\varepsilon} Q + \bar{S}\xi + \xi_k^m K_m + \lambda_D D + \lambda_A A. \qquad (14.7.2)$$

The covariant derivative D_μ and the curvature $R_{\mu\nu}^A$ can be written following the rules outlined in the previous section once the quantum numbers of a particular field are known under conformal transformation. However, before we do that, we have to give the constraint equations involving $R_{\mu\nu}$

$$R_{\mu\nu}(P) = 0, \qquad (14.7.3a)$$

$$R_{\mu\nu}(Q)^{\Gamma^\nu} = 0, \qquad (14.7.3b)$$

$$R_{\mu\lambda}^{mn}(M) e_m^\lambda e_{n\nu} - \tfrac{1}{2} R_{\lambda\nu}(Q)\gamma_\mu \psi^\lambda - \frac{i}{4} e\varepsilon_{\mu\nu\rho\sigma} R^{\rho\sigma}(A) = 0. \qquad (14.7.3c)$$

The first and third constraints are, of course, the same as in the non-supersymmetric case whereas the second constraint is specific to the supersymmetric theories. These are necessary in order to convert the P_m-transformations to generalized coordinate transformations as discussed earlier. We also define the covariant derivative in a manner similar to the one given in eqn. (14.6.17), except that A' goes over to new gauge degrees of freedom of the theory.

Moreover, we note that eqn. (14.7.3) can be used to reexpress ω_μ^{mn}, ϕ_μ, and f_μ^m gauge fields in terms of other fields of the theory and the Q- and P-transformation rules get changed. We do not go into these details here and refer the reader to the review article by Van Nieuwenhuizen [2] and a recent review article by Kugo and Uehara [5]. (Important earlier references are given in Ref. [6].)

(a) Transformation Rules for Matter Fields

To study matter couplings to supergravity we need transformation rules of matter fields under the conformal group. Let us consider the chiral multiplet $\Sigma \equiv (\mathscr{A}, \chi_L, \mathscr{F})$. It transforms under the various superconformal transformations as follows:

Q-supersymmetry

$$\delta_Q \mathscr{A} = \tfrac{1}{2}\bar{\varepsilon}\chi_L,$$

$$\delta_Q \chi_L = -\frac{i}{2}(1 + \gamma_5)(\mathscr{F} + \gamma^m D_m^C \mathscr{A})\varepsilon,$$

$$\delta_Q \mathscr{F} = \frac{i}{2}\bar{\varepsilon}\lambda^m D_m^C \chi_L. \qquad (14.7.4)$$

Note that the only difference from global supersymmetry is that ∂_m is replaced by the conformal covariant derivative D_m^C.

S-supersymmetry
Let the corresponding parameter be ξ

$$\begin{aligned} \delta_S \mathscr{A} &= 0, \\ \delta_S \chi_L &= -i\lambda \mathscr{A}(1 + \lambda_5)\xi, \\ \delta_S \mathscr{F} &= i(\lambda - 1)\bar{\xi}\chi_L, \end{aligned} \tag{14.7.5}$$

where λ is the scale dimension of the $\mathscr{A}$-component of the superfield Σ.

D-(scale) transformations

$$\begin{aligned} \delta_D \mathscr{A} &= \lambda \lambda_D \mathscr{A}, \\ \delta_D \chi_L &= (\lambda + \tfrac{1}{2})\lambda_D \chi_L, \\ \delta_D \mathscr{F} &= (\lambda + 1)\lambda_D \mathscr{F}. \end{aligned} \tag{14.7.6}$$

K_m-(conformal) transformations

$$\delta_K \mathscr{A} = \delta_K \chi_L = \delta_K \mathscr{F} = 0. \tag{14.7.7}$$

Chiral-transformations

$$\begin{aligned} \delta_A \mathscr{A} &= \frac{i\lambda}{2} \lambda_A \mathscr{A}, \\ \delta_A \chi_L &= i\left(\frac{2\lambda - 3}{4}\right)\lambda_A \chi_L, \\ \delta_A \mathscr{F} &= i\left(\frac{\lambda - 3}{2}\right)\lambda_A \mathscr{F}. \end{aligned} \tag{14.7.8}$$

Similar rules can be written for the vector multiplet with component (C, Z, H, K, V_μ, Λ, D). Now we are ready to write the rules for constructing the matter-coupled Lagrangian for the case of supergravity [5], [6]. In complete analogy with the case of global supersymmetry, there are two kinds of Lagrangians for which the action is invariant under supersymmetry transformations: (i) F-type and (ii) $\hat{D}$-type. But we see from eqn. (14.7.4) that, due to local symmetries, $\delta_Q F$ is no more a full divergence but has additional terms which must be compensated. Furthermore, to maintain scale (D) and chiral (A) invariance we must have scale dimension $\lambda = 3$; and to maintain S-invariance something must be added. Taking all this into account, it has been noted that the conformal invariant F-type Lagrangian is given by [6]

$$S_F = \int d^4x\, e[F + \tfrac{1}{2}\bar{\psi}_\mu \gamma^\mu \chi_L + \tfrac{1}{2}\bar{\psi}_\mu \sigma^{\mu\nu}(A - iB\gamma_5)\psi_\nu], \tag{14.7.9}$$

where we have rewritten $\mathscr{A} = A + iB$, A, B being real fields. Similarly, the locally conformal invariant D-type Lagrangian is

$$S_D = \int d^4x\, e\left[D + \Box^c C - \frac{i}{2}\bar{\psi}\cdot\gamma\gamma_5(\not{D}^c Z + \Lambda) - \tfrac{1}{2}\bar{\psi}_\mu\sigma^{\mu\nu}(H + i\gamma_5 K)\phi_\nu\right], \tag{14.7.10}$$

where $\Box^c$ is the conformal d'Alambertain which is a supersymmetric generalization of eqn. (14.6.22) and is given by

$$\Box^c = \partial_m D^{mc} - \sum_{A'} \delta_{A'}(h_m^{A'})D_m^c \tag{14.7.11}$$

and

$$D_m^c \mathscr{A} = \partial_m \mathscr{A} - \frac{i}{2}\bar{\psi}_m \chi_L - \lambda b_m \mathscr{A} - \frac{i\lambda}{2} A_m \mathscr{A}. \tag{14.7.12}$$

We saw that in eqn. (14.6.22) the Hilbert action arose because of the conformal gauge variation of the dilatation gauge field b_μ. In a similar manner, the variation of the term $\bar{\psi}_m\chi_L$ yields the gravitino counterpart of the Hilbert action, i.e.,

$$\Box^c = -\tfrac{1}{6}|\mathscr{A}|^2\left(R + \frac{1}{e}\bar{\psi}_\mu\gamma_5\gamma_\rho\partial_\sigma\psi_\lambda\varepsilon^{\mu\rho\sigma\lambda}\right) + D_\mu\mathscr{A}^* D^\mu\mathscr{A} + \cdots. \tag{14.7.13}$$

We recognize the coefficient of $\mathscr{A}^*\mathscr{A}$ as the Lagrangian for pure Poincaré supergravity.

If we start with chiral fields of arbitrary scale dimension, we must multiply them until we get their total scale dimension (to be three) to construct S_F. To construct S_D from chiral fields we must multiply them with antichiral fields. We also often have to multiply several vector multiplets. We therefore need the conformal generalization of the multiplication rules for global supersymmetry. For chiral fields there is no change. For vector fields

$$V_1 \times V_2 = V_3,$$

where

$$\begin{aligned} V_3 = \Big[& C_1C_2,\, C_1Z_2 + C_2Z_1,\, C_1H_2 + C_2H_1 - \tfrac{1}{2}Z_1^T C^{-1} Z_2, \\ & C_1K_2 + C_2K_1 + \frac{i}{2}Z_1^T C^{-1}\gamma_5 Z_2,\, C_1V_m^2 + C_2V_m^1 + \frac{i}{2}Z_1^T C^{-1}\gamma_5\gamma_m Z_2, \\ & \{C_1\Lambda_2 + \tfrac{1}{2}(H_1 - i\gamma_5 K_1 + i\gamma_5 \not{V}_1 - D^c Z_2) + (1 \leftrightarrow 2)\}, \\ & C_1D_2 + C_2D_1 + H_1H_2 + K_1K_2 - V_M^1 V^{m2} - D_m^c C_1 D^{cm} C_2 \\ & - Z_1^T C^{-1}\Lambda_2 - Z_2^T C^{-1}\Lambda_1 - \tfrac{1}{2}Z_1^T C^{-1} D^c Z_2 - \tfrac{1}{2}Z_2^T C^{-1}\not{D}^c Z_1 \Big]. \end{aligned} \tag{14.7.14}$$

Similarly, we give the formula for the vector multiplet obtained from the product of a chiral and antichiral multiple Φ and Φ^* where $\Phi = (A + iB, \chi_L, F + iG)$

$$
\begin{aligned}
\Phi_1\Phi_2^* = \Bigg[& A_1A_2 + B_1B_2, -\{(B_1 + i\gamma_5 A_1)\chi_2 + 1 \leftrightarrow 2\}, \\
& -\{(A_1F_2 + B_1G_2) + 1 \leftrightarrow 2\}, \{(B_1F_2 - A_1G_2) + 1 \leftrightarrow 2\}, \\
& \left\{B_1 D_m^c A_2 - A_1 D_m^c B_2 + \frac{i}{2}\bar{\chi}_{1L}\gamma_m\chi_{2L} + 1 \leftrightarrow 2\right\}, \\
& \{(G_1 + i\gamma_5 F_1)\chi_{2L} + \not{D}^c(B_1 + i\gamma_5 A_1)\chi_2 + 1 \leftrightarrow 2\}, \\
& \{2F_1F_2 + 2G_1G_2 - 2D_m^c A_1 D_m^c A_2 - 2\not{D}_m^c B_1 D_m^c B_2 - \chi_{1L}^T C^{-1}\not{D}^c\chi_{2L} \\
& - \chi_{2L}^T C^{-1}\not{D}^c\chi_{1L}\}\Bigg].
\end{aligned}
\tag{14.7.15}
$$

(b) The Pure Supergravity Lagrangian

We illustrate the use of the action formula by constructing the Lagrangian for pure supergravity. We consider an auxiliary chiral multiplet Σ with conformal and chiral weight 1 and construct a real vector multiple out of it taking $\Sigma^*\Sigma$. This has weight 2 and therefore its D-term has got weight 4 and we can use the S_D formula to obtain (writing $\mathscr{F} = F + iG$)

$$
\begin{aligned}
S_D = \tfrac{1}{2}\int d^4x\, e\Bigg[& F^2 + G^2 + \Box^c(A^2 + B^2) - \frac{i}{2}\bar{\psi}\cdot\gamma\gamma_5(D^c(\mathscr{A}\chi_L) + \chi\mathscr{F}) \\
& + \tfrac{1}{2}\bar{\psi}_\mu\sigma^{\mu\nu}(1 + \gamma_5)(\mathscr{A}\mathscr{F} + \cdots)\phi_\nu\Bigg].
\end{aligned}
\tag{14.7.16}
$$

Now, we can fix the conformal gauge by setting $b_m = 0$ and the S-gauge by setting $\chi = 0$; $\mathscr{A} = \kappa^{-1}$ by using chiral and scale transformations. The resulting action, then, is of the form (using eqn. (14.7.14))

$$
S_D = \tfrac{1}{2}\int d^4x\, e\left[-\frac{1}{6\kappa^2}\mathscr{L}_{SG} + \frac{A_\mu^2}{4\kappa^2} + F^2 + G^2\right]. \tag{14.7.17}
$$

We thus obtain the supergravity Lagrangian in a very intuitive manner using familiar methods of gauge theories. Furthermore, we see that the auxiliary fields F, G, A_μ of supergravity are related to an underlying conformal symmetry.

Another important thing to be noted is that the ψ_μ-field appearing in the connection formula has dimension $+1/2$ rather than $+3/2$ which is to be expected for a propagating spinor. However, after the conformal gauges are fixed, a new spin 3/2 field can be defined such that $\psi_\mu^{\text{new}} = (1/\kappa)\phi_\mu$; the ψ_μ^{new} has dimension 3/2 and is the actual dynamical field.

§14.8. Matter Couplings and the Scalar Potential in Supergravity

To discuss the application of $N = 1$ supergravity we must couple a system of the Yang–Mills matter system to supergravity. The rules for the construction are given in the previous section. We illustrate it for the case of a matter field assigned to a chiral superfield $S(z, \chi_L, h)$ where components are denoted in increasing order of the power of θ. We ignore the gauge fields to start with.

The starting point is to assign zero conformal weight ($\lambda = 0$) to the superfield S and choose an auxiliary multiplet $\Sigma(\mathscr{A}, \chi_L', \mathscr{F})$ with $\lambda = +1$. In global (or rigid) supersymmetry the superpotential part denotes the self-interactions. Let us take a superpotential $g(S)$. Let us choose another function $f(S, S^\dagger)$ to obtain the kinetic energy part. It is clear that $g(s)$ will become part of the F-type action whereas $f(S, S^\dagger)$ will be used in constructing the D-type action. Note that we have chosen an arbitrary function $f(S, S^\dagger)$; each term in its power series expansion must have at least one S and one $S^\dagger$.

To construct the F-type Lagrangian we consider the $(g(S)\Sigma^3)$ superfield since it has scale dimension 3 and, remember that by using the s-transformation, D- and A-transformations, we can write

$$\Sigma = \left(\kappa^{-1}, 0, \frac{u}{3}\right), \tag{14.8.1}$$

so that

$$\Sigma^3 = (\kappa^{-3}, 0, u), \tag{14.8.2}$$

$$g(s)\Sigma^3 = (g(z)\kappa^{-3}, \chi_L g'(z), ug(z)\kappa^{-1} - g''\kappa^{-2}\chi_L^T C^{-1}\chi_L^\dagger + g'(z)h\kappa^{-3}). \tag{14.8.3}$$

So we have

$$S_F = \tfrac{1}{2}\int d^4x\, e\left[\kappa^{-1}ug(z) + hg'(z)\kappa^{-3} - g''\chi_L^T C\chi_L\kappa^{-2} + \tfrac{1}{2}\bar{\psi}\cdot\gamma\chi_{Li}g'(z)\kappa^{-2} + \frac{\kappa^{-2}}{2}g(z)\bar{\psi}_\mu\sigma^{\mu\nu}(1-\gamma_5)\psi_\nu\right] + \text{h.c.} \tag{14.8.4}$$

To discuss the kinetic energy part we write down the real vector multiplet $\Sigma^*\Sigma$ using the formula in eqn. (14.7.6)

$$\Sigma^*\Sigma = (\kappa^{-2}, 0, -\tfrac{2}{3}u, -\tfrac{2}{3}A_m\kappa^{-1}, 0, \tfrac{2}{9}(uu^* - A_m A^m\kappa^{-2})). \tag{14.8.5}$$

Note that the origin of the A_m term is the covariant conformal derivative $D_m^c \cdot 1$. Using formula (14.7.5) we can also expand $f(S, S^\dagger)$ into its component fields

$$
\begin{aligned}
f(S, S^\dagger) = [\,& f(z, z^*), -2if'^i\chi_{iL}, -2f'^i h_i + 2f''^{ij}\chi_{Li}^T C^{-1}\chi_{Lj}, \\
& if'^i D_m z_i - if_i' D_m z^{*i} - 2if''^j \bar{\chi}_L^{\,i}\gamma_m \chi_{Lj}, \\
& -2if_i''^j h_j \chi_L^i + 2if_k'''^{ij}\chi_L^k \chi_{Li}^T C^{-1}\chi_{Lj} + 2if_i''^j \not{D} z^{*i}\chi_{Lj}, \\
& 2f_j''^i h_i h^{*j} - 2f_k'''^{ij}\chi_{Li}^T C^{-1}\chi_{Lj} h^{*k} - 2f_{ij}'''^k \chi_{Lj}^T C \chi_{Li}^* h_k \\
& + 2f_{kl}''''^{ij}\chi_{Lk}^\dagger C \chi_{iL}^* - 2f_i''^j D_m z^{*i} D_m z_j + \chi_{Li}^T C^{-1}\chi_{Lj} \\
& - 2f_i''^j \bar{\chi}_{Lj} D\chi_{Li} - 2(f_{ji}'''^i D_\mu z^{*k} + f_j'''^{ik} D_\mu z_k \bar{\chi}_{Li}\gamma^\mu \chi_{Lj})]. \qquad (14.8.6)
\end{aligned}
$$

To finally write down the action we have to multiply $f(S, S^\dagger)\Sigma\Sigma^*$ and use the formula in eqn. (14.7.10). The Lagrangian has a long complicated form, but let us write down some of the most important terms

$$
\begin{aligned}
\kappa^{-2}e^{-1}\mathscr{L}_{\text{K.E.}} = & -\tfrac{1}{6} f(z, z^*) e^{-1}\mathscr{L}_{SG} \\
& + f_j''^i(-\tfrac{1}{2}D_\mu z_i D^\mu z^{j*} - \bar{\chi}_{Li}\not{D}\chi_L^j + \tfrac{1}{2}h_i h^{*j}) \\
& + \tfrac{1}{3}u^*(f'^i h_i - f''^{ij}\chi_{Li}^T C^{-1}\chi_{Lj}) \\
& + \frac{i}{3}A^\mu(\tfrac{1}{2}f_j''^i \bar{\chi}_L^j \gamma_\mu \chi_{Li} + f'^i(D_\mu z_i - \bar{\psi}_{\mu L}\chi_{Li})), \qquad (14.8.7)
\end{aligned}
$$

where

$$
\begin{aligned}
\mathscr{L}_{SG} &= (R + R_\psi + \tfrac{1}{3}(u^* u - A_m A_m)), \\
e^{-1}\mathscr{L}_{\text{P.E.}} &= -\tfrac{1}{2}g''^{ij}\chi_{Li}^T C^{-1}\chi_{Lj} + \tfrac{1}{2}g'^i h_i + \tfrac{1}{2}gu \\
&\quad + \tfrac{1}{2}\bar{\psi}_L \cdot \gamma\chi_{Li} g'^i + \tfrac{1}{2}g\bar{\psi}_{\mu L}\sigma^{\mu\nu}\psi_{\nu R}. \qquad (14.8.8)
\end{aligned}
$$

An important point to note is that even though we wrote for the action

$$
S = \int d^4\theta\, (f(s, s^\dagger)\Sigma\Sigma^*) + \int d^2\theta\, (g(s)\Sigma^3) + \text{h.c.} \qquad (14.8.9)
$$

we can rescale the auxiliary field by $\Sigma g^{-1/3}$ to argue that the action will depend only on the function $f(z, z^*)/g^{2/3}$. We can define a new function for this

$$
e^{\mathscr{G}(z, z^*)/3} = \frac{f(z, z^*)}{g^{2/3}(z)}. \qquad (14.8.10)
$$

We would now like to discuss two important implications of the supergravity Lagrangian written as: (i) the supergravity modifications of the effect scalar potential; and (ii) the spontaneous breaking of supersymmetry and the super-Higgs effects.

Effective Scalar Potential in Supergravity Theories

To address this question we have to eliminate the auxiliary fields and cast the kinetic energy terms in canonical form. Let us therefore isolate the relevant parts of the Lagrangian from eqns. (14.8.7) and (14.8.8). First we focus on the part of the Lagrangian that involves only the auxiliary fields and bosonic fields

$$e^{-1}\mathscr{L}_{\text{aux}} = \tfrac{1}{18}f(z, z^*)(uu^* - A_m A^m) + \tfrac{1}{2}f_j''^i h_i h^{*j}$$
$$+ \tfrac{1}{2}g'^i h_i + \tfrac{1}{3}u^* f'^i h_i + \tfrac{1}{2}g^* u^* + f'^i D_\mu z_i + \text{h.c.} \qquad (14.8.11)$$

Let us define a new expression

$$T = 3\ln\left(-\left(\frac{f}{3}\right)\right) \quad \text{and} \quad \tilde{u} = u + T'^i h_i. \qquad (14.8.12)$$

We can then rewrite $e^{-1}\mathscr{L}_{\text{aux}}$ as follows:

$$e^{-1}\mathscr{L}_{\text{aux}} = \tfrac{1}{18}f(\tilde{u}\tilde{u}^* - A_m A^m) + \tfrac{1}{6}f T_j''^i h_i h^{*j}$$
$$+ \tfrac{1}{2}g\left[\tilde{u} - h_i\left(T'^i - \frac{g'^i}{g}\right)\right] + f'^i D_\mu Z_i + \text{h.c.}, \qquad (14.8.13)$$

this leads to the following field equations:

$$\tilde{u} = \frac{9g^*}{2f},$$
$$\frac{f}{3}h_i T_k''^i = -\tfrac{1}{2}g^*\left(\frac{g_k^{*'}}{g^*} - T_k'^*\right),$$
$$\tfrac{2}{3}fA_\mu = if'^I D_\mu Z_i. \qquad (14.8.14)$$

Substituting back into (14.8.13) we find

$$e^{-1}\mathscr{L}_{\text{aux}} = -\frac{9}{4f}|g|^2 - \frac{3}{f}(T''^{-1})_l^k\left[\frac{g^*}{2}\left(\frac{g_k'^*}{g^*} - T_k'\right) \times \frac{g}{2}\left(\frac{g'^l}{g} - T'^l\right)\right]$$
$$-\frac{1}{4f}[f'^i D_\mu z_i - f_i' D_\mu z^{*i}]^2. \qquad (14.8.14a)$$

Equation (14.8.14a) has no auxiliary fields and gives the form of the effective potential in terms of the dynamical fields and their functions.

Let us now look at other terms in the Lagrangian and try to bring them to canonical form by adjusting the arbitrary function $f(z, z^*)$. First, we see that to get the Hilbert action for gravity, we have to do Weyl rescaling of the vierbein field $e_{m\mu}$ as follows:

$$e_{m\mu} \to e^\sigma e_{m\mu}$$

and

$$e_m^\mu \to e^{-\sigma} e_m^\mu, \qquad e \to e^{4\sigma} e,$$

$$\chi \to e^{-\sigma/2}\chi,$$

$$\psi_\mu \to e^{\sigma/2}\psi_\mu,$$

where

$$\sigma = \tfrac{1}{2}\ln\left(-\frac{3}{f\kappa^2}\right) = -\frac{T}{6}. \tag{14.8.15}$$

Under this transformation

$$\tfrac{1}{6}efR \to -\frac{1}{2\kappa^2}eR - \tfrac{3}{4}e(\partial_\mu \ln f)^2 + \text{four-divergence},$$

$$e\varepsilon^{\mu\nu\rho\sigma}\bar{\psi}_\mu\gamma_5\gamma_\rho\partial_\sigma\psi_\nu \to -\frac{1}{2\kappa^2}\bar{\psi}_\mu\gamma_5\gamma_\rho D_\sigma\psi_\nu. \tag{14.8.16}$$

We now call $\kappa^{-1}\psi_\mu \equiv \psi_\mu^g$ the dynamical gravitino field with dimension 3/2. This gives us the correct form of the gravitational part of the Lagrangian. Next we wish to get the canonical form for the kinetic energy part of the scalar fields. For that purpose, first note that using the $(\partial_\mu \log f)$ term in eqn. (14.8.16), we can cast the kinetic energy part of the z-field in a suitable form. We can now rewrite the entire bosonic part of the Lagrangian as follows:

$$e^{-1}\mathscr{L}_B = \frac{\kappa^2}{4}e^{-T}[3gg^* + T_l''^{-1k}(g_k^{*\prime} - T_k' g^*)(g_l' - T_g' g)]$$

$$-\frac{1}{2\kappa^2}R + \frac{2}{\kappa^2}T_j''^i D_\mu z_i D_\mu^* z^j. \tag{14.8.17}$$

It is now clear that, if we choose $T = -(\kappa^2/2)z^*z$, the effective potential takes the form

$$V_{\text{eff}} = \exp\left(+\frac{\kappa^2}{2}z^*z\right)\left[\left|\frac{\partial g}{\partial z} + \frac{\kappa^2}{2}z^*g\right|^2 - 3\kappa^2|g|^2\right]. \tag{14.8.18}$$

We we include gauge fields we must add D-terms to this potential. Equation (14.8.18) is the fundamental equation for model building in $N = 1$ supergravity. The particular form of the potential is intuitively understandable, referring to our equation which says that the Lagrangian can only be a function of $(f/g^{2/3})$; therefore, the effective potential must depend not just on g as in the case of global supersymmetry but on both f and g.

Exercise

First redefining $\Sigma \to \Sigma g^{-1/3}$ we can recast the kinetic energy term in the form

$$\int d^4\theta\left(\frac{f}{|g|^{2/3}}\Sigma^*\Sigma\right).$$

Show that this form also leads to the same effective potential.

Another point of major importance to cosmology is the fact that the potential has negative terms in it so that subsequent to spontaneous breaking of the value of the potential need not be positive definite unlike that in the case of global supersymmetry. This point will be elaborated upon in the next chapter.

§14.9. Super-Higgs Effect

Since supersymmetry is now a gauge symmetry it is important to ask whether the analog of the Higgs mechanism exists in this case. If it does, subsequent to the spontaneous breakdown of supersymmetry of the massless Goldstone fermion, the Goldstino and the massless gravitino should combine to give a massive spin 3/2 field. We wish to demonstrate in this section that this indeed happens. Our discussion will be very schematic and we will leave out most of the technical details which can be found in the work of Cremmer *et al.* [6]. Let us look at the fermionic part of the F-type Lagrangian in eqn. (14.8.8)

$$S_F = \int d^4x\, e[-\tfrac{1}{2}g''^{ij}\chi_{Li}^T C^{-1}\chi_{Lj} + \tfrac{1}{2}\bar{\psi}_L\gamma\chi_{Li}g'^i + \tfrac{1}{2}g\bar{\psi}_{\mu L}\sigma^{\mu\nu}\psi_{\nu R}]. \tag{14.9.1}$$

When supersymmetry is broken, $g'^i \neq 0$. Therefore, there are bilinears involving $\bar{\psi}\cdot\gamma\chi$, the latter field being the Goldstino. Thus, gravitino acquires a mass whose magnitude is proportional to

$$M_{3/2}^2 = \exp\left(-\frac{\kappa^2}{2}z^*z\right)\frac{\kappa^4|g|^2}{4}. \tag{14.9.2}$$

An important point to note about the gravitino mass is that, before its value can be fixed, the cosmological constant must be set equal to zero to that order. This is due to the fact that a combination of terms of the term

$$\mathscr{L}' = 3em^2\kappa^{-2} - \frac{i}{2}m\varepsilon^{\mu\nu\lambda\sigma}\bar{\psi}_\mu\gamma_5\Sigma_{\nu\lambda}\psi_\sigma \tag{14.9.3}$$

is invariant under local supersymmetry. This means that a cosmological constant is equivalent to a gravitino mass. Thus, the physical mass of the gravitino is obtained when the cosmological constant is set equal to zero.

For simplicity, we have ignored the Yang–Mills couplings in the above discussion. They can easily be included by replacing $S \to e^{gV}S$ in the function f and including the following generalized kinetic energy term for the gauge fields

$$\mathscr{S}_g = \int d^4x\, \mathrm{Re}\int d^2\theta\, (f_{\alpha\beta}(s)W_a^\alpha\varepsilon^{ab}W_b^\beta) + \text{h.c.} \tag{14.8.8}$$

If $f_{\alpha\beta}(S) = \delta_{\alpha\beta}$, this gives the canonical kinetic energy term for the gauge multiplet. It is also clear that if we choose $f_{\alpha\beta} = (1 + S)\delta_{\alpha\beta}$ then, on expanding the Lagrangian, we will have a term of the form $h\lambda^T C^{-1}\lambda$, i.e., a mass term for the gaugino, after rescalings, and spontaneous supersymmetry breakings which can give rise to a gaugino mass term.

§14.10. Different Formulations of Supergravity

We saw in Section 14.4 that to match the bosonic and fermionic degrees of freedom for the gravitino–graviton supermultiplet we need six bosonic auxiliary fields. In the formulation just presented, those auxiliary fields are identifiable with the complex F-component of an auxiliary chiral superfield which is used to fix conformal gauges and the gauge field corresponding to chiral symmetry. However, supergravity can be formulated by different choices of auxiliary fields. We list below the various formulations and their authors.

(a) The New Minimal Formulation developed by Sohnius and West [7] uses a real linear multiplet L, i.e., a multiplet satisfying $D^2L = \bar{D}^2L = 0$. Its components can be written as (C, η, B_m) where B_m is a transverse vector field, η is a Majorana field, and C is a real field. The auxiliary fields in this case are two transverse vector fields B_m and the transverse chiral field A_m. The chiral field becomes transverse (and therefore has only three components) because the real auxiliary field has zero chiral quantum number and cannot be used to fix chiral gauge. The matter coupling to supergravity in this gauge has been studied and straightforward techniques lead to a scalar potential that is different from the one obtained in the old minimal case [8].

(b) The Breitenlohner Formulation [9]. In this formulation a complex linear multiplet is used as the auxiliary field. The components of a complex linear multiplet are $(\mathscr{A}, \eta_1 + i\eta_2, H, B_m, \Lambda)$. After fixing the gauge the multiplet becomes $(1, \eta_1, H, B_{m_1} + iB_{m_2}, \Lambda)$. We have fourteen auxiliary bosonic fields (i.e., H_1, H_2, B_{m_1}, B_{m_2}, and A_m) and eight fermionic auxiliary fields (η_1, Λ), leaving six unpaired bosonic fields to match degrees of freedom in the graviton–gravitino multiplet. There also exists a tensor calculus for this multiplet [5].

(c) The de Wit–Van Niuenhuizen [10] Formulation uses a real vector multiplet as the auxiliary multiplet. Here again, after conformal gauge fixing, six unmatched bosonic fields remain as required.

We can give new formulations taking different sets [8], [11] (chiral, vector, etc.), of multiplets as auxiliary fields, but the most interesting for application to particle physics seems to be the old minimal formulation or any formulation that contains a chiral field in its auxiliary field set. In this case only an effective scalar potential, which is nontrivially different from that of global supersymmetry, emerges.

References

[1] D. Freedman, S. Ferrara, and P. Van Nieuwenhuizen, *Phys. Rev.* **D13**, 3214 (1976);
S. Deser and B. Zumino, *Phys. Lett.* **62B**, 335 (1976).
[2] For a review see
P. Van Nieuwenhuizen, *Phys. Rep.* **68**, 189 (1981).
[3] C. Misner, K. S. Thorne, and J. Wheeler, *Gravitation*, Freeman, San Francisco, 1970;
S. Weinberg, *Gravitation and Cosmology*, Wiley, New York, 1972.
[4] M. Kaku, P. K. Townsend, and P. Van Nieuwenhuizen, *Phys. Rev.* **D17**, 3179 (1978).
[5] T. Kugo and S. Uehara, *Nucl. Phys.* **B222**, 125 (1983).
[6] K. S. Stelle and P. C. West, *Nucl. Phys.* **B145**, 175 (1978);
P. Van Nieuwenhuizen and S. Ferrara, *Phys. Lett.* **B76**, 404 (1978);
E. Cremmer, S. Ferrara, L. Girardello and A. van Proeyen, *Phys. Lett.* **116B**, 231 (1982); *Nucl. Phys.* **B212**, 413 (1983);
E. Cremmer, B. Julia, J. Scherk, S. Ferrara, L. Girardello, and P. Van Nieuwenhuizen, *Nucl. Phys.* **B147**, 105 (1979);
A. H. Chamseddine, R. Arnowitt, and P. Nath, *Phys. Rev. Lett.* **49**, 970 (1982).
[7] M. Sohnius and P. West, *Phys. Lett.* **105B**, 353 (1981).
[8] C. S. Aulakh, M. Kaku, and R. N. Mohapatra, *Phys. Lett.* **126B**, 183 (1983);
For recent discussion of theories with multiple compensators, see
K. T. Mahanthappa and G. Stabler, VPI preprint, 1985.
[9] P. Breitenlohner, *Phys. Lett.* **67B**, 49 (1977).
[10] B. de Wit and P. Van Nieuwenhuizen, *Nucl. Phys.* **B139**, 216 (1978).
[11] V. O. Rivelles and J. G. Taylor, *Phys. Lett.* **113B**, 467 (1982).

CHAPTER 15

Application of Supergravity ($N = 1$) to Particle Physics

§15.1. Effective Lagrangian from Supergravity

In this chapter we would like to build models of elementary particle interactions using the ideas developed in the previous chapter. To summarize the discussion we note that the general action that couples supergravity ($N = 1$) to the Yang–Mills matter field system can be written as follows:

$$S_{\text{eff}} = \int d^4x \, e(f(S, e^{gV}s^\dagger)\Sigma^*\Sigma)_D + \int d^4x \, e(g(S)\Sigma^3)_F + \text{h.c.}$$

$$+ \int d^4x \, e(f_{\alpha\beta}(S) W^{\alpha,a} W_a^\beta)_F + \text{h.c.}, \tag{15.1.1}$$

where S and V denote the matter and gauge fields, respectively, and D and F denote the generalized D- and F-terms invariant under local supersymmetry transformations defined in eqns. (14.7.9) and (14.7.10), respectively.

In terms of these function we can write the Lagrangian for the matter Yang–Mills system as follows (denoting $S_i = (z_i, \chi_i)$ and $V = (V_\mu, \lambda)$)

$$\mathscr{L}(z_i, \chi_i, V_\mu, \lambda, \psi_\mu) = \mathscr{L}_{\text{K.E.}} - V + \mathscr{L}_g + \mathscr{L}', \tag{15.1.2}$$

where $\mathscr{L}_{\text{K.E.}}$ denotes the kinetic energy for the matter Yang–Mills fields, $\mathscr{L}_g$ for the graviton–gravitino Lagrangian is given by

$$\mathscr{L}_g = -\frac{e}{2\kappa^2} R - \varepsilon^{\mu\nu\rho\sigma} \bar{\psi}_\mu \gamma_5 \gamma_\nu D_\rho \psi_\sigma - m_g \bar{\psi}_\mu \sigma^{\mu\nu} \psi_\nu, \tag{15.1.3}$$

where

$$m_g^2 = \kappa^{-2} e^{-\mathscr{G}}, \tag{15.1.4}$$

where

$$\mathcal{G} = 3\ln\left(-\frac{\kappa^2 f}{3}\right) - \ln\frac{|g|^2}{3}\kappa^6$$

$$= 3\ln\left(-\frac{f}{|g|^{2/3}}\right) - \ln 9. \qquad (15.1.5)$$

We note that this is the form suggested by rescaling arguments given in the previous chapter.

The effective scalar potential V is given by

$$V(z, z^*) = -e^{-\mathcal{G}}(3 + \mathcal{G}'^i \mathcal{G}_i''^{-1j} \mathcal{G}_j'). \qquad (15.1.6)$$

To obtain the canonical form for the kinetic energy we choose

$$\mathcal{G}(z, z^*) = -\kappa^2 z_i z^{*i} - \ln\frac{|g|^2}{4}\kappa^6, \qquad (15.1.7)$$

in which case we get

$$V = \exp\left(\kappa^2 \sum_i z^{*i} z_i\right)\left[\sum_i \left|\frac{\partial g}{\partial z_i} + \frac{\kappa^2}{2} z^{*i} g\right|^2 - 3\kappa^2 |g|^2\right] \qquad (15.1.8)$$

and

$$m_g^2 = \frac{\kappa^4}{4}\exp\left(-\kappa^2 \sum_i z^{*i} z_i\right)|g|^2. \qquad (15.1.9)$$

$\mathcal{L}'$ consists of terms involving powers of κ multiplying $f'_{\alpha\beta}$, etc., and for the simplest choice $f_{\alpha\beta} = \delta_{\alpha\beta}$ such terms are small. A particularly interesting term in $\mathcal{L}'$ is of the form $\kappa^{-1} f' \lambda^T c^{-1} \lambda$ which can be nonzero if $f_{\alpha\beta} = (1 + a\kappa z)\delta_{\alpha\beta}$. This will generate a finite tree level mass for the gauginos without affecting anything else. In the subsequent section we would like to apply these formulas to building models.

An important property of eqn. (15.1.6) worth emphasizing is that the potential is invariant under the transformation

$$g(z, z^*) \to g(z, z^*) + f(z) + f'(z^*).$$

This property is known as Kahler invariance. As a first application of this formalism let us choose $g(z)$ as follows:

$$g(z) = \mu z^2 + m^3. \qquad (15.1.10)$$

It is then easy to see that $g'(z) = \mu z$ and minimum occurs at $z = 0$ which preserves supersymmetry. Thus, we should expect the gravitino mass to vanish. But eqn. (15.1.9) tells us that

$$m_g^2 = \frac{\kappa^4}{4} m^6. \qquad (15.1.11)$$

Equation (15.1.8) gives for V at the minimum

$$V_{\min} = -3\kappa^2 m^6. \tag{15.1.12}$$

This gives the relation

$$V_{\min} = -12\kappa^{-2} m_g^2, \tag{15.1.13}$$

which is an explicit verification of the connection between the cosmological constant and the gravitino mass given in eqn. (14.9.3) provided we write $m_g = m/2$. Thus, in this case, the gravitino is actually massless and the appearance of the mass is due to the cosmological constant being nonzero, as was explained in the previous chapter. This point was emphasized by Deser and Zumino [1].

§15.2. The Polonyi Model of Supersymmetry Breaking

As in the case of global supersymmetry models, the physics discussion of supergravity models starts with the superpotential function $g(z)$. A particularly useful form for this was proposed in an unpublished paper by Polonyi [2] which splits g into a sum of two functions

$$g = g_1(z_a) + g_2(z), \tag{15.2.1}$$

where z_a denotes the usual matter fields (or the scalar member of the matter chiral multiplets) and z denotes a gauge singlet scalar field which is the scalar member of a chiral multiplet. The $g_2(z)$ is used to break supersymmetry and involves parameters which are of the order $\kappa^{-1} \equiv M_p$. Polonyi chose

$$g_2(z) = \mu^2(z + \beta). \tag{15.2.2}$$

To simplify the study of its implications we choose $g_1(z_a) = 0$. It is easy to see that, if the gravitational effects are ignored (i.e., $\kappa = 0$), eqn. (15.2.2) leads to a flat potential independent of z (see Fig. 15.1). Once gravitational effects are

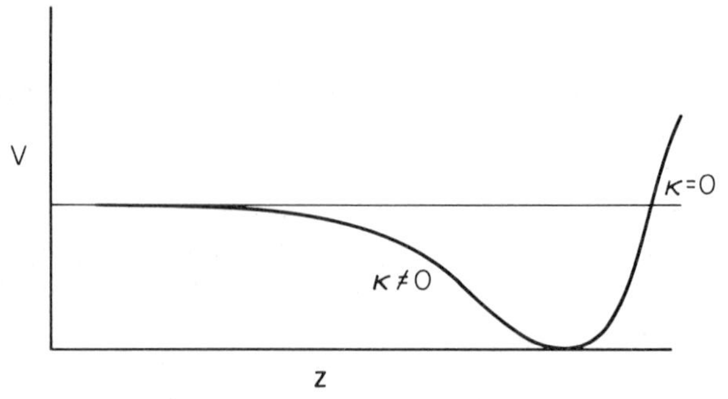

Figure 15.1. Form of the Polonyi potential in the absence of gravitational interactions.

turned on, the presence of the negative terms produces a minimum of the potential at $z \sim O(M_p)$. We would like to find this minimum and we would also require that the value of the potential at this minimum vanish. This is to ensure that the cosmological constant vanishes. The vanishing of the cosmological constant is an additional attractive feature of supergravity not shared by global supersymmetry models and is, in fact, needed to give a consistent definition of the physical gravitino mass.

Using eqn. (14.8.17) we can write the effective scalar potential as

$$V(z, z_a) = \exp\left(\frac{\kappa^2}{2}\left[|z|^2 + \sum_a |z_a|^2\right]\right)\left[\left|\mu^2 + \frac{\kappa^2}{2}\mu^2 z(z + \beta)\right|^2 + \sum_a \frac{\kappa^4}{4}\mu^4|z_a|^2|z + \beta|^2 - 3\kappa^2\mu^4|z + \beta|^2\right]. \quad (15.2.3)$$

For the choice of the parameter $\mu^2\kappa^2 < 1$, the minimum of V to leading order is given by minimizing the potential ignoring the second term in eqn. (15.2.3). We then have the following constraint equations:

$$\left.\frac{\partial V}{\partial z}\right|_{z=z_0} = 0,$$

$$V(z_0) = 0, \quad (15.2.4)$$

choosing real z we obtain

$$1 + \frac{\kappa^2}{2}z(z + \beta) = \pm\sqrt{3}\kappa(z + \beta) \quad (15.2.5)$$

and

$$\frac{\kappa^2}{2}(2z + \beta)\left(1 + \frac{\kappa^2}{2}z(z + \beta)\right) = 6\kappa^2(z + \beta). \quad (15.2.6)$$

This leads to the value of z at the minimum, i.e., z_0 being

$$z_0^{(\mp)} = 2\kappa^{-1}(-\sqrt{3} + 1)(\pm),$$

$$\beta^{(\mp)} = 2\kappa^{-1}(-2 + \sqrt{3})(\pm). \quad (15.2.7)$$

The corresponding value of the gravitino mass, denoted by m_{g_0}, is (choosing the $z_0 = z_0^-$)

$$m_{g_0} = e^{2(\sqrt{3}-2)}\mu^2\kappa. \quad (15.2.8)$$

The numerical value for the lowest order gravitino mass m_{g_0} is an interesting one if we choose the supersymmetry breaking scale $\mu \simeq 10^{11}$ GeV, i.e., $m_g \simeq 10^3$ GeV. This is of the same order of magnitude as the weak interaction symmetry breaking scale m_W/g. This raises the interesting possibility that the scale of weak interactions could be induced by supergravity. In fact, this suggestion appears one step closer to realization once we evaluate the potential in eqn. (15.2.3) and keep terms to zeroth power in (κm_g)

$$V(z_0, z_a) = m_{g_0}^2 \sum_a |z_a|^2. \tag{15.2.8a}$$

Note that z_a are the Lorentz scalar components of the light matter superfields such as quark, leptons, etc., and that among the light fields is included the doublet Higgs field of the Weinberg–Salam models, responsible for the electro-weak symmetry breaking whose gravitationally induced mass is also m_{g_0}. If this mass term appeared with a negative sign the electro-weak symmetry breaking would be induced by supergravity and thus would solve a major problem of unified gauge theories.

The next step in trying to explore this possible connection between supergravity and electro-weak-symmetry breaking is to include $g_1(z_a)$ and see whether it can trigger symmetry breaking, in spite of the positive (mass)^2 term for the Higgs field. Let us write down the effective potential involving the light fields z_a in the presence of nonzero $g_1(z_a)$ at $z = z_0$. To zeroth order in (κm_g) we have [3], [4]

$$V(z_a) = m_g^2 \sum z_a^* z_a + \sum_a \left| \frac{\partial g_1}{\partial z_a} \right|^2 + m_g \left[\sum_a z_a \frac{\partial g_1}{\partial z_a} + (A - 3) g_1 + \text{h.c.} \right] + D\text{-terms}, \tag{15.2.9}$$

where $A = 3 - \sqrt{3}$. It is worth pointing out at this stage that there is no reason to implement supersymmetry breaking by the Polonyi-like superpotential. We can choose other forms. The interesting point is that the effective potential in eqn. (15.2.9) retains its form except that the value of A changes provided there are no other intermediate scales in the theory between M_p and μ^2/M_p ($M_p = \kappa^{-1}$). In the next section we will explore the question of relating electro-weak symmetry breaking to m_g for arbitrary A.

Another important point regarding the effective potential is that the structure of the supersymmetry breaking terms is uniquely fixed by supergravity, thus removing a certain degree of arbitrariness from supersymmetry model building. A curious feature of the supersymmetry breaking terms in eqn. (15.2.9) is that, even though the scale of supersymmetry breaking is of order 10^{11}–10^{12} GeV, the mass splitting between the components of a supermultiplet is only of order m_g. We may then ask how this can be reconciled with the mass formula in a supermultiplet which says that

$$m_\psi^2 - m_A^2 = f_{\chi\psi A} M_{\text{SUSY}}^2. \tag{15.2.10}$$

The answer is that the Goldstino χ-coupling to matter in this case is extremely small, i.e.,

$$f_{\chi\psi A} = \mu^2 \kappa^2. \tag{15.2.11}$$

The diagram responsible for the Goldstino coupling to matter is shown in Fig. 15.2.

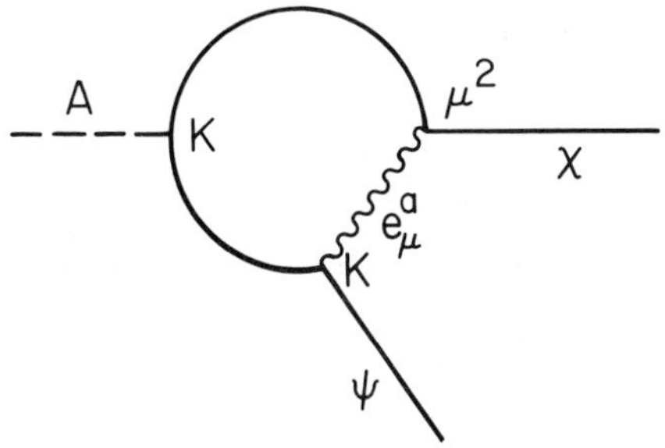

Figure 15.2. The one-loop graph that induces Goldstino coupling to matter.

§15.3. Electro-Weak Symmetry Breaking and Supergravity

In this section we pursue the idea of connecting the scale of electro-weak symmetry breaking with the gravitino mass. In the absence of the light-particle superpotential, i.e., $g_1(z_a) = 0$, we saw that (mass)^2 for the Higgs multiplet is positive leading to no electro-weak breaking. Let us include a nontrivial superpotential involving two Higgs doublets H_u, H_d and a singlet Y

$$g_1(H_u, H_d, Y) = \lambda_1 H_u H_d Y + \lambda_2 Y^3. \tag{15.3.1}$$

Recall that H_u and H_d are the two $SU(2)_L$ doublet superfields introduced in globally supersymmetric $SU(2)_L \times U(1)$ model building. The singlet Y is introduced and the form of g_1, in eqn. (15.3.1) without any dimensional parameters, is chosen so that the weak symmetry breaking (if it occurs) will automatically be proportional to m_g. Using eqn. (15.2.9) we can write the low-energy effective potential as

$$V(z_a) = m_g^2 \sum_{a=H_\mu, y_d, Y} z_a^* z_a + \sum_a \left|\frac{\partial g_1}{\partial z_a}(z_a)\right|^2 + m_g \frac{A}{3} \sum_{z_a} \left(z_a \frac{\partial g_1}{\partial z_a} + \text{h.c.}\right) \tag{15.3.2}$$

$$= \left(1 + \frac{A}{3}\right) \sum_a \left| m_g z_a + \frac{\partial g_1}{\partial z_a}\right|^2 + \left(1 - \frac{A}{3}\right) \sum_a \left| m_g z_a - \frac{\partial g_1}{\partial z_a}\right|^2. \tag{15.3.3}$$

It then follows that for $|A| < 3$ the absolute minimum of V is at $\langle H_u \rangle = \langle H_d \rangle = \langle Y \rangle = 0$. Thus, a necessary condition for electro-weak symmetry breaking is that $|A| \geq 3$ [5]. Clearly the Polonyi potential does not satisfy this condition. Thus, we must go beyond this simple analysis to realize the dream of connecting m_W to m_g. It has been further pointed out that, even if we assume $|A| \geq 3$, the symmetry breaking solution in realistic examples leads to the breakdown of electric charge. Therefore, alternative approaches must be sought.

An ingeneous alternative method has been proposed by Alvarez-Gaume, Polchinski, and Wise [6], who note that the effective potential in eqn. (15.2.9) has been defined at the Planck scale, i.e., $m = M_p$. To study their behavior at lower energies the parameters must be extrapolated down to the TeV scale. Since the nature of the extrapolation is determined by radiative corrections, the various parameters in $V(z_a)$ will change from their values at the Planck scale so, at the TeV scale, the potential should be written as

$$V(z_a)|_{\text{TeV}} = \sum_a M_a^2 Z_a^* Z_a + \sum \mu_{abc} Z_a Z_b Z_c + \text{h.c.}$$
$$+ \sum \lambda_{abcd} Z_a^* Z_b^* Z_c Z_d + \text{h.c.} + D\text{-terms}, \qquad (15.3.4)$$

with M_a, μ_{abc}, and λ_{abcd} satisfying the following boundary conditions:

$$M_a^2(M_p) = m_g^2,$$
$$\mu_{abc}(M_p) = Am_g,$$
$$\lambda_{abcd}(M_p) = \frac{\partial^3 g_1^*}{\partial Z_a^* \, \partial Z_b^* Z_c^*} \frac{\partial^3 g_1}{\partial Z_c \, \partial Z_d \, \partial Z_e}. \qquad (15.3.5)$$

If the process of extrapolation makes the (mass)^2 of the Higgs doublets H_u and H_d negative then, conspiring with the D-terms, it will give rise to electro-weak symmetry breaking. To see that this can indeed be realized for certain values of the parameters in the Lagrangian, we look at a realistic $SU(2)_L \times U(1)_Y \times SU(3)_C$ model. As mentioned in Chapter 12 it will have the matter fields as shown in Table 15.1.

The most general gauge invariant superpotential involving these fields can be written as

$$W = h_{u,ab} Q_a H_u U_b^c + h_{d,ab} Q_a H_d D_b^c + f_{e,ab} L_a H_d E_b^c + \mu H_u H_d. \qquad (15.3.6)$$

The last term in eqn. (15.3.6) introduces a mass parameter but we will keep it small and will not adjust it to get electro-weak breaking. The significance of this parameter is that it provides a solution to the strong CP-problem which

Table 15.1

Fields	$SU(2)_L \times U(1)_Y \times SU(3)_C$ quantum number
Q_a ($a = 1, 2, 3$ for generation)	$(\frac{1}{2}, \frac{1}{3}, 3)$
U_a^c ($a = 1, 2, 3$ for generation)	$(0, -\frac{4}{3}, 3^*)$
D_a^c ($a = 1, 2, 3$ for generation)	$(0, +\frac{2}{3}, 3^*)$
L_a ($a = 1, 2, 3$ for generation)	$(\frac{1}{2}, -1, 1)$
E_a^c ($a = 1, 2, 3$ for generation)	$(0, +2, 1)$
H_u ($a = 1, 2, 3$ for generation)	$(\frac{1}{2}, +1, 1)$
H_d ($a = 1, 2, 3$ for generation)	$(\frac{1}{2}, -1, 1)$

we do not describe here [7]. It is now possible to write the renormalization group equations for the various parameters in order to study their extrapolation from the Planck mass down to the TeV scale. The general analysis of the renormalization group equations for softly-broken supersymmetric theories has been carried out in the literature and can be used to write the equaions for m_a^2 and other parameter equations for m_a^2. Other parameters of the potential are given below

$$V = \sum_a \left| \frac{\partial W}{\partial Z_a} \right|^2 + \sum_a M_a^2 Z_a^* Z_a + m_g A_c^{ab} f_{e,ab} L_a H_d E_b^c + A_u^{ab} h_{u,ab} Q_a H_i U_b^c$$
$$+ A_d^{ab} h_{d,ab} Q_a H_d D_b^c + \tilde{A} \mu M_g H_u H_d + \text{h.c.}, \tag{15.3.7}$$

with ($t = \ln m$) [6], [8]

$$\frac{\partial M_{\bar{H}_u}^2}{\partial t} = \frac{1}{8\pi^2} [3h_{u,33}^2 (M_{\bar{H}_u}^2 + M_{\bar{U}_3^c}^2 + M_{\bar{Q}_3}^2 + M_g^2 |A_u^{33}|^2)$$
$$- (3|\tilde{M}_2|^2 g_2^2 + |\tilde{M}_1|^2 g_1^2)], \tag{15.3.8}$$

$$\frac{\partial M_{\bar{U}_3^c}^2}{\partial t} = \frac{1}{8\pi^2} [2h_{u,33}^2 (M_{\bar{H}_u}^2 + M_{\bar{U}_3^c}^2 + M_{\bar{Q}_3}^2 + M_g^2 |A_u^{33}|^2)$$
$$- (\tfrac{16}{3}|\tilde{M}_3|^2 g_3^2 + \tfrac{16}{9}|\tilde{M}_1|^2 g_1^2)], \tag{15.3.9}$$

$$\frac{\partial M_{\bar{Q}_3}^2}{\partial t} = \frac{1}{8\pi^2} [(h_{u,33})^2 (M_{\bar{H}_u}^2 + M_{\bar{U}_3^c}^2 + M_{\bar{Q}_3}^2 + M_g^2 |A_u^{33}|^2)$$
$$- (\tfrac{16}{3}|\tilde{M}_3|^2 g_3^2 + 3|\tilde{M}_2|^2 g_2^2 + \tfrac{1}{9}|\tilde{M}_1|^2 g_1^2)], \tag{15.3.10}$$

where $\tilde{M}_a$ are the gaugino masses for the SU(3), SU(2), and U(1) groups, respectively, g_a's being the corresponding gauge couplings. There are other equations that must also be analyzed. However, to illustrate the basic point, we need only the above three equations.

From eqn. (15.3.8) we see that, if $\tilde{M}_2$ and $\tilde{M}_1$ are chosen small, the slope $\partial M_{\bar{H}}^2/\partial t$ is positive which means that as we come down from the Planck scale $M_{\bar{H}}^2$ decreases and, therefore, if $h_{u,33}$ is sufficiently large, $m_{\bar{H}}^2$ can be negative at the TeV scale and the SU(2) × U(1) symmetry will break down. We must, however, be careful since in the same approximation $m_{\bar{U}_3^c}^2$ and $m_{\bar{Q}_3}^2$ also decrease with energy and *could* become negative leading to the lead result that the $SU(3)_C \times U(1)_{em}$ symmetries will break down. This can be prevented since we note that, due to the coefficients 3, 2, and 1 in the slopes of eqns. (15.3.8), (15.3.9), and (15.3.10) (arising, respectively, from three-colors and the doublet and singlet nature of Q and U), the $m_{\bar{H}}^2$ decreases faster than $m_{\bar{Q}_3}^2$ and $m_{\bar{U}_3^c}^2$. So for h_u^{33}, within a certain limited range, at the TeV scale only $m_{\bar{H}}^2$ becomes negative and $m_{\bar{Q}_3}^2$ and $m_{\bar{U}_3^c}^2$ remain positive. Thus, for only $SU(2)_L \times U(1)$ to occur, $h_{u,33}$ and hence m_t must lie within certain limits. It is interesting that m_t lies in the range 100 GeV $\leq m_t \leq$ 190 GeV. The recent discovery of the t-quark in the mass range of 40–60 GeV therefore rules out the simple-minded analysis carried out here. The idea is, however, quite profound and in some variations may provide an understanding of the electro-weak scale.

§15.4. Grand Unification and $N = 1$ Supergravity

In this section we would like to study several applications of supergravity to grand unified models. We saw in Chapter 12 that, while global supersymmetry ameliorates somewhat the gauge hierarchy problem, it gives rise to several other problems, such as vacuum degeneracy and the associated cosmological difficulties. It would, therefore, be interesting to see if supergravity affects these considerations. More importantly, since the general grand unification scales are of the order of 10^{15}–10^{19} GeV, we may hope that embedding grand unified models into $N = 1$ supergravity models may explain the grand unification scale in terms of the scale of gravity, i.e., the Planck scale. Combining this with the discussion of the previous section, we may visualize a scenerio where the only independent scales are the scales of supersymmetry breaking μ_S and the Planck mass, and all other scales are consequences of the unification of gravity with other forces of nature.

It is with the hope of realizing this dream that a number of grand unified models have been embedded into $N = 1$ supergravity theories. While no convincing model of this kind exists, we review in this section some of the attempts in this direction and the difficulties that arise in trying to realize this goal. We will work with the SU(5) model for the purpose of illustration and discuss the following points: (i) supergravity has the potential to solve the vacuum degeneracy problem; and (ii) to maintain the tree level hierarchy of mass scales the superpotential must satisfy certain constraints.

(a) Lifting Vacuum Degeneracy with Supergravity

To study this problem we consider the superpotential of the SU(5) model given in Chapter 13 involving only the {24}-dimensional Higgs superfield Φ

$$W = z \operatorname{Tr} \Phi + x \operatorname{Tr} \Phi^3 + y \operatorname{Tr} \Phi^3. \tag{15.4.1}$$

We recall our discussion in Chapter 13 that, if we keep supersymmetry unbroken at the grand unification scale, i.e., $F_\phi \doteq 0$, we get three degenerate vacuums corresponding to the unbroken symmetries SU(5), SU(4) × U(1), and SU(3) × SU(2) × U(1)

$$\text{SU(5)}: \quad \langle \Phi \rangle = 0,$$

$$\text{SU(4)} \times \text{U(1)}: \quad \langle \Phi \rangle = b \begin{pmatrix} 1 & & & & \\ & 1 & & & \\ & & 1 & & \\ & & & 1 & \\ & & & & -4 \end{pmatrix},$$

$$\text{SU}(3) \times \text{SU}(2) \times \text{U}(1): \quad \langle \Phi \rangle = a \begin{pmatrix} 1 & & & & \\ & 1 & & & \\ & & 1 & & \\ & & & -3/2 & \\ & & & & -3/2 \end{pmatrix}, \tag{15.4.2}$$

where

$$b = \frac{10}{33} \frac{x}{y}$$

and

$$a = \frac{4}{3} \frac{x}{y}.$$

The inclusion of supergravity modifies the form of the effective potential at low energies and to the lowest order in powers of κ we have

$$V \simeq \left| \frac{\partial W}{\partial \phi} \right|^2 + \kappa^2 \left(\phi W \frac{\partial W}{\partial \phi} + \text{h.c.} \right) + \kappa^2 \operatorname{Tr} \phi^2 \left| \frac{\partial W}{\partial \phi} \right|^2 - 3\kappa^2 |W(\phi)|^2 + \mathrm{O}(\kappa^4). \tag{15.4.3}$$

Since $\partial W / \partial \phi = F_\phi = 0$ for the minima in eqn. (15.4.2) we find at the minimum

$$V \simeq -3\kappa^2 |W(\phi)|^2. \tag{15.4.4}$$

Using eqn. (15.4.2) we find that

$$V_{\text{SU}(5)} = 0, \qquad V_{\text{SU}(4) \times \text{U}(1)} = -\left(\frac{7200}{35937} \right)^2 \frac{x^6}{y^4} 3\kappa^2$$

and

$$V_{\text{SU}(3) \times \text{SU}(2) \times \text{U}(1)} = -3 \left(\frac{112}{9} \right)^2 \frac{x^6}{y^4} \kappa^2.$$

Thus the vacuum degeneracy is lifted by gravitational interaction as was pointed out by Weinberg [8]. A new problem arises. The physical vacuum corresponding to the symmetry SU(3) × SU(2) × U(1) has a large cosmological constant. This can, however, be set equal to zero by subtracting from $W(\phi)$ the constant $\sqrt{3}(\frac{112}{9})(x^3/y^2)\kappa$, but it is easy to see that this makes the minima with SU(4) × U(1) and SU(5) symmetries lower than the desired minimum. It has, however, been argued by Weinberg [9] that, if the universe is in the SU(3) × SU(2) × U(1) minimum, the tunneling probability to the lower minima is not significant. The question could be raised as to why the universe chose the highest of the three minima in the first place. It is, however, important that supergravity does remove the utterly high degree of vacuum degeneracy associated with global supersymmetry.

(b) Maintaining the Hierarchy of Mass Scales at the Tree Level

We saw in the previous section that coupling the electro-weak model to $N = 1$ supergravity can produce a mass scale m_g of order 10^3–10^4 GeV through supersymmetry breaking. If, instead of an electro-weak symmetry whose breaking scale is of order m_g, we consider a grand unified symmetry (such as SU(5), SO(10), etc.) or a partial unified symmetry (such as $SU(2)_L \times SU(2)_R \times SU(4)_C$) which manifest themselves at a scale $\mu \gg m_g$, we can ask whether the gravitino mass can still be considered as inducing the W-boson mass. One technical problem that arises is the following: if we assume $\mu = M_U \approx 10^{15}$ GeV, $\kappa M_U \approx 10^{-4}$ and *a priori* the gravitino mass could receive tree level corrections of order $M_U(\kappa M_U)^n$, $n = 1, 2, 3, \ldots$. For the tree level mass hierarchy to remain the theory must be such that $n > 3$. This problem has been analyzed in several papers and it has been observed that, in general, the superpotential has to satisfy certain constraints. To understand these constraints let us classify the fields entering the superpotential according to their masses and vacuum expectation values as (ϕ_α, T_i, D_a), where the subscript α denotes fields which have v.e.v. as well as mass of order M_U, i for fields with v.e.v. of order m_g and mass of order M_U, and a for fields with both v.e.v. and mass of order m_g. The constraints to be satisfied by the superpotential in order to maintain the gauge hierarchy are [3]:

$$\left.\frac{\partial^2 g}{\partial T_i\, \partial\phi_\alpha}, \frac{\partial^2 g}{\partial D_a\, \partial\phi_\alpha}, \frac{\partial^2 g}{\partial\phi_\alpha\, \partial\phi_\beta}\right|_{\text{minimum}} \sim O(m_g). \tag{15.4.5}$$

The next question is: For a superpotential that satisfies eqn. (15.4.5), what is the low-energy form of the effective potential? Using the form in eqn. (15.2.1), where z_a goes over z_α, z_i, z_a, we can write down the effective potential as

$$V(z_a, z_a^\dagger) = \exp\left(\frac{\kappa^2}{2}\sum_n z_n^* z_n\right)[|\tilde{g}_{1,a}|^2 + m_1^2 z_a^* z_a$$
$$+ (\omega + \omega^\dagger) + m_g^2(\tilde{g}_{1,\alpha} G_\alpha^{(0)} + \tilde{g}_{1,i} G_i^{(0)} + \text{h.c.})], \tag{15.4.6}$$

where

$$\omega = m_2 \tilde{g}_1 + m_3 z^a \tilde{g}_{1,a} \tag{15.4.7}$$

$$\tilde{g}_1 = g_1(z_\alpha, z_i, z_a) - g_1(z_\alpha, z_i, 0), \tag{15.4.8}$$

$$G_m = \frac{\partial g_1}{\partial z^m} + \frac{\kappa^2}{2} z_m g_1, \tag{15.4.9}$$

$$m_1^2 = \tfrac{1}{2} m_g^2 [|G_z^{(0)}|^2 - |g_2^{(0)}|^2],$$
$$m_2 = \tfrac{1}{2} m_g [Z^{(0)} \bar{G}_z^{(0)} - 3 g_2^{(0)}],$$
$$m_3 = \tfrac{1}{2} m_g \bar{g}_2^{(0)}, \tag{15.4.10}$$

where

$$\bar{g}_2 = \frac{\kappa}{\mu^2} g_2, \qquad \bar{G} = \frac{\kappa}{\mu^2} G, \quad \text{etc.,}$$

and the superscript zero implies the value of the function at the zeroth order minimum.

As an example of a grand unified model coupled to $N = 1$ supergravity we consider the SU(5) model discussed in Ref. [3], with the following structure for the superpotential, in terms of the Higgs superfields $\Phi\{24\}$, $H\{5\}$, $H\{\bar{5}\}$, $U\{1\}$ and the matter fields $\bar{F}\{\bar{5}\}$ and $T\{10\}$

$$g_1(\Phi, H, H', F, T) = \lambda_1 \left(\tfrac{1}{3} \operatorname{Tr} \Phi^3 + \frac{M}{2} \operatorname{Tr} \Phi^2 \right) + \lambda_2 H(\Phi + 3M')H' + \lambda_3 UH'H + Tf_1 TH + H'Tf_2 \bar{F}. \tag{15.4.11}$$

The role of the various pieces of g_1 is clear. The terms proportional to λ_1 are responsible for SU(5) breaking while keeping supersymmetry unbroken. The λ_2 coupling is the one that keepts the Weinberg–Salam Higgs doublet massless; the f_1 and f_2 terms lead to fermion masses. The supersymmetry breaking is achieved by the usual Polonyi term. We leave it as an exercise to the reader to work out the symmetry breaking chain as well as other consequences of the model. These ideas have been applied in various papers for the discussion of grand unified models with $N = 1$ supergravity [9].

References

[1] S. Deser and B. Zumino, *Phys. Rev. Lett.* **38**, 1433 (1977).
[2] J. Polonyi, Budapest preprint no. KFKI-1977-93, 1977.
[3] A. H. Chamseddine, R. Arnowitt, and P. Nath, *Phys. Rev. Lett.* **49**, 970 (1982); *Phys. Lett.* **121B**, 33 (1983); *Phys. Lett.* **120B**, 145 (1983).
[4] R. Barbieri, S. Ferrara, and C. A. Savoy, *Phys. Lett.* **119B**, 343 (1982).
[5] H. P. Nilles, M. Srednicki, and D. Wyler, *Phys. Lett.* **124B**, 337 (1983); *Phys. Lett.* **120B**, 346 (1983).
[6] L. Alvarez-Gaume, J. Polchinski, and M. Wise, *Nucl. Phys.* **B221**, 495 (1983).
[7] R. N. Mohapatra, S. Ouvry, and G. Senjanovic, *Phys. Lett.* **126B**, 329 (1983).
[8] S. Weinberg, *Phys. Rev. Lett.* **48**, 1303 (1982).
[9] L. Hall, J. Lykken, and S. Weinberg, *Phys. Rev.* **D27**, 2359 (1983);
L. Ibanez, *Nucl. Phys.* **B218**, 514 (1983).

CHAPTER 16

Beyond $N = 1$ Supergravity

§16.1. Beyond Supergravity

So far, in this book, we have described the philosophical motivations, the mathematical foundations, and working principles for locally supersymmetric unification that described all four forces of nature, weak electromagnetic, and strong as well as gravitational, within one theoretical framework. A theorist's dream is, however, more ambitious and rightly so, since, even the very elegant $N = 1$ supergravity leaves many questions unanswered: a partial list includes:

(i) The gauge symmetry describing electro-weak unification has to be put in by hand. Thus, we really need two fundamental principles to derive the laws of physics: first, the equivalence principle to derive gravitational forces; and second, the local Yang–Mills symmetry to derive the rest of the interactions. It certainly would be more satisfying if both these principles could be combined into one.

(ii) The matter fields are chosen to fit phenomenology rather than being an outcome of the theoretical principles. It would certainly be more desirable if the basic principle that yields the physical laws could also yield the matter multiplets.

(iii) Finally, the age-old problem of divergences that beset the local field theories since their introduction to physics does not get resolved by the $N = 1$ supergravity theories. In fact, this problem is worse for $N = 1$ supergravity theories than either globaly supersymmetric or non-supersymmetric theories. In that sense, it could be construed as a step backward in the quest for the ultimate theory.

The search for answers to such questions has led to many interesting theoretical advances. In this chapter, we provide a brief introduction to and overview of the extended supersymmetries and higher dimensional supergravities which have emerged as interesting candidates for this ultimate unification, some of which also seem to have a more fundamental origin in string theories of matter.

§16.2. Extended Supersymmetries ($N = 2$)

We have seen in previous chapters that $N = 1$ global supersymmetry not only improves the divergence structure of the theory but also constrains the parameters in an interesting manner even after soft breakings are introduced. It is, therefore, tempting to go beyond $N = 1$ supersymmetry. The simplest extension is to consider $N = 2$ supersymmetry. Here, instead of considering a superspace with one two-component complex Weyl spinor coordinate θ as the fermionic coordinates, we will consider two sets of θ-coordinates θ_a, $a = 1, 2$. This theory has been studied in detail both for global [1] and local supersymmetry [2], [3] and it turns out that their structures are very different.

Before discussing the representations of $N = 2$ supersymmetry, it is worth pointing out that $N = 2$ supersymmetry algebra can support an additional element, the central charge denoted by Z in addition to the usual supersymmetry generators: Q_a, $a = 1, 2$, i.e., the $N = 2$ SUSY algebra is written as

$$\{Q_{a,\alpha}, Q_{b,\beta}\} = \varepsilon_{ab}\varepsilon_{\alpha\beta}Z, \tag{16.2.1}$$

$$\{Q_{a,\alpha}, \bar{Q}_{b,\beta}\} = \delta_{ab}(\sigma^\mu)_{\alpha\dot{\beta}}P_\mu. \tag{16.2.2}$$

It is obvious that for $N = 1$ supersymmetry, the right-hand side of eqn. (16.2.1) is zero. In physical applications, we generally restrict ourselves to the $Z = 0$ sector of the algebra. For this sector, the simplest irreducible multiplets are the hypermultiplet and the Yang–Mills multiplet. Their representation content is given in terms of $N = 1$ superfields as follows:

Hypermultiplet:	(ψ, χ),
Yang–Mills multiplet:	(W, ϕ),

where ψ, χ, and ϕ are chiral multiplets and W is the chiral multiplet constructed out of the $N = 1$ gauge multiplet in the Weiss–Zumino gauge. An important point, however, is that, under an internal symmetry group, ψ and χ transform like $\{N\}$- and $\{\bar{N}\}$-representations respectively under the group, whereas W and ϕ both transform like adjoint representations. This property is important for physical applications since it implies that, if quarks and leptons are assigned to the hypermultiplet, there must necessarily be mirror multiplets with $V + A$ interactions [4].

Before discussing the supersymmetric action for this case, we wish to analyze the structure of the hypermultiplet and the Yang–Mills multiplet. For this purpose, we write ψ, χ, W and ϕ in terms of components

$$\begin{aligned} \psi &= (A_\psi, \tilde{\psi}), \\ \chi &= (A_\chi, \tilde{\chi}), \\ W &= (\lambda, F_{\mu\nu}), \\ \phi &= (A_\phi, \tilde{\phi}). \end{aligned} \tag{16.2.3}$$

We can exhibit action of the SUSY generators Q^1 and Q^2 on the components as follows:

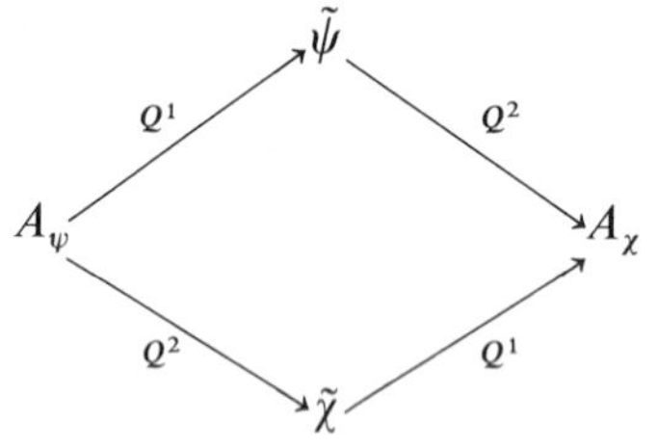

and

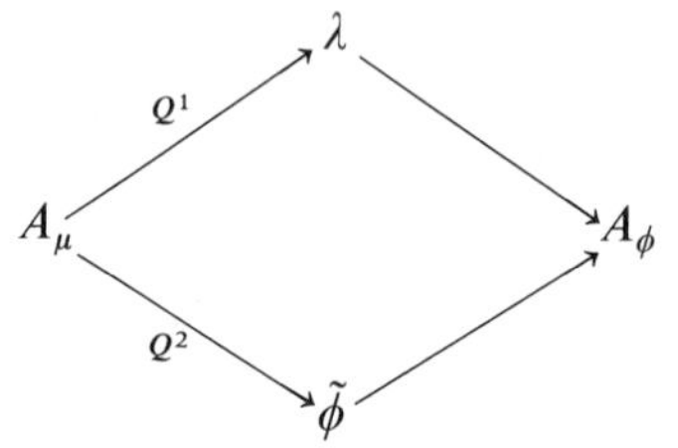

Note that the internal index a of the supersymmetry generator is shared only by the fermions $(\tilde{\psi}, \tilde{\chi})$ and $(\lambda, \tilde{\phi})$. If we define an SU(2) group on these indices, then $(\tilde{\psi}, \tilde{\chi})$ and $(\lambda, \tilde{\phi})$ transform as doublets and all other fields transform as singlets under the group.

The $N = 2$ supersymmetric action can be written in terms of the $N = 1$ supermultiplets: the matter multiplet (ψ, χ) and the gauge multiplet (W, ϕ) as follows:

$$\begin{aligned} S = &\int d^4x \, d^4\phi \, [\mathrm{Tr} \, \phi^\dagger e^{gV} \phi + \bar{\psi} e^{gV} \psi + \bar{\chi}_e{}^{-gV} \chi] \\ &+ \frac{1}{64g^2} \int d^4x \, d^2\theta \, WW + g \int d^4x \, d^2\theta \, (\psi\phi\chi) + \text{h.c.} \end{aligned} \tag{16.2.4}$$

The first point to note about this action is that it is invariant under the internal SU(2) group defined earlier.

It is clear that the structure of the above action is too restrictive to be of much use in model building. However, in the hope that this problem may receive a cure when $N = 2$ breaking terms are included, we proceed to point out one important property of this theory. This has to do with its divergence structure. We look at the β-function of this theory

$$\beta(g) = \frac{g^3}{8\pi^2}\left[\sum_\sigma n_\sigma T(R_\sigma) - C_2(G)\right], \tag{16.2.5}$$

where n_σ is the number of matter multiplets and T is defined as follows:

$$\mathrm{Tr}(\theta_a\theta_b) = \delta_{ab}T(R_\sigma), \tag{16.2.6}$$

where θ_a are the generators of the gauge group in the representation to which the matter multiplets belong. We find that, $\beta(g)$ vanishes, if

$$\sum_\sigma n_\sigma T(R_\sigma) = C_2(G). \tag{16.2.7}$$

Since the theory has no other coupling construct, it is finite if $\beta(g) = 0$. Thus, by choosing the number of matter multiplets appropriately, we may hope to construct a fully finite mode of elementary particle interaction. However, no such model has emerged because of the highly restrictive nature of eqn. (16.2.6) for known simple groups.

§16.3. Supersymmetries with $N > 2$

The search for the ultimate unification has led to the study of properties of extended supersymmetries with N larger than 2 in four space–time dimensions. The algebra for these supersymmetries reads

$$\begin{aligned}
[Q^i_\alpha, T_K] &= i(S^{ij})_K Q^j_\alpha, \\
\{Q^i_\alpha, Q^i_\beta\} &= i\varepsilon_{\alpha\beta}Z^{ij}, \\
\{Q^i_\alpha, Q^{*j}_\beta\} &= 2\delta^{ij}(\sigma^\mu)_{\alpha\beta}P_\mu,
\end{aligned} \tag{16.3.1}$$

where i denotes the index whose maximum value denotes the number of supersymmetries in the theory. T_K denotes the generator of the internal symmetry group G of the extended theory, analogous to the SU(2) symmetry of the $N = 2$ theory. For the massless case, the group G has to be U(N) except for $N = 4$ where it could be either U(4) or SU(4) [5].

To study the particle content of N-extended supersymmetric theories, we consider a massless particle with maximum helicity $|\lambda|$ and apply the operator Q^i_α to it successively in which case we find $|\lambda|$, $|\lambda| - \frac{1}{2}$, $|\lambda| - 1, \ldots, |\lambda| - N/2$. There is no helicity with $|\lambda| - (N + 1)/2$ as multiplying $(N + 1)Q^i_\alpha$'s $(i = 1, \ldots, N)$ gives zero since the Q^i_α's are anticommuting spinorial objects. It is, therefore, clear that if we want to consider matter multiplets with only spin 0 and 1/2, we can have at most $N = 2$. Similarly, for $N = 4$, the smallest irreducible multiplet must contain a spin 1 field along with spin 0 and spin 1/2 fields. Since the highest spin elementary particle of interest is the graviton with spin 2, the above counting argument restricts the N to be ≤ 8.

Table 16.1

	Number of supersymmetries			
Helicity λ	$N = 1$	$N = 2$	$N = 3$	$N = 4$
1	1	1	1	1
1/2	1	2	3 + 1	4
0		1 + 1	3 + 3	6
−1/2	1	2	1 + 3	4
−1	1	1	1	1

In Tables 16.1 and 16.2 respectively, we give the particle content of the various extended supersymmetric theories for highest spin 1 (or the gauge multiplet), highest spin 2 (supergravity multiplet), and zero mass. Note that $N = 3$ and $N = 4$ Yang–Mills multiplets as well as $N = 7$ and $N = 8$ supergravity multiplets have essentially the same number of fields; as a result those theories are identical. Supersymmetric Lagrangians for these theories have been constructed. We do not discuss these here, except to note that as we go to $N = 8$ supergravity neither gravity nor the Yang–Mills multiplet nor the matter multiplet can exist in isolation; they exist together and there mutual interaction is dictated by symmetry. This attractive feature was the basis for the hope that extended supergravity may provide true unification of all interactions as well as constituents. The problem, however, is that SO(8), which is the largest gauge group of these theories is not big enough to include the realistic strong and electro-weak gauge group $SU(3)_c \times SU(2)_L \times U(1)_Y$. Moreover, all $N > 1$ supersymmetry theories are vectorlike, i.e., the interactions are symmetrical with respect to both left- and right-handed chiralities. On the other hand, we know that low-energy weak interactions are left-

Table 16.2

	Number of supersymmetries						
λ	$N = 1$	$N = 2$	$N = 3$	$N = 4$	$N = 5$	$N = 6$	$N = 1$
2	1	1	1	1	1	1	1
3/2	1	2	3	4	5	6	7 + 1
1		1	3	6	10	15 + 1	21 + 7
1/2			1	4	10 + 1	20 + 6	35 + 21
0				1 + 1	5 + 5	15 + 15	35 + 35
−1/2			1	4	1 + 10	6 + 20	21 + 35
−1		1	3	6	10	1 + 15	7 + 21
−3/2	1	2	3	4	5	6	1 + 7
−2	1	1	1	1	1	1	1

handed. Thus these models, though very interesting, cannot be realistic models of quark–lepton interactions.

§16.4. Higher Dimensional Supergravity Theories

Soon after the discovery of extended supersymmetric theories, it was realized that we can obtain these theories from $N = 1$ supersymmetric theories constructed in higher space–time dimensions. As a heuristic example of how this comes about, let us look at $N = 2$ supersymmetry. As discussed in the previous section, it involves two sets of fermionic coordinates θ^1 and θ^2 where θ^i are four-component Majorana spinors. If we define a new spinor ξ such that

$$\xi = \begin{pmatrix} \theta^1 \\ \theta^2 \end{pmatrix} \tag{16.4.1}$$

the new eight-component spinor can be thought of as a single spinor in six-dimensional space–time ($d = 6$). A general property of SO(N) spinors is that they decompose into a number of spinors when SO(N) is reduced to one of its subgroups. This is the intuitive way to see how extended supersymmetric theories arise after dimensional reduction of higher dimensional supersymmetric theories. Of course, we have to be careful since Majorana and Weyl constraints cannot always be imposed in arbitrary dimensions.

Another reason to consider higher dimensional supersymmetry is that the chirality problem of extended supersymmetric theories may be evaded in the process of dimensional reduction, since in higher dimensions we have to consider only $N = 1$ supersymmetry.

An important ingredient in the discussion of higher dimensional supersymmetric theories is whether the Majorana spinors can be defined in arbitrary dimensions. We now discuss this question in arbitrary even dimension $d = 2M$. Define the d Γ-matrices Γ^n, $n = 0, 1, \ldots, d-1$ satisfying the Clifford algebra

$$\{\Gamma^n, \Gamma^m\} = 2\eta^{mn}, \tag{16.4.2}$$

where $\eta^{mn} = (+, ---\cdots)$. We also choose Γ^m such that $\Gamma^0 = \Gamma^{0\dagger}$ and $\Gamma^i = -\Gamma^{i\dagger}$, $i = 1, \ldots, d-1$. Let us write down the Dirac equation

$$(i\Gamma^n(\partial_{n-ieA_n} + m)\psi = 0. \tag{16.4.3}$$

For $m = 0$, we find from eqn. (16.4.3)

$$(-i\Gamma^{n^T}\partial_n)\bar{\psi}^T = 0. \tag{16.4.4}$$

Since Γ^{n^T} also satisfy Clifford algebra, there exists a unitary matrix C such that

$$C\Gamma^{n^T}C^{-1} = \eta\Gamma^n. \tag{16.4.5}$$

Here η can be either $+1$ or -1, $C\bar{\psi}^T$ satisfies the Dirac equation, and C is the familiar charge conjugation matrix. The Majorana condition can therefore be written as

$$C\bar{\psi}^T = \psi. \tag{16.4.6}$$

However, eqn. (16.4.6) can be imposed provided

$$(C\Gamma_0^T)(C^*\Gamma^0) = 1, \tag{16.4.7}$$

or using eqn. (16.4.5),

$$CC^* = \eta. \tag{16.4.7a}$$

However, we show below that eqn. (16.4.7a) need not be satisfied for all values of d. For simplicity, let us choose $\eta = -1$. Let us now use the following identity for the Γ-matrices and their products:

$$2^m \sum_A (\Gamma_A)_{ij}(\Gamma_A^{-1})_{kl} = \sum_A 1 \cdot \delta_{jk}\delta_{il}, \tag{16.4.8}$$

where

$$\Gamma_A = \Gamma^m, \tfrac{1}{2}[\Gamma^m, \Gamma^n], \ldots.$$

It is easy to convince oneself that $A = 1, \ldots, 2^{2m}$. From eqn. (16.4.8), we obtain

$$2^m \sum_A \mathrm{Tr}(C\Gamma_A^T C^{-1}\Gamma_A^{-1}) = \sum_A 1 \cdot \mathrm{Tr}(C^T C^{-1}). \tag{16.4.9}$$

Let

$$C^T = \lambda C. \tag{16.4.10}$$

Using $\eta = -1$ in eqn. (16.4.5) we obtain

$$\lambda = \cos\frac{\pi m}{2} - \sin\frac{\pi m}{2}. \tag{16.4.11}$$

It follows from this that, for $m = 2$ (i.e., $d = 4$), $\lambda = -1$; therefore eqns. (16.4.10) and (16.4.7) are consistent with each other. But for $m = 3$ (or $d = 6$), $\lambda = +1$; thus eqn. (16.4.7) cannot hold. This means in $d = 6$, mod 8 we cannot define Majorana spinors, whereas in $d = 2, 4$ mod 8, we can.

In $d = 2, 4$ mod 8, we can show that the Γ-matrices can be chosen pure imaginary in which case, we can set $C = \Gamma^0$, which on using eqn. (16.4.6), leads to $\psi = \psi^*$ as the Majorana condition.

The next question we may ask is: In which dimension can we define a Majorana Weyl spinor?, i.e., where both the following conditions are satisfied

$$\psi = \psi^*, \qquad \Gamma_{\mathrm{FIVE}}\psi = \psi, \tag{16.4.12}$$

where

$$\Gamma_{\mathrm{FIVE}} = \eta\Gamma^0\Gamma^1, \ldots, \Gamma^{d-1}, \tag{16.4.13}$$

clearly Γ_{FIVE} must be a real diagonal matrix. This implies $\eta^2 = 1$ which can be true only if $d = 2$ mod 8.

We can now determine the number of supersymmetries that will result from dimensional reduction of a D-dimensional theory. In D-dimensions the spinor is $2^{D/2}$ dimensional. However, the Majorana or Majorana–Weyl condition reduces the number of two or four respectively. We denote this factor by r. Thus the number of supersymmetries N in four dimensions is

$$N = \tfrac{1}{2}r \cdot 2^{D/2}. \tag{16.4.14}$$

Now, coming back to dimensional reduction from $d = 6$ to $d = 4$ to obtain $N = 2$ supersymmetry, we see that we cannot naively define a Majorana spinor. However, a symplectic spinor can be defined by the condition

$$(\psi^i)^* = -\Sigma^{ij}C^{-1}\Gamma^0\psi^j. \tag{16.4.15}$$

We can use them to complete dimensional reduction [6] and using our counting we get $N = \frac{1}{2}\cdot\frac{1}{2}\cdot 2^3 = 2$.

Having outlined an important subtlety in implementing dimensional reduction, we now pass on to discuss some specific higher dimensional theories. Two which have received greatest attention are the $d = 11$ and $d = 0$ supergravity theories. We present these theories now.

$d = 11$ Supergravity

We assume the signature of this theory to be $(+ - - - - \cdots)$. The fields of this theory are:

Elfbein E^N_B, N = curved space index,

B = flat tangent space index.

Gravitino ψ_N

Antisymmetric Three Index Bosonic Tensor Field A_{MNP}

The Lagrangian invariant under the following supersymmetry transformations

$$\begin{aligned} \delta E^A_M &= -i\kappa\bar{\varepsilon}\Gamma^A\psi_M, \\ \delta A_{MNP} &= \tfrac{3}{2}\bar{\varepsilon}\Gamma_{[MN;}\psi_{P]}, \\ \delta\psi_M &= \frac{1}{\kappa}\hat{D}_\mu\varepsilon \end{aligned} \tag{16.4.16}$$

is given by [6]

$$\begin{aligned} \mathscr{L} = &-\frac{E}{4\kappa^2}R(\omega) - \frac{i}{2}E\bar{\psi}_M\Gamma^{MNP}D_N\left(\frac{\omega+\hat{\omega}}{2}\right)\psi_P - \frac{E}{48}F_{MNPQ}F^{MNPQ} \\ &+ \frac{2\kappa}{(144)^2}\varepsilon^{M_1,\ldots,M_{11}}F_{M_1,\ldots,M_4}F_{M_5,\ldots,M_8}A_{M_9,\ldots,M_{11}} \\ &+ \frac{\kappa E}{192}(\bar{\psi}_M\Gamma^{MNPQRS}\psi_N + 12\bar{\psi}^P\Gamma^{QR}\psi^S)(F_{PQRS} + \hat{F}_{PQRS}), \end{aligned} \tag{16.4.17}$$

where

$$E = \text{Det } E_M^A,$$

$$D_M = \partial_M - \tfrac{1}{4}\omega_{M,AB}\Gamma^{AB},$$

and

$$\omega_{MAB} = \tfrac{1}{2}(-\Omega_{MAB} + \Omega_{ABM} - \Omega_{BMA}) + K_{MAB},$$

$$K_{MAB} = \frac{i}{4}[-\bar{\psi}_N\Gamma^{NP}_{MAB}\psi_P + 2(\bar{\psi}_M\Gamma_B\psi_A - \bar{\psi}_M\Gamma_A\psi_B + \bar{\psi}_B\Gamma_M\psi_A)],$$

$$\Omega^A_{MN} = 2\partial_{[N}E^A_{M]},$$

$$\tilde{\omega}_{MAB} = \omega_{MAB} + \frac{i}{4}\bar{\psi}_N\Gamma^{NP}_{MAB}\psi_P,$$

$$F_{MNPQ} = 4\partial_{[M}A_{NPQ]},$$

$$\tilde{F}_{MNPQ} = F_{MNPQ} - 3\bar{\psi}_{[M}\Gamma_{NP}\psi_{Q]},$$

$$\tilde{D}_M(\tilde{\omega})\psi_N = D_M(\tilde{\omega})\psi_N + T_M^{PQRS}\tilde{F}_{PQRS}\psi_N, \tag{16.4.18}$$

$$T_{SMNPQ} = \frac{1}{144}[\Gamma^{SMNPQ} - 8\Gamma^{[MNP}g^{Q]S}], \tag{16.4.19}$$

Under coordinate transformation by a parameter ξ^M the fields transform as follows:

$$\delta E^A_M = E^A_N\partial_M\xi^N + \xi^N\partial_N E^A_M,$$

$$\delta\psi_M = \psi_N\partial_M\xi^N + \xi^N\partial_N\psi_M,$$

$$\delta A_{MNP} = 3A_{Q[NP}\partial_{M]}\xi^Q, \tag{16.4.20}$$

An interesting result in this model was noted by Freund and Rubin [7] who showed that by choosing a ground state where

$$F_{\mu\nu\rho\sigma} = 3m\varepsilon_{\mu\nu\rho\sigma}$$

$$F_{mnpq} = 0 \tag{16.4.21}$$

where $\mu, \nu = 0, \ldots, 3$ and $m, n, \ldots = 4, \ldots, 11$, we obtain a compactification to the $M_4 \times M_7$ space where M_4 can be identified with space–time. This kind of ground state is however not unique [8], and there exist other kinds of solutions of this model where the four-dimensional theory corresponds to $N = 8$ supergravity. In any case, none of these theories have been shown to lead to realistic models of electro-weak interactions.

§16.5. $d = 10$ Super-Yang–Mills Theory

Another higher dimensional theory of great interest and model building potential is the ten-dimensional super-Yang–Mills theory. This is of particular interest in the superstring models [9], [10], which have recently been proposed as the ultimate unified theory of all forces. The zero slope limit of the superstring theories are also supposed to lead to ten-dimensional super-Yang–Mills theories. Therefore, in this section, we present the particle content and Lagrangian for the $d = 10$ theory [11], [12].

Particles

$$\text{Vielbein: } E_M^A,$$
$$B_{MN} = -B_{NM},$$
$$\text{Spinor } \lambda,$$
$$\text{Real Scalar } \phi,$$
$$\text{Yang–Mills Field } A_M^\alpha,$$
$$\text{Gaugino } \chi^\alpha,$$

α is the index of the gauge group.

The supersymmetric Lagrangian for this model is

$$\begin{aligned} E^{-1}\mathscr{L} = &-\frac{1}{2\kappa^2}R - \frac{i}{2}\bar{\psi}_M\Gamma^{MNP}D_N(\omega)\psi_P + \tfrac{3}{4}\phi^{-3/2}H_{MNP}H^{MNP} + \frac{i}{2}\bar{\lambda}\Gamma^M D_M(\omega)\lambda \\ &+ \tfrac{9}{16}(\partial_M\phi/\phi)^2 + \tfrac{3}{8}\sqrt{2}\bar{\psi}_M(\not{\partial}\phi/\phi)\Gamma^M\lambda - \frac{\sqrt{2}}{16}\phi^{-3/4}H_{MNP}(i\bar{\psi}_Q\Gamma^{QMNPR}\psi_R \\ &+ \sqrt{2}\bar{\psi}_Q\Gamma^{MNP}\Gamma^Q\lambda + 6i\bar{\psi}^M\Gamma^N\psi^P - \bar{\chi}^\alpha\Gamma^{MMP}\chi^\alpha) - \tfrac{1}{4}\phi^{-3/4}F_{MN}^\alpha F^{\alpha MN} \\ &+ \frac{i}{2}\bar{x}^\alpha\Gamma^M(D_M(\omega)x)^\alpha - \frac{i\kappa}{4}\phi^{-3/8}(\bar{x}^\alpha\Gamma^M\Gamma^{NP}F_{NP}^\alpha)\left(\psi_M + \frac{i\sqrt{2}}{12}\Gamma_{M\lambda}\right) \\ &+ \text{four Fermi interactions,} \end{aligned} \tag{16.5.1}$$

where

$$D_N\psi_P = (\partial_N - \tfrac{1}{2}\omega_{N[RS]}\Gamma^{RS})\psi_P - 2\omega_{NP}^Q\psi_Q, \qquad \kappa \text{ is the gravitational constant.} \tag{16.5.2}$$

$$\begin{aligned} D_N\lambda &= (\partial_N - \tfrac{1}{2}\omega_{N[RS]}\Gamma^{RS})\lambda \\ (D_N\chi)^\alpha &= (\partial_N - \tfrac{1}{2}\omega_{N[RS]}\Gamma^{RS})\chi^\alpha - f^{\alpha\beta\delta}A_N^\beta\chi^\delta. \end{aligned} \tag{16.5.3}$$

The Yang–Mills field strength

$$F_{MN}^\alpha = \tfrac{1}{2}\partial_{[M}A_{N]}^\alpha + f^{\alpha\beta\delta}A_M^\beta A_N^\gamma, \tag{16.5.4}$$

$f^{\alpha\beta\gamma}$ are structure constants of the gauge group

$$H_{MNP} = \partial_{[M}B_{NP]} - \frac{\kappa}{\sqrt{2}}(\Omega_{MNP}^{\text{YM}} - \Omega_{MNP}^L), \tag{16.5.5}$$

where Ω^{YM} and Ω^L are the Yang–Mills and Lorentz Chern–Simons forms given by

$$\Omega^{\mathrm{YM}}_{MNP} = \mathrm{Tr}(A_{[M}F_{NP]} - \tfrac{2}{3}A_{[M}A_N A_{P]})$$

and

$$\Omega^{L}_{MNP} = \mathrm{Tr}(\omega_{[M}R_{NP]} - \tfrac{2}{3}\omega_{[M}\omega_N\omega_{P]}) \tag{16.5.6}$$

It has recently been considered as a serious candidate for the ultimate unification group of quark–lepton interactions including gravity for the following reasons:

(a) This theory is supposed to arise in the zero-slope limit of the superstring theories.
(b) For the case where the Yang–Mills group is chosen to be either SO(32) or $E_8 \times E'_8$ [14], this theory is free [9] of gauge, as well as gravitational, anomalies [13]. This is a unique feature of this model. The anomaly cancellation requires the presence of the Ω-terms in eqn. (16.5.5). Their presence, however, breaks the apparent supersymmetry of the Lagrangian but it is expected that once higher order terms are included, supersymmetry invariance will be restored.

We now comment briefly on the compactification of this model to obtain four-dimensional theories: the physically relevant compactification of this model is of $M_4 \times K_6$ type where M_4 is the four-dimensional Minkowski space and K_6 is a six-dimensional compact manifold. An interesting compactification scheme that is believed to preserve $N = 1$ supergravity arises when K_6 is a Calabi–Yau manifold with SU(3) holonomy. In this case, the $E_8 \times E'_8$ gauge group breaks down to $E_8 \times E_6$. Since the $\{248\}$-dimensional representation of E'_8 when decomposed under E_6, contains $\{27\}$-dimensional representations, some of the gauginos of the E'_8 group turn into quarks and leptons. However, in general, we obtain hundreds of massless $\{27\}$ and $\{\overline{27}\}$ representations. To obtain the number of generations to be less than four, K cannot be simply connected. If K admits a discrete symmetry group that acts freely, then compactifying the ten-dimensional theory to $M_4 \otimes (K/G)$ leads to five $\{27\}$-dimensional and one $\{\overline{27}\}$-dimensional massless superfields (for $G = Z_5 \times Z_5$) which can then be identified with quarks and leptons and Higgs superfields. This model has many features of realistic quark–lepton models because the $\{27\}$-dimensional superfield decomposes under the SO(10) subgroup of E_6 [16] as follows:

$$\{27\} = \{16\} + \{10\} + \{1\}. \tag{16.5.7}$$

The same representation can also serve as a Higgs superfield and can be used to implement gauge symmetry breaking. Two important phenomenological problems of this model are the following:

(a) Since the model has a right-handed neutrino and it has no $\{126\}$-dimensional Higgs representation of the SO(10) group (which will be part of the $\{351\}$-dimensional representation of E_6), the conventional "see-saw" mechanism used in Chapter 6 to understand tiny neutrino mass does not exist. However, alternative mechanisms [17] have been proposed for this purpose.

(b) *A priori*, there is the possibility of proton decaying rapidly.

Another problem, which may require deeper understanding of the compactification spaces, is how to introduce *CP*-violation into these models. Thus, even though ten-dimensional $E_8 \times E_8'$ super-Yang–Mills theory has raised expectations for a realistic unification model, a lot of work remains to be done.

References

[1] P. Fayet, *Nucl. Phys.* **B113**, 135 (1976); **B149**, 137 (1979);
M. F. Sohnius, K. Stella, and P. West, *Nucl. Phys.* **B173**, 127 (1980).

[2] J. Bagger and E. Witten, *Nucl. Phys.* **B222**, 1 (1983).

[3] B. deWit, P. G. Lauwers, R. Philippe, S. Q. Su, and A. Van Proyen, preprint NIKHEF-H/83-13 (1983);
P. Breitenlohner and M. Sohnius, *Nucl. Phys.* **B165**, 483 (1980).

[4] F. del Aguila, B. Brinstein, L. Hall, G. G. Ross, and P. West, HUTP 84/A001, 1984;
S. Kalara, D. Chang, R. N. Mohapatra, and A. Gangopadhyaya, *Phys. Lett.* **145B**, 323 (1984);
J. P. Deredings, S. Ferrara, A. Masiero, and A. Van Proyen, *Phys. Lett.* **140B**, 307 (1984).
For physical applications see
J. M. Frére, I. Meznicescu, and Y. P. Yao, *Phys. Rev.* **D29**, 1196 (1984);
A. Parkes and P. West, **127B**, 353 (1983).

[5] R. Haag, J. T. Lopuszanski, and M. Sohnius, *Nucl. Phys.* **B88**, 257 (1975).

[6] J. Koller, Cal. Tech. preprint 68-975, 1982;
G. Sierra and P. K. Townsend, Ecole Normale preprint, LPTENS 83/26, 1983;
J. Sherk, Ecole Normale preprint, 1979;
M. Duff, Lectures at the GIFT Summer School, San Feliu de Guixols, Spain, 1984.

[7] P. G. O. Freund and M. Rubin, *Phys. Lett.* **B97**, 233 (1980).

[8] M. Duff and C. N. Pope, in *Supersymmetry and Supergravity* '82 (edited by S. Ferrara *et al.*), World Scientific, Singapore, 1983.

[9] M. Green and J. H. Schwarz, *Phys. Lett.* **149B**, 117 (1984).

[10] P. Candelas, G. T. Horowitz, A. Strominger, and E. Witten, *Nucl. Phys.* **B258**, 46 (1985);
K. Pilch and A. N. Schellekens, Stony Brook preprint, 1985;
E. Witten, Princeton preprint, 1985.

[11] E. Bergshoeff, M. de Roo, B. de Wit, and P. Van Niuewenhuizen, *Nucl. Phys.* **B195**, 97 (1982).

[12] G. F. Chapline and N. S. Manton, *Phys. Lett.* **120B**, 105 (1983).

[13] L. Alvarez-Gaume and E. Witten, *Nucl. Phys.* **B234**, 269 (1983).

[14] D. Gross, J. Harvey, E. Martinec, and R. Rohm, *Phys. Rev. Lett.* **54**, 502 (1985).

[15] M. Dine, V. Kaplunovsky, M. Mangano, C. Nappi, and N. Seiberg, *Nucl. Phys.* B (1985) (in press);
J. Breit, B. Ovrut, and G. Segre, *Phys. Lett.* **B158**, 33 (1985);

A. Sen, *Phys. Rev. Lett.* **55**, 33 (1985);
J. P. Deredings, L. Ibanez, and H. P. Nilles, *Phys. Lett.* **155B**, 65 (1985); and CERN preprint CERN-TH-4228/85, 1985;
V. S. Kaplunovsky, *Phys. Rev. Lett.* **55**, 1036 (1985);
M. Dine and N. Seiberg, *Phys. Rev. Lett.* **55**, 366 (1985);
V. S. Kaplunovsky and Chiara Nappi, *Comments on Nuclear and Particle Physics* (1986) (to appear).

[16] F. Geursey, P. Ramond, and P. Sikivie, *Phys. Lett.* **60B**, 117 (1976);
Y. Achiman and B. Stech, *Phys. Lett.* **77B**, 389 (1987);
Q. Shafi, *Phys. Lett.* **79B**, 301 (1978);
P. K. Mohapatra, R. N. Mohapatra, and P. Pal, *Phys. Rev.* D (1986).

[17] R. N. Mohapatra, *Phys. Rev. Lett.* **56**, 561 (1986);
U. Sarkar and S. Nandi, *Phys. Rev. Lett.* **56**, 564 (1986).

"I thought that my voyage had come to its end at the last limit of my power—that, the path before me was closed, that provisions were exhausted and the time come to take shelter in a silent obscurity.

But I find that thy will knows no end in me. And when old words die out on the tongue, new melodies break forth from the heart; and where the old tracks are lost, new country is revealed with its wonders."

GITANJALI, RABINDRA NATH TAGORE

Index